Towards Water Secure Societies

Lars Ribbe • Andreas Haarstrick •
Mukand Babel • Sudeh Dehnavi •
H. K. Biesalski
Editors

Towards Water Secure Societies

Coping with Water Scarcity and Quality Challenges

Springer

Editors
Lars Ribbe
TH Köln
Cologne, Germany

Andreas Haarstrick
TU Braunschweig
Braunschweig, Germany

Mukand Babel
Water Engineering and Management
Asian Institute of Technology
Bangkok, Thailand

Sudeh Dehnavi
TH Köln
Cologne, Germany

H. K. Biesalski
University of Hohenheim
Stuttgart, Baden-Württemberg, Germany

ISBN 978-3-030-50652-0 ISBN 978-3-030-50653-7 (eBook)
https://doi.org/10.1007/978-3-030-50653-7

This Springer imprint is published by the registered company Springer Nature Switzerland AG
The registered company address is: Gewerbestrasse 11, 6330 Cham, Switzerland

About This Issue

This book is a collection of papers, that were presented in the framework of the International Conference Series: Water Security and Climate Change (WSCC) (https://watersecurity.info/). The thematic scope of the volume is on water scarcity including droughts and water quality challenges.

About Conference Series: Water Security and Climate Change (WSCC)

Water Security as a 21st Century Challenge for Science, Politic and Society

Water resources form a pivotal base for economic development, social wellbeing and environmental integrity. However, achieving water security in the face of rapid change—in populations, economies, geopolitics and climate—is one of the major challenges faced by society today, at both a global and local level. More efficient use and integrated management of water are critical to addressing the growing demand for water and the increasing frequency of droughts and floods resulting from climate change. However, according to the *"Sustainable Development Goals Report 2019"* of the United Nations most countries are unlikely to reach full implementation of integrated water resources management by 2030 (United Nations, 2019).
Furthermore, attaining water security is a multidimensional challenge that implies a range of issues from protection of the environment and water quality, coping with scarcity to flood control. These diverse facets of water management require interdisciplinary approach including environmental, engineering and social sciences as well as cross sectoral cooperation between business, civil society and government.
In response to these requirements, the Conference Series on "Water Security and Climate Change" (WSCC—https://watersecurity.info/) was initiated in 2016 by the international network of universities comprising the Centers for Natural Resources and Development—CNRD (TH Köln—https://www.cnrd.info/), the International Network on Sustainable Water Management in Developing Countries—SWINDON (TU Braunschweig—

http://www.exceed-swindon.org/) and the Food Security Center—FSC (University Hohenheim—https://fsc.uni-hohenheim.de/en/home).

The WSCC aims to bring together scientists, policy-makers and stakeholders from around the world and from various sectors and disciplines to discuss the diverse facets of water security in dynamic environments and its relation to climate variability and change. The first WSCC took place in 2016 in Bangkok (Thailand), and subsequently conferences were organized in 2017 in Cologne (Germany), 2018 in Nairobi (Kenya) and in 2019 in San Luis Potosí (Mexico).

Contents

Editors and Contributors

About the Editors

Lars Ribbe is a Professor for integrated land and water resources management and Dean of the Faculty of Spatial Development and Infrastructure Systems at the TH Köln (University of Applied Sciences). Holding a PhD in the field of hydro-informatics (Friedrich Schiller University Jena), his current work in research and teaching is river basin assessment, modeling, and management and he is specifically interested in developing knowledge systems helping decision-makers to cope with prevailing water resource-related challenges. He is actively involved in promoting a global network of Integrated Water Resources Management (IWRM) master programs and is the lead of the Exceed project Centers for Natural Resources and Development (CNRD).

Andreas Haarstrick has studied Chemistry at the Technische Universität Braunschweig in Germany (1983–1989) and received his doctorate in 1992 in the field of biotechnological up- and downstreaming processes of biopolymers. Since 2006, he is Professor of Bioprocess Engineering at the TU Braunschweig. His teaching and research interests cover modeling biological and chemical processes in heterogeneous systems, development of models predicting pollutant reduction in and emission behavior of landfills, growth kinetics at low substrate concentrations under changing environmental conditions, Advanced Oxidation Processes (AOP), and groundwater management. Since 2012, he is the Managing Director of the exceed-Swindon project dealing with the sustainable water management in developing countries (www.exceed-swindon.org).

Mukand Babel is a Professor of Water Engineering and Management and the Chair of Climate Change Asia initiative at the Asian Institute of Technology (AIT), Thailand. He was the Chair of Academic Senate of AIT. He has been a Visiting/Honorary Professor at the University of Exeter, UK; Kyoto University, Japan; University Teknologi Mara, Malaysia; and Sam Higginbottom University of Agriculture, Technology and Sciences (SHUATS), India. His professional experience in teaching, research, and consultancy spans over 35 years mainly in Asia. He has been a Member of the Executive Committee, Asia Pacific Division of IAHR. Similarly, he was a member of the Executive Board of IWRA (2016-18) and a Board

Member of the Asia Water Council. With over 250 publications in international refereed journals, books, and book chapters and conference proceedings, he conducts interdisciplinary research relating hydrology and water resources with economic, environmental, and socioeconomic aspects of water to address diverse water problems and issues including water security assessment, water-energy-food nexus, and climate change impact and adaptation in water sector.

Sudeh Dehnavi is a researcher and lecturer in the sustainable development of food and water systems. Her interdisciplinary research and educational work's core topic is Natural Resources and Agricultural Economics, Food and Water Security and Governance, and Sustainability Entrepreneurship. Her methodical expertise lies especially in the areas of water and food security assessment and policy analysis; social and economic analysis; knowledge system design for decision support and market-based research. She has years of experience in strategic planning and scientific coordination of international development projects and networks; design and implementation of international scientific conferences; science-policy dialogue workshops; and educational excursions.

Prof. Dr. med. H. K. Biesalski was the head of the Department of Biological Chemistry and Nutrition at University of Hohenheim in Stuttgart, Germany, and director of the food security center of the University Stuttgart Hohenheim until his retirement in 2017. He is member of the WHO micronutrient initiative and the FAO/WHO expert group nanomaterials. He is the elected member of the high-level expert panel of the global forum on food security and nutrition. In 2007/2008, he was the elected member of the Institute of advanced studies in Berlin. He was the principal investigator of a couple of clinical studies and of studies dealing with malnutrition and micronutrients in South Asia and different countries in Africa. Biesalski is a member of the advisory board of the German ministry of agriculture and nutrition and further executive boards and scientific advisory boards of German health societies and academies. He is the editor of nutrition and metabolism in oncology, European editor of Nutrition (until 2017), and European editor of nutrition and metabolism and board member of several scientific journals. He has edited nine books related to nutrition medicine, clinical nutrition, and ageing edited in 6 different languages, and has published more than 250 papers in peer reviewed scientific journals. He has chaired several international conferences (among others, the WHO Conference on Nutrition and Cancer in 1997 and the First Congress Nutrition Teams). He is the chair and organizer of the International conference "Hidden Hunger". The upcoming fifth one will be held on September 16–19, 2022. He has got a couple of awards, e.g., the Konrad Lang Medal of the German Society of Nutrition Medicine, the Vitamin Award of the Society for applied Vitamin research, and the prestigious Justus von Liebig Price for World Nutrition in 2017.

Contributors

Suha Al-Madbouh Institute for Technology and Resources Management in the Tropics and Subtropics (ITT)/TH-Köln, Cologne, Germany

Victor Alcaraz-Gonzalez Departamento de Ingeniería Química, Universidad de Guadalajara, Guadalajara Jalisco, Mexico

Mukand Babel Asian Institute of Technology (AIT), Klong Luang, Pathumthani, Thailand

Mwenda Borona Intergrative Research Institute on Transformations of Human-Environment Systems (IRI THESys), Humboldt-Universität zu Berlin, Berlin, Germany

Alexander Chusov Peter the Great St. Petersburg Polytechnic University, St. Petersburg, Russia

Thomas Claßen StadtLandschaft & Gesundheit, Bielefeld, Germany

Nguyen Mai Dang Thuyloi University (TLU), Hanoi, Vietnam

Eyüp Debik Environmental Engineering Department, Yildiz Technical University, Istanbul, Turkey

M. Elshemy Faculty of Engineering, Tanta University, Tanta, Egypt

Jannatul Ferdous Climate Change Lab, Department of Civil Engineering, Military Institute of Science and Technology (MIST), Dhaka, Bangladesh

Jorge Gironás Centro Interdisciplinario de Cambio Global. Pontificia, Universidad Católica de Chile, Santiago, Chile;
Departamento de Ingeniería Hidráulica y Ambiental, Pontificia Universidad Católica de Chile, Santiago, Chile;
Centro de Desarrollo Urbano Sustentable CEDEUS CONICYT/FONDAP/15110020, Santiago, Chile;
Centro de Investigación Para La Gestión Integrada del Riesgo de Desastre CIGIDEN CONICYT/FONDAP/15110017, Santiago, Chile

Victor Gonzalez-Alvarez Departamento de Ingeniería Química, Universidad de Guadalajara, Guadalajara Jalisco, Mexico

Md. Arman Habib Climate Change Lab, Department of Civil Engineering, Military Institute of Science and Technology (MIST), Dhaka, Bangladesh

A. A. Hassan Faculty of Engineering, Ain Shams University, Cairo, Egypt

Md. Maruful Hoque Climate Change Lab, Department of Civil Engineering, Military Institute of Science and Technology (MIST), Dhaka, Bangladesh

Claudia Hornberg Faculty of Health Sciences, Bielefeld University, Bielefeld, Germany

G. M. Tarekul Islam Institute of Water and Flood Management, Bangladesh University of Engineering and Technology, Dhaka, Bangladesh

Elaine T. Lawson Institute for Environment and Sanitation Studies (IESS), College of Basic and Applied, Sciences, University of Ghana, Legon, Accra, Ghana

Julia Lednova Peter the Great St. Petersburg Polytechnic University, St. Petersburg, Russia

Trung Tuan Luong Vietnam Academy for Water Resources, Hanoi, Vietnam

Cheikh Mbow Future Africa, University of Pretoria, Pretoria, South Africa

Francisco Javier Meza Centro Interdisciplinario de Cambio Global. Pontificia, Universidad Católica de Chile, Santiago, Chile;
Departamento de Ecosistemas Y Medio Ambiente. Pontificia, Universidad Católica de Chile, Santiago, Chile;
Aquasec. IAI Center of Excellence for Water Security, Tucson, USA

A. Nada Faculty of Engineering, Tanta University, Tanta, Egypt

Lam Xuan Nguyen Vietnam Academy for Water Resources, Hanoi, Vietnam

Melanie Oertel Centro Interdisciplinario de Cambio Global. Pontificia, Universidad Católica de Chile, Santiago, Chile;
Departamento de Ecosistemas Y Medio Ambiente. Pontificia, Universidad Católica de Chile, Santiago, Chile

Issa Ouedraogo The CGIAR Research Program on Climate Change, Agriculture and Food Security (CCAFS), ICRISAT West & Central Africa Regional Office, BP 320, Bamako, Mali

M. Tauhid Ur Rahman Department of Civil Engineering, Military Institute of Science and Technology (MIST), Dhaka, Bangladesh

Mashrekur Rahman Institute of Water and Flood Management, Bangladesh University of Engineering and Technology, Dhaka, Bangladesh

Md. Munsur Rahman Institute of Water and Flood Management, Bangladesh University of Engineering and Technology, Dhaka, Bangladesh

Md. Rasheduzzaman Climate Change Lab, Department of Civil Engineering, Military Institute of Science and Technology (MIST), Dhaka, Bangladesh

Lars Ribbe Faculty of Spatial Development and Infrastructure Systems, TH Köln - University of Applied Sciences, Cologne, Germany

Frank Rögener TH Köln, Cologne, Germany

Abdul Rauf Zanya Salifu Institute for Environment and Sanitation Studies (IESS), College of Basic and Applied, Sciences, University of Ghana, Legon, Accra, Ghana

Devesh Sharma Central University of Rajasthan, Kishangarh, Ajmer, India

Victor Shinde Asian Institute of Technology (AIT), Klong Luang, Pathumthani, Thailand

Justyna Sycz Faculty of Spatial Development and Infrastructure Systems, TH Köln - University of Applied Sciences, Cologne, Germany

Sven Theus DAR - Deutsche, Abwasser-Reinigungs-GmbH, Wiesbaden, Germany

Trinh Duc Tran Vietnam Academy for Water Resources, Hanoi, Vietnam

Vu Thanh Tu Thuyloi University (TLU), Hanoi, Vietnam

Charlotte Wrigley-Asante Department of Geography and Resource Development, College of Humanities, University of Ghana, Legon, Accra, Ghana

B. A. Zeidan Faculty of Engineering, Tanta University, Tanta, Egypt

Introduction

Justyna Sycz and Lars Ribbe

1 Water Scarcity in the International Development Discourse

1.1 Water Resources Under Pressure

The United Nations Water (UN-Water) defines water security as "the capacity of a population to safeguard sustainable access to adequate quantities of and acceptable quality water (…)" (UN-Water 2013). In terms of quantities, this definition encompasses both water scarcity and water abundance. The presented book focuses on water scarcity where the water demand excesses the water availability.

Stable distribution systems to serve the water supply has been a central aspect of all development and urbanization processes since ancient times. The first civilizations were formed on the banks of rivers: Ancient Egyptians on the Nile, the Mesopotamians in the Fertile Crescent on the Tigris/Euphrates Rivers, the Ancient Chinese on the Yellow River, and the Ancient Indian on the Indus (Terje and Coopey 2010). At present, 158 of 239 national capitals are located along rivers with access to surface water (own study). The continuing availability of water is a basis for food and energy and, industry, economic growth, as well as human health and environment. Hence, sustainable access to required quantities of acceptable quality water has been and will continue to be a pivotal condition for human development and welfare.

However, currently, the world's water resources are under considerable pressure due to both climatic as well as anthropogenic factors. Global water withdrawal had increased at a rate of 15% per decade between 1960 and 2010 as a function of population growth, economic development, and changing consumption patterns (Wada et al. 2016; WWDR 2020). Moreover, water abstractions are expected to increase by some 55% between 2000 and 2050, mainly in conjunction with growing demands from manufacturing (+400%), thermal power generation (+140%), and domestic use (+130%) (OECD 2012).

As demand rises, freshwater becomes increasingly scarce. Organic waste, fertilizers, pesticides, and emerging contaminants significantly influence water quality reducing the availability of usable resources (WWDR 2020). While the hydraulic infrastructure developed over the late 19th and 20th century have allowed economic development and enabled large-scale agriculture (Schmandt et al. 2013) it has led to the overexploitation of surface and groundwater resources with continuously declining water levels (Savenije et al. 2014). Climate change is expected to aggravate the situation with predicted

J. Sycz (✉) · L. Ribbe
Faculty of Spatial Development and Infrastructure Systems, TH Köln - University of Applied Sciences, Cologne, Germany
e-mail: justyna.sycz@th-koeln.de

L. Ribbe et al. (eds.), *Towards Water Secure Societies*,
https://doi.org/10.1007/978-3-030-50653-7_1

increases in temperature causing sea level rise and saltwater intrusion as well as changes in the frequency and severity of precipitation and associated droughts (IPCC 2014). According to FAO increasing drought events have affected more people worldwide in the last 40 years than any other natural hazard (FAO date not defined nd).

Taking into account the monthly variation of water availability, about two-thirds of the global population—"4.0 billion people"—are currently living in areas under conditions of severe blue water scarcity for at least for 1 month of the year (Mekonnen and Hoekstra 2016, p. 1). In these areas, the ratio of the blue water consumption to total blue water availability is in the range of 2.0 or higher (Mekonnen and Hoekstra 2016). The demand, which more than two-fold exceeds the supply results in conflicts between users, the irreversible depletion of groundwater, dry rivers, and negative impacts on the environment. The number of people living under conditions of severe blue water scarcity for at least 1 month of the year is expected to increase to 4.8–5.7 billion people by 2050 (Burek et al. 2016).

1.2 Water Scarcity Gains Attention in the International Development Discourse

In the global development discourse, the concern about water as a pivotal base for economic growth, social well-being, and environmental integrity that may become scarce, dates back to the first intergovernmental conference on water: the United Nations Water Conference in Mar del Plata in 1977. Although the emphasis was placed on water supply and sanitation while water scarcity itself was not a major concern, the final conference report called for urgent action to avoid expected water crisis in the face of "the alarming increase in world population, the enhanced requirements of industry, the needs of urban concentrations and the often irrational utilization of natural resources, and increasing pollution." (UN 1977, p. 96).

The next milestone in the discourse was marked by the Dublin Statement on Water and Sustainable Development. The statement was adopted on January 31, 1992 during a meeting of water experts that took place at the International Conference on Water and the Environment (ICWE) in Dublin, Ireland (ICWE 1992). In contrast to the conference in Mar del Plata the key message from Dublin was not a warning but rather a recognition that scarcity and misuse of freshwater already pose a serious threat to sustainable development and protection of environment. To address the emerging water crisis, the Dublin Statement set out recommendations for action at local, national, and international levels based on four guiding principles: "Fresh water is finite and vulnerable (…)"; "Water development and management should be based on a participatory approach (…)"; "Women play a pivotal role as providers and users of water (…)."; Water has an economic value (…) (ICWE 1992, p. 4).

The Dublin Principles led the foundation for the concept of "integrated water resources management" (IWRM) that offers solutions to the water crisis by linking the social and economic development with protection of natural ecosystems (ICWE 1992).

Consequently, water and its governance have gained increased attention as a policy concern resulting in numerous initiatives to support integrated water management for sustainable development.

At the UN Conference on Environment and Development held in 1992 in Rio de Janeiro, world leaders adopted the Agenda 21, a global action plan for sustainable development. The Chapter 18 of the Agenda 21 set a framework for integrated water management with the overall objective to "make certain that adequate supplies of water of good quality are maintained for the entire population of this planet, while preserving the hydrological, biological and chemical functions of ecosystems, adapting human activities within the capacity limits of nature and combating vectors of water-related diseases" (UN 1992).

In 1996, two important global organizations were established with the mission to coordinate implementation of integrated water resources management: the World Water Council (WWC) and the Global Water Partnership

(GWP). While the WWC focuses on awareness promotion and building of political commitment (WWC 2021), concentrates the GWP on the implementation of IWRM concepts at the operational levels (GWP 2021).

In 1997 the WWC launched the first World Water Forum (WWF) bringing together participants from all levels and areas, including politics, academia, civil society and the private sector around the world to collaborate and mobilize action on critical water issues. Since 1997, the WWF takes place regularly every three years.

The next significant milestones in the global water discourse placed the United Nations (UN). In 2000 the organization launched the Millennium Development Goals (MDGs) for the year 2015 addressing water aspects in the Target C of the Goal 7: "Ensuring environmental sustainability". The target aimed to reduce until 2015 by half the proportion of worlds population without access to drinking water and sanitation (United Nations, 2000). In 2015, the UN recognized water as an integral part of all human development and ecosystem needs by formulating the Sustainable Development Goal 6 (SDG 6): "To ensure availability and sustainable management of water and sanitation for all by 2030" (UN 2015).

The progress toward SDG 6 was recognized by the UN in 2020 as being "way off track" (UN-Water 2020). In response to this concern, *The SDG 6 Global Acceleration Framework* was launched to speed up the progress (UN-Water 2020).

The water governance and the evolving challenge of water scarcity were intensively discussed in an academic and political context after the meeting in Mar del Plata in 1977. However, it has been the only United Nations Water Conference until today. Indeed, the first conference since 1977 is scheduled for 2023 in New York (UN-Water 2020).

1.3 Paradigm Shift in Water Management

The milestones of the global development discourse on water mark a paradigm shift in the way to manage and think about water. Until the 1970s, the field of water management was known as "water resources development" with the guiding paradigm that "water is a resource to be exploited" (Savenije and van der Zaag 2008; Savenije et al. 2014). The main challenges of water management were to enforce water supplies. Consequently, the efforts to administer water and access aimed at developing technologies and large-scale, physical infrastructures, such as dams and reservoirs. After the conference in Mar del Plata in the 1980s, the term "water resources development" was progressively replaced by the term "water resources management" (Savenije et al. 2014). The change of terms reflects the paradigm shift from the notion "water is a resource to be exploited" to the recognition that "water can be overexploited" and becomes scarce. Accordingly, the challenges of water management changed from enforcing water supplies to avoiding overexploitation and establishing a balance between water availability and demand to fulfill all needs of humans and ecosystems.

To tackle the new challenges, in the 1990s emerged the concept of "integrated water resources management" (IWRM). The IWRM concept offers solutions to the current water crisis by promoting the coordinated development and management of water. This is ensured through linking water with other vital resources such as land, considering the entire hydrological water cycle together with human interventions, and promoting a balance between water use for socio-economic well-being and for maintaining the ecosystem functions (Global Water Partnership, 2000).

The changing challenges of water management are also reflected in the internationally agreed global goals on water. The Target C of the Millennium Development Goal 7 aimed to reduce until 2015 by half the proportion of the worlds population without access to drinking water and sanitation (United Nations 2000). While MDG 7, Target C, focused only on sanitation and drinking water, the 2030 Agenda for Sustainable Development adopted by the United Nations in 2015 formulated the Sustainable Development Goal 6 "Clean Water and Sanitation" with eight global targets covering the complete water cycle: From sanitation and drinking water supply (Targets 6.1 and 6.2), water quality (6.3), water-use efficiency and scarcity (6.4), IWRM (6.5) to the protection of ecosystems (6.6) with two additional targets on capacity-building (6.a) and participation (6.b) (United Nations 2015).

Figure 1 summarizes the above-described milestones of addressing the water crisis in the development discourse and the parallel paradigm shift in the way in which to manage and think about water.

1.4 Water Scarcity is Local

Since 2015, the World Economic Forum has regularly listed water crisis as one of the five main threats to livelihoods in the current times in its annual Global Risks Report (WEF 2020).

Although water crisis is identified as a high global risk, water scarcity is a local phenomenon. At the global level, the "planetary boundaries" for freshwater availability have not been crossed (Steffen et al. 2015) and on annual basis sufficient

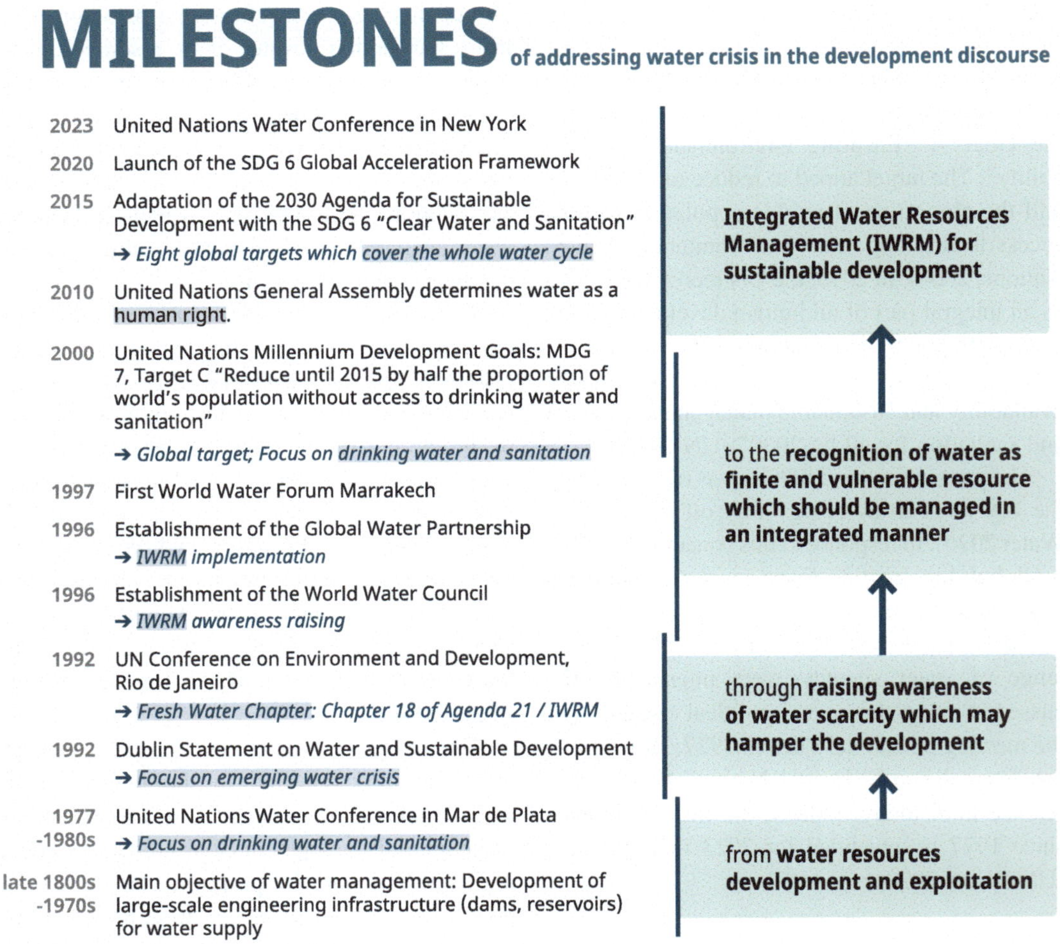

Fig. 1 Milestones of addressing water crisis in the development discourse

freshwater is available to meet the increasing water demand. However, the demand for and availability of water fluctuate in spatial and temporal variations as a result of natural hydrological variability (Mekonnen and Hoekstra 2016; Liu et al. 2017), socioeconomic changes, or as a consequence of inadequate planning and management approaches with lacking institutional capacity to provide the necessary water services and develop infrastructure to control the storage, distribution, and access to water resources (FAO 2012; Jaeger et al. 2013). This leads to water scarcity in several parts of the world during specific times of the year.

Hence, understanding water scarcity in the context of its regional dynamics is crucial for formulating adequate national policies and applying suitable local management strategies as well as technical and infrastructural solutions.

1.5 Research Demand at Local Scales: Water Scarcity Addressed in This Volume

In response to the evolving perception of water crisis in the development discourse the topic has also gained increasing attention in research. This is demonstrated in the rapidly growing number of publications after the 1992 conference in Dublin. For the overarching term of "water crisis" Google Scholar records 289.000 publications between 1900 and 1992 and 1.780.000 publications for the time between 1993 and 2020. However although the number of scientific publications has significantly increased since 1992, there remains an urgent demand to conduct extensive research on emerging water scarcity challenges at regional scales, document these studies, compare the results looking for the best solutions and disseminate them.

In response to this demand, this volume aims to bring regional cases to the center. It provides a balanced overview of contributions that reflect the interdisciplinary and context-specific characteristics of water scarcity and quality including its environmental, technological, and socioeconomic aspects. The book compromises thirteen studies from Africa, South-East Asia, Middle East, Chile, Russia, and Germany conducted by hydrologists, environmental and chemical engineers, agronomists, and social scientists. The thematic spectrum ranges from water security assessment framework, through water treatment technologies, wastewater reuse, adaptation to water scarcity in agriculture to the monitoring of sediment concentration in river ecosystems.

The following section includes a description of the main content of the chapters.

2 Main Content of the Chapters

2.1 Water Security Assessment Framework

The first chapter sets the scene and introduces a methodology for water security assessment. Achieving water security is a complex pathway that requires a holistic understanding and treatment of the interdisciplinary nature of water resources. This covers hydro-climatic, institutional, and socioeconomic aspects as well as spatial and temporal variability in water demand and availability determined by natural boundaries. Considering this interdisciplinary nature of water, Nguyen Mai Dang et al. in Chap. 2 argue that water security assessment has to be guided by an integrated approach applied not at local or national level but rather at the river basin scale. Accordingly, the authors develop an operational framework for assessing water security with five dimensions: water availability, water productivity, water-related disasters, watershed health, and water governance. Furthermore, they validate the framework through the application for the Red—Thai Binh River Basin in Vietnam.

2.2 Wastewater Treatment and Reuse

Safe wastewater reuse is recognized as a sustainable and efficient solution to reduce water stress (FAO 2020; Rodriguez et al. 2020). However, practical implementation of this method is still lacking due to both: sociocultural and political barriers as well as missing technologies to deal with the challenging treatment of emerging pollutants (UN 2018). The authors of the chapters from 3 to 8 investigate how to overcome these barriers.

Suha Al-Madbouh et al. in Chap. 3 address sociocultural dimensions of wastewater reuse for agriculture. Although the practice can significantly alleviate water stress especially in the arid and semi-arid zones, its application often meets strong opposition from farmers (UN 2018). The authors explain that understanding farmers reasons for opposition is crucial for developing solutions addressing implementation barriers. Accordingly, they design a conceptual model for examining farmers' opinion of and willingness to reuse wastewater and apply it in the Wadi Al-Far`a Watershed in Palestine, one of the most important catchments and agricultural areas in the West Bank (EQA 2006).

Chapter 4 addresses technological challenges faced by the treatment of agro-industrial wastewater. Alcaraz-Gonzalez et al. describe the process of anaerobic digestion—a frequently applied biological method to treat agro-industrial effluents with the production of biogas as a by-product. Currently, the search for optimal operating conditions to enhance the production of biogas and ensure the best treatment quality lies at the center of research interests (Karlsson et al. 2014). The chapter contributes to this research by exploring how the latest instrumentation, control and automation approaches in wastewater systems can increase the efficiency of anaerobic digestion processes.

Chapter 5 addresses the water treatment challenges in the Rosetta Branch in the Nile Delta in Egypt. Several studies have documented poor water quality caused by the discharge from agricultural drainage transported to the Rosetta Branch through Tala Drain (Mostafa 2015; El-Sayed et al. 2020). The most applied wastewater treatment method in the Tala Drain is the treatment using dilution—a method that involves mixing freshwater with drainage water. A. A. Hassan et al. in Chap. 5 investigate how the dilution process affects water quality, identify quality gaps, and based on this knowledge they recommend additional water treatment methods.

While the previous chapters focus on agricultural wastewaters, chapters 6-8 bring municipal and industrial effluents to the center. Rögener in Chap. 6 provides a general overview of emerging challenges for the treatment of municipal and industrial wastewaters and describes the most applied treatment methods. Chaps. 7 and 8 trace these challenges from a regional perspective. In Chap. 7, Rögener offers a holistic comparison of the municipal wastewater treatment sectors in Germany and Russia comprising technological, institutional, and climatic factors. Based on the comparison results, the author indicates the *Membrane Bioreactor Technology* as an adequate method for both countries to tackle emerging municipal wastewater treatment challenges.

In Chap. 8 Eyüp Debik brings industrial pollutants to the center and examines the most adequate wastewater treatment methods for pollutions coming from the dominant industries in Turkey.

2.3 Evidence-Based Adaptation to Water Scarcity in Agriculture

As the largest water consumer accounting for 70% of withdrawals globally (FAO 2020) and for up to 90% in many developing countries (UN 2018), agriculture is especially vulnerable to water scarcity and droughts requiring urgent adaptation measures. Currently, thanks to research and practical experience a wide portfolio of adaptation strategies is available ranging from water harvesting, storage, and reuse through efficient irrigation technologies or increasing crop productivity. However, the selection of an adequate strategy from the portfolio always depends on local conditions and should be evidence-based. This includes a profound understanding of causes for water scarcity, adaptation capacities of local communities, and information about resources availability and demand. The authors in Chaps. 9–13 contribute to gathering this knowledge through five case studies from Ghana, Burkina Faso, Bangladesh, Chile, and Vietnam.

Abdul Rauf Zanya Salifu et al. in Chap. 9 investigate vulnerability and adaptation capacities to water scarcity of groundnut farmers in West Ghana depending on age, gender, economic status, and ethnic origins. The authors argue that understanding such socioeconomic determinants is crucial for selecting demand-oriented adaptation measures. They find that high land tenure insecurity and limited access to credit facilities faced by women and non-natives within the study area as well as a lack of knowledge about technological adaptation possibilities are the main barriers to water scarcity adaptation.

One of the main reasons for reduced water availability for agriculture in the coastal zones of Bangladesh is not the decreasing precipitation but rather the saltwater intrusion. Hence, Md Arman Habib et al. in Chap. 10 evaluate the progress of saltwater intrusion in the southwest coast of Bangladesh and its impact on freshwater availability for agricultural production. The chapter concludes by recommending integrated adaptation methods like the cultivation of saline-resistant crops combined with sullage water reuse and mulching technique for reducing the effects of soil salinity.

Mwenda Borona et al. in Chap. 11 investigate how staple cereals cultivated in the Sahel Region of Burkina Faso respond to climatic variability and changes in water availability at inter-annual and inter-decadal levels. The authors conclude that the study results are a crucial knowledge source for adapting to water scarcity through changes in the planting calendar and cultivation of water stress-resilient crops.

We close this section with two contributions dedicated to drought forecasting and monitoring. Drought events are a recurrent phenomenon not only in water scare arid and semi-arid climates but also in traditional water-rich tropical zones (Erfanian et al. 2017; FAO 2019) affecting more people globally than any other natural hazard (Wilhite and Glantz 1985; FAO nd). Agriculture as the main water user is the most strongly affected sector absorbing up to 80% of all direct impacts (FAO nd).

Experience shows that the impacts of drought can be mitigated through timely drought forecasting and pro-active drought management (Wilhite et al. 2014; FAO 2019). However, contrary to visible and immediate flood, drought is a creeping phenomenon resulting from complex interactions among meteorological anomalies, land surface processes, and human activities that make it very difficult to forecast (Andreu et al. 2015; Hao et al. 2018). Currently, the development of new methods and use of technological innovations, e.g., satellite imagery or land surface and crop model simulations for drought forecasting and monitoring lie in the center of drought research.

In line with the above, we selected two articles that contribute to this research focusing on two different climatic zones: one in the semi-arid Chile and another one in tropical Vietnam. Oertel et al. in Chap. 12 explore how results gained by standardized drought indices can be used for operational drought forecast in the Maipo River Basin in semi-arid Chile. Trung Tuan Luong et al. in Chap. 13 evaluate the application of remote sensing images for water balance monitoring in small reservoirs of Central Vietnam and describe how to use the water storage data to forecast water shortages for irrigated areas serviced by monitored reservoirs.

2.4 River Ecosystems: Monitoring of Sediment Concentration and Transport Using Satellite-Based Products

While previous chapters discuss the water availability challenges for agriculture and human use, the last chapter brings the river ecosystems to the center and covers the topic of sediments concentration and transport. Suspended sediments in rivers are an important parameter for sustainable river management. However, traditional in situ measurement is very challenging in terms of time, cost, and spatial coverage. New technologies such as satellite-based products

Table 1 Main content of the Chaps. 2–14

Thematic block	Content	Spatial scale
Water security	*Chapter* 2: Development of a holistic water security assessment framework	*River Basin*: The Red—Thai Binh River Basin, Vietnam
Wastewater treatment and reuse for agriculture	*Chapter* 3: Development of a conceptual model to assess farmers perceptions of wastewater reuse	*River Basin*: Wadi Al-Far`a Watershed, Palestine
	Chapter 4: Maximize biogas production in anaerobic degradation processes applying instrumentation, control, and automation approaches	General overview
	Chapter 5: Assessing the impacts of wastewater treatment using dilution on water quality; development of a water quality assessment framework	*River Basin*: Rosetta Branch and Tala Drain in the Nile Delta, Egypt
Municipal and industrial wastewater treatment and reuse	*Chapter* 6: Overview of upcoming challenges for industrial and municipal wastewater treatment	General overview
	Chapter 7: Municipal wastewater treatment and evaluation of *Membrane Bioreactor Technology* as a treatment method to remove microplastics, micropollutants, and multiresistant germs	*Country level*: Russia, Germany
	Chapter 8: Identifying the most adequate wastewater treatment methods for pollutions coming from the dominant industries in Turkey: the textile, poultry slaughterhouses, and olive processing industries	*Country level*: Turkey
Water scarcity and agriculture	*Chapter* 9: Exploring interlinkages between socioeconomic characteristics of farmers and their vulnerabilities and adaptation capacities to water scarcity	*Local level*: Lawra and Nandom Districts, Ghana
	Chapter 10: Evaluating the impacts of salinity driven freshwater scarcity on agricultural production and livelihoods	*Local level*: Tala and Shyamnagar Upazilas of Satkhira district in the southwest part of Bangladesh
	Chapter 11: Investigating how different staple cereals respond to climate variability at an inter-annual and inter-decadal level	*Local level*: Cassou, Ziro Province in Southern Burkina Faso
	Chapter 12: Using multivariate standardized drought indices to identify drought events	*River Basin*: Maipo River Basin in semi-arid Chile
	Chapter 13: Using remote sensing images to estimate water balance in irrigation reservoirs and assess water availability for agriculture	*Local level*: Central Vietnam
River ecosystems and sediment concentration	*Chapter* 14: Monitoring sediment transport and concentration using satellite-based products	*River Basin*: Lower Ganges River in Bangladesh

offer a range of alternative methods to collect this data. The last chapter of the book investigates the applicability of open-access Landsat Enhanced Thematic Mapper (ETM+) images to investigate the sediment concentration in the *Lower Ganges (Padma) River*.

2.5 Key Recommendations for Addressing Water Scarcity

Water scarcity is not only a natural phenomenon beyond human agency. Hydrological and water conflicts studies show that water scarcity is mostly caused by a combination of high hydrological variability, a lack of investment in water infrastructure, and weak water governance (Mehta 2003; FAO 2012; Pahl-Wostl et al. 2016; UN 2018, 2019).

In this context, there is a range of opportunities to successfully address water scarcity but it requires adequate knowledge, investments, and cooperation. With the WSCC conference series and this volume, we aim to create a platform for cross-sectoral cooperation and knowledge exchange on water secure societies. In the concluding chapter, we bring together key recommendations for addressing water scarcity designed by around 200 experts from around the world who contribute to the dialogue of WSCC and are confronted with water challenges in their daily working lives.

Table 1 summarizes the main content of the thematic Chaps. 2–14 presented in this volume.

References

Andreu J, Solera A, Haro-Monteagudo D, Paredes-Arquiola J, Van Lanen HAJ, Drought: Research and Science-Policy Interfacing, London 2015. ISBN: 978–1–138–02779–4

Burek P, Satoh Y, Fischer G, Kahil M.T, Scherzer A, Tramberend S, Nava L.F, Wada Y, Eisner S, Flörke M, Hanasaki N, Magnuszewski P, Cosgrove B, Wiberg D (2016) Water Futures and Solution—Fast Track Initiative (Final Report). IIASA Working Paper WP 16–006. http://pure.iiasa.ac.at/13008/

El Sayed SM, Hegab MH, Mola HRA, Ahmed NM, Goher ME (2020) An integrated water quality assessment of Damietta and Rosetta branches (Nile River, Egypt) using chemical and biological indices. Environ Monit Assess 192(4):228. https://doi.org/10.1007/s10661-020-8195-4

Environmental Quality Authority—EQA (2006) The Far`a and Jerash integrated watershed management project. Baseline study report for Wadi Al-Far`a, EQA

Erfanian A, Wang G, Fomenko L (2016) Unprecedented drought over tropical South America in 2016: significantly under-predicted by tropical SST. Sci Rep 7:5811, 2017. https://doi.org/10.1038/s41598-017-05373-2, http://www.fao.org/3/a-i7584e.pdf

FAO—Food and Agriculture Organization of the United Nations (nd) Drought and agriculture. http://www.fao.org/land-water/world-water-day-2021/drought/en/. Accessed 09th Mar 2021

FAO (2012) Coping with water scarcity. An action framework for agriculture and food security. http://www.fao.org/3/i3015e/i3015e.pdf

FAO (2019) Proactive Approaches to Drought Preparedness. Where are we now and where do we go from here? http://www.fao.org/3/ca5794en/ca5794en.pdf

FAO (2020) The State of Food and Agriculture 2020. Overcoming water challenges in agriculture, Rome. https://doi.org/10.4060/cb1447en

Global Water Partnership (2000) Integrated Water Resources Management. TAC Background Papers No. 4. Stockholm, Sweden https://www.gwp.org/en/learn/KNOWLEDGE_RESOURCES/Global_Resources/background-papers/

Global Water Partnership (2021) About. Available at: https://www.gwp.org/en/

Hao Z, Singh VP, Xia Y (2018) Seasonal drought prediction: advances, challenges, and future prospects. Rev Geophys 56:108–141. https://doi.org/10.1002/2016RG000549

ICWE (1992) The Dublin statement on water and sustainable development. Dublin: International Conference on Water and the Environment (ICWE). https://www.wmo.int/pages/prog/hwrp/documents/english/icwedece.html

IPCC (2014) Climate Change: Synthesis Report. Contribution of Working Groups I, II and III to the Fifth Assessment Report of the Intergovernmental Panel on Climate Change [Core Writing Team, Pachauri RK, Meyer LA (eds)]. https://www.ipcc.ch/report/ar5/syr/

Jaeger WK, Plantinga AJ, Chang HK, Dello G, Grant D, Hulse J, McDonnell J, Lancaster S, Moradkhani H, Morzillo AT, Mote P, Nolin A, Santelmann M, Wu J (2013) Toward a formal definition of water scarcity in natural-human systems. Water Res Res 49:4506–4517. https://doi.org/10.1002/wrcr.20249

Karlsson A, Björn A, Yekta SS, Svensson Bo H (2014) Improvement of the biogas production process. https://www.diva-portal.org/smash/get/diva2:776575/FULLTEXT01.pdf

Liu J, Yang H, Gosling SN, Kummu M, Flörke M, Pfister S, Hanasaki N, Wada Y, Zhang X, Zheng C, Alcamo J, Oki T (2017) Water scarcity assessments in the past, present and future. Earth's Futu 5(6):545–559. https://doi.org/10.1002/2016EF000518

Mehta L (2003) Contexts and constructions of water scarcity. Econ Pol Wkly 38(48):5066–5072

Mekonnen MM, Hoekstra AY (2016) Four billion people facing severe water scarcity. Sci Adv 2(2):e1500323. https://doi.org/10.1126/sciadv.1500323

Mostafa MK (2015) Impact of improving water quality at the tala drain on the rosetta branch water quality. J Environ Prot 6:1149–1157. https://doi.org/10.4236/jep.2015.610102

OECD (Organisation for Economic Co-operation and Development) (2012) Environmental outlook to 2050: the consequences of inaction. OECD Publishing, Paris. https://www.oecd.org/g20/topics/energy-environment-green-growth/oecdenvironmentaloutlookto2050the-consequencesofinaction.htm
Rodriguez DJ, Serrano HA, Delgado A, Nolasco D, Saltiel G (2020) From waste to resource. Shifting paradigms for smarter wastewater interventions in Latin America and the Caribbean. https://www.unwater.org/world-bank-launch-report-on-the-value-of-wastewater/
Pahl-Wostl C, Gupta J, Bhaduri A (2016) Water security: a popular but contested concept. In: Pahl-Wostl C, Gupta J, Bhaduri A (eds) Handbook on water security. ISBN: 9781782548003
Savenije HHG, Van der Zaag P (2008) Integrated water resources management: concepts and issues. Phys Chem Earth 33(2008):290–297. https://doi.org/10.1016/j.pce.2008.02.003
Savenije HHG, Hoekstra AY, Van der Zaag P (2014) Evolving water science in the Anthropocene. In: Hydrol Earth Syst Sci 18(2014):319–332. https://doi.org/10.5194/hess-18-319-2014
Schlosser CA, Strzepek K, Gao X, Fant C, Blanc É, Paltsev S, Jacoby H, Reilly J, Gueneau A (2014) The future of global water stress: an integrated assessment. Earth's Futur 2:341–361. https://doi.org/10.1002/2014EF000238
Schmandt J, North GR, Ward GH (2013) How Sustainable are Engineered Rivers in Arid Lands?, J. sustain. dev. energy water environ syst 1(2):78-93. https://doi.org/10.13044/j.sdewes.2013.01.0006
Steffen W, Richardson K, Rockström J, Cornell SE, Fetzer I, Bennett EM, Biggs R, Carpenter SR, de Vries W, de Wit CA, Folke C, Gerten D, Heinke J, Mace GM, Persson LM, Ramanathan V, Reyers B, Sörlin S (2015) Planetary boundaries. Guiding human development on a changing planet. Science 347:6223, 1259855. https://science.sciencemag.org/content/347/6223/1259855
Terje T, Coopey R (2010) Rivers and society: from early civilizations to modern times, series II, vol II. IB Tauris, London, New York
United Nations Water (nd) Water scarcity. https://www.unwater.org/water-facts/scarcity/. Accessed 09th Mar 2020
United Nations—UN (1977) Report of the United Nations Water Conference Mar del Plata, 14–25 Mar 1977. https://digitallibrary.un.org/record/724642.
UN (1992) United Nations Conference on Environment & Development Rio de Janerio, Brazil, 3 to 14 June 1992 AGENDA 21, Available at: https://sustainabledevelopment.un.org/content/documents/Agenda21.pdf
United Nations (2000) Millennium development goals. https://www.un.org/millenniumgoals/
UN (2010) The human right to water and sanitation. Resolution adopted by the general assembly on 28, United Nations General Assembly Document A/RES/64/292. New York: United Nations. Available at: https://undocs.org/A/RES/64/292
UN-Water (2013) Water Security & the Global Water Agenda: a UN-Water Analytical Brief. https://www.unwater.org/app/uploads/2017/05/analytical_brief_oct2013_web.pdf
UN (2015) Transforming our world: the 2030 Agenda for sustainable development, 21 Oct 2015, A/RES/70/1. https://www.refworld.org/docid/57b6e3e44.html
UN (2018) Sustainable Development Goal 6. Synthesis Report on Water and Sanitation. https://www.unwater.org/publication_categories/sdg-6-synthesis-report-2018-on-water-and-sanitation/
UN (2019) The Sustainable Development Goals Report. https://unstats.un.org/sdgs/report/2019/The-Sustainable-Development-Goals-Report-2019.pdf
UN World Water Development Report (2020) (WWDR). Water and Climate Change. Paris, UNESCO. https://www.unwater.org/
UN-Water (2020) United Nations launches framework to speed up progress on water and sanitation goal, 9 July 2020. https://www.unwater.org/un-water-launch-the-sdg-6-global-acceleration-framework/
Wada Y, de Graaf IEM, van Beek LPH (2016) High-resolution modeling of human and climate impacts on global water resources. J Adv Model Earth Syst 8:735–763
Wilhite DA, Glantz M (1985) Understanding the drought phenomenon: the role of definitions. Water Int 10:111–120
Wilhite DA, Sivakumar MV, Pulwarty R (2014) Managing drought risk in a changing climate: the role of national drought policy. Weather Clim Extrem 3:4–13. https://doi.org/10.1016/j.wace.2014.01.002
World Economic Forum—WEF (2020) The global risks report. https://www.weforum.org/reports/the-global-risks-report-2020
World Water Council (2021) About us. https://www.worldwatercouncil.org/en

Water Security Assessment for the Red River Basin, Vietnam

Nguyen Mai Dang, Vu Thanh Tu, Mukand Babel, Victor Shinde, and Devesh Sharma

Abstract

Water scarcity is becoming one of the major concerns for water management on global and catchment scales. This requirs efficient and sustainable management practicies. As water availability varies depending on local hydro-climatic and socio-economic conditions, sustainable water management has to rely on profound information about water security on local sacle and consider the interdisciplinary character of the resource. Hence, this paper aims to develop a general framework for water security assessment at local scales with five dimensions: water availability, water productivity, water-related disaster, watershed health, and water governance. Furthermore, the framework is validated by employing a case study design using the Red River Basin in Vietnam. The climate in Vietnam has tropical characteristics with uneven distribution of rainfall and fluctuated flows across all parts of the territory with significant differences in water availability along the country. In this context, the assessment of water security at local scale is essential for efficient distribution and decision-making. The results indicate that the Red River had a moderate level of water security during a period of 5 years from 2010 to 2015 but with significant spatial differences according to water availability and quality. The average ratio of water use over water availability varies in sub-basins such as in Da, Lo, and Thao Rivers in a range of 0.02–0.04, while those in Red River Delta and Thai Binh River Delta are 0.44 and 0.25, respectively. The average ratio of industrial revenue and agricultural revenue per water supply in overall is 40 USD/m^3 and 2.37 USD/m^3, respectively. It can be explained that the rainfall patterns change in both space and time, and the catastrophic floods on the Red River Basin have occurred frequently with severe impact on socio-economic activities leading to the differences of water availability. We suggest to develop a new master plan for the management of water resources in the Red River Basin based on the concept of Integrated Water Resource Management and considering spatial and temporal varieties at water availability.

N. M. Dang (✉) · V. T. Tu
Thuyloi University (TLU), 175 Tay Son, Dong Da, Hanoi, Vietnam
e-mail: dang@tlu.edu.vn; dang@wru.vn

M. Babel · V. Shinde
Asian Institute of Technology (AIT), Km 42 Paholyothin Road, Klong Luang, Pathumthani 12120, Thailand

D. Sharma
Central University of Rajasthan, Kishangarh, Ajmer, India

L. Ribbe et al. (eds.), *Towards Water Secure Societies*,
https://doi.org/10.1007/978-3-030-50653-7_2

Keywords

Water security index · Water available · Water productivity · Water-related disasters · Water health · Water governance

1 Introduction

Water is a vital source of life and always plays a very important role in sustainability, socio-economic development, and diversity of ecosystems (Grey and Sadoff 2007). Water resources are crucial for economic development, poverty reduction, social welfare, and environmental protection. However, due to the negative effects of climate change, extreme weather can lead to an exacerbation of adverse impact on people's health and productivity. Water quality is emerging concern for the future due to the growth of population and environmental degradation by rapid economic development. Therefore, ensuring sufficient and safe water supplies for inhabitants is becoming increasingly challenging. The numbers of solutions have been employed in recent decades to raise public awareness of the significance of water resources and better management of water resources. An integrated watershed management is a key initiative to overcome the problems regarding water availability, water demand, and water quality (Siegfried et al. 2014). To address this massive challenge, the international community needs to look at water issues more comprehensively in terms of water security and makes efforts to better identify as well as assess the linkages to other sectors and issues such as equity (The World Bank 2016).

According to the definition of International Hydrological Programme of UNESCO in 2012, the intimidation of water security is not only caused by floods, droughts, and pollution to human societies, but also includes the impacts of insufficient water supply, in both quantity and quality, for food and energy production, for domestic and industrial purposes, and for sustaining ecosystem productivity.

The Asian Water Development Outlook (AWDO 2013) provides a robust, pragmatic, and readily understood framework for assessing water security. The report proposed five key indicators, which denote the inherent tensions among the different water uses as water resources come under increasing stress. These, when aggregated, provide an indicator of national and regional water securities. The indicators (referred to as dimensions) had seen as the means of measuring the outcomes of integrated water resource management. Based on that, Bangladesh evaluated the overall water security status in their country through five individual indices including Household Water Security Index, Economic Water Security Index, Urban Water Security Index, Environmental Water Security Index, and Resilience to Water-Related Disasters Index. When we consider the necessary water security index for economic development of Bangladesh, it can be seen that water security indexes for food production, industry, and energy are 4.89, 5.56, and 3.78, respectively, on a scale of 1–10. Therefore, necessary water security index is 14.22 out of 30 for the economic development of Bangladesh, which gives us an Economic Water Security Index value of 3 on a scale of 5. For urban water security index, the first sub-index value is 1 in a scale of 5 as only 20% (i.e., less than 90%) urban people get safe water supply in Bangladesh. As only 17% (less than 60%) urban people get the facility of wastewater treatment in Bangladesh the second sub-index value also becomes 1. Moreover, considering the property loss due to flood or disasters, which is $127 for an urban area of Bangladesh, the third sub-index becomes 4. Averaging all of these three values and also applying some adjustment factors, the Urban Water Security Index for Bangladesh is 1 in a scale of 5. Based on human interventions and its possible threat on floodplain and disturbance of natural flow due to various infrastructural developments on floodplain, the Environmental Water Security Index for Bangladesh becomes 1. Furthermore, based on population density and rate of population growth, rate of poverty, telecommunication, and literacy, the value of Resilience to Water-Related Disasters Index is also 1 for Bangladesh on a scale of 5.

The addressing water security in 5-year plan 2016–2020 in the People Republic of China points out the current situation and recommendation for increasing water security in the next few years. Water used for agriculture is the biggest, accounting for 65% of freshwater supplied for all of water uses; however, the water productivity is quite low at 3.6 USD/m^3 (Asian Development Bank 2016; Yong 2015). The pollution of rivers is a big problem due to weak management. The non-point pollution from agriculture and sediment loads is damaging the health of rivers and adding stress on the availability of water resources. The sewer and drainage pipes are under-developed. From 2008 to 2010, over 60% of the 351 surveyed cities suffered from urban flooding. Every year, water-related disasters affect 280 million people and cause a loss of 38 billion USD.

The concept of water security index introduced by Asian Development Bank (ADB) & Asian Water Development Outlook was accepted in Bhutan in a meeting of the Technical Advisory Committee in December 2014. The indicators were adapted to Bhutan conditions in a national IWRM workshop held in March 2015. The TAC endorsed the results of the workshop in May 2015. The BWSI system has been developed to accommodate information collected at district level. Currently, data is not available at the district level for many indicators. For enabling proper functionality of the system, the available national-level data have been deduced for the districts, which is most likely different from the actual level. User discretion is advised for the use of such district-level information for the year 2015. However, designated agencies are expected to gradually over time improve reliability of the information for the indicators identified against them.

In the framework of water security assessment of Canada, 2015, an assessment and management of water security involves four core elements: (1) Evaluate the current status of water quality and quantity; (2) Determine thresholds above which water is insecure; (3) Assess risk, taking into account stressors like development and climate change; and (4) Integrate monitoring and assessment results in decision-making and policies. The Water Security Assessment Framework tools have been developed for small communities to assess and manage water security in their watershed, but many of the concepts and issues are applicable at larger scales (Bakker and Allen 2015).

Tuan (2011) indicated that Vietnam has been affected by five factors that can influence its water security as follows: (1) Vietnam has about 208 rivers, but more than 60% of flows from outside Vietnam. (2) The annual rainfall is high, but is unevenly allocated in terms of space and time; the surface flows change clearly by season, accounting for 75–85% in the rainy season. (3) Vietnam is one of the top countries, which is the most affected by climate change and global warming. (4) Due to the increasing pressure of population, industrial growth, urbanization, food demand, and the reduction of land and upstream forests, water sources are being exploited more and more. It is hard to control the decrease in water quality; many industrial parks, residential areas, and hospitals do not have good wastewater treatment systems. (5) The demand for water rises while the volume and quality of water go down further is a big challenge for Vietnam's water security.

The Red–Thai Binh River System is a transboundary river basin, flowing through three countries including Vietnam, China, and Laos, with a total area of about 169,000 km^2. In the territory of Vietnam as can be seen in Fig. 1, the area is around 87,840 km^2, accounting for 51.3% of total area. In the territory of Vietnam, the Red River was formed by three major tributaries including the Da, Lo, and Thao Rivers. The Thai Binh River also was formed by three major tributaries including Cau, Thuong, and Luc Nam Rivers. The Red and Thai Binh Rivers are connected to each other by Duong and Luoc Rivers. In this paper, the water security assessment was done for the Red–Thai Binh Rivers in the Vietnam part only.

Fig. 1 The Red–Thai Binh River Basin in the territory of Vietnam (from left to right and up to down: Da sub-basin, Thao sub-basin, Lo sub-basin, Red River Delta, and Upper Thai Binh sub-basin)

Due to the climate change, the Red River was ranked in top river basins, which are strongly affected in terms of not only the submergence but also the hydrological extremes, which can pose a burden to local people as well as the government AECOM Asia Co. Ltd. and GFD (2012). Thus, in order to improve the understanding and awareness of water resources affecting socio-economic conditions in the area, water security assessment should be done for the Red River. That provides a better concept of integrated water management that balances resource protection and resource use. It takes a wider awareness of all water availability and demands in both quality and quantity, water-related disasters, the aquatic ecosystem health, and adaptive governance. In the second section, the methodology is outlined, and the third section results on water security are presented.

2 Methodology

This study has the objectives to apply common methodology for water security assessment. A general framework (Table 1) is developed as a composite index, with seven indicators and five dimensions including water availability, water productivity, water-related disaster, watershed health, and water governance. The indicators will be done in two ways: (1) through an extensive literature review of past and on-going water security studies/projects, and (2) by consulting and interacting with local water resource managers, decision-makers, and other experts through workshops. Hence, the quantifiable indicators were established for evaluating the dimensions including availability index, quality index, water-induced disaster index, environment

Table 1 The dimensions and indicators for water security assessment

Dimension	Parameter	Indicator	Data
Water availability	Sustainable basin exploitation	Water stress	Water use (MCM)
			Average annual rainfall (mm)
			Drainage area (m^2)
			Water availability (MCM)
			Water use/water availability
Water productivity	Economic value of water	Industry water productivity factor	Industrial revenue ($US) (GPP of each province in basin)
			Industrial water use (MCM)
			Industrial revenue/industrial water demand (US$/$m^3$)
		Agriculture water productivity factor	Agricultural revenue ($US) (GPP of each province in basin)
			Agricultural water use (MCM)
			Agricultural revenue/agricultural water demand (US$/$m^3$)
Water-related disasters	Drought factor	Ratio of the area with water-saving irrigation to the total area of arable land	Area of irrigation (ha)
			Area of arable (ha)
			Area of irrigation/area of arable
	Flood factor	Flood damage	Flood loss (million US$)
			Flood-affected people (persons)
			Total population (persons)
			Flood-affected people/total population %
Watershed health	River health	Surface water quality factor	BOD_5 (mg/l)
			BOD_5 permission(mg/l)
			BOD_5 concentration/permissible limit
Water governance	Overall management of the water sector	Institution factor	Questionnaire
	Potential to adapt to future changes	Adaptability factor	Questionnaire

protection index, public consciousness index, governance index, and adaptability index.

3 Data Collection and Discussions

3.1 Data Collection

In general, most of the province has not an abundant data due to some reasons such as data were not recorded properly, continuously, and in detail. The data mostly were collected from the kinds of reports, projects, research, and journals that published in the reliable sources such as the Department of Water Resources Management of Vietnam, Department of Agriculture and Rural Development, Department of Natural Resources and Environment, People's Committee of Province, Statistical Offices, and Center of Meteo-Hydrological Stations. Field survey and Questionnaire are also implemented as shown in Table 2. Most of the data are collected from the typical reports such as the summary of flood prevention activities and planning in each year, from 2010 to 2015, the Planning and Management of Water Resources in 21 provinces, Meteo-Hydrological data, and Statistical Yearbook.

3.2 Water Availability

In the period of 2010–2015, the average of total rainfall in the Red River Basin is 1800 mm, the water availability is ranging from 82 to 118 billion m^3, of which 61.1% of water produced in the territory of Vietnam. Due to the terrain and uneven distribution of rainfall, the flow in different parts of the basin also varies. The water availability in Da River accounts for 42% of the total water in the whole basin. The Thao River Basin covers an area approximately as big as the Da River Basin but has the smallest flow, accounting for 19%. The Lo River Basin has the smallest area but has the second most significant flow after the Da River, accounting for 25.4%.

Within the Red–Thai Binh River Basin, there are four largest reservoirs such as Hoa Binh, Son La, and Lai Chau on the Da River with the total volume of 13.7 billion m^3, Thac Ba reservoir on Chay River and Tuyen Quang reservoir on Lo River with the total volume of 2.1 billion m^3 and 1.7 billion m^3, respectively. Most of them have important roles in water supply for hydropower, agricultural activities, drinking, and flood control.

With the population growth of around 1.3% per year and the increasing water demand for

Table 2 The questionnaire for the assessment of water governance and adaptation

Questions	Yes	No
Governance index		
1. Does the basin have an Integrated Water Resources Management (IWRM) plan?		
2. Does the basin have a River Basin Organization (RBO)?		
3. Is there a provision to incentivize water conservation and/or water source protection?		
4. Is public opinion sought when developing water-related plans?		
5. Is there a mechanism to monitor pollution offences?		
Adaptability index		
1. Is there a centralized database of water-related information?		
2. Is there an early warning system in place?		
3. Are future drivers of change (e.g., climate change) taken in consideration when developing long-term DWR master plans?		
4. Is there a mechanism for staff to upgrade knowledge?		
5. Is there a flexibility to change water allocation quota for different users, whenever there is need to do so?		
Additional comments ……………………………………………………………………		

Fig. 2 The ratio of water use per water available in sub-basins of Red–Thai Binh River Basin

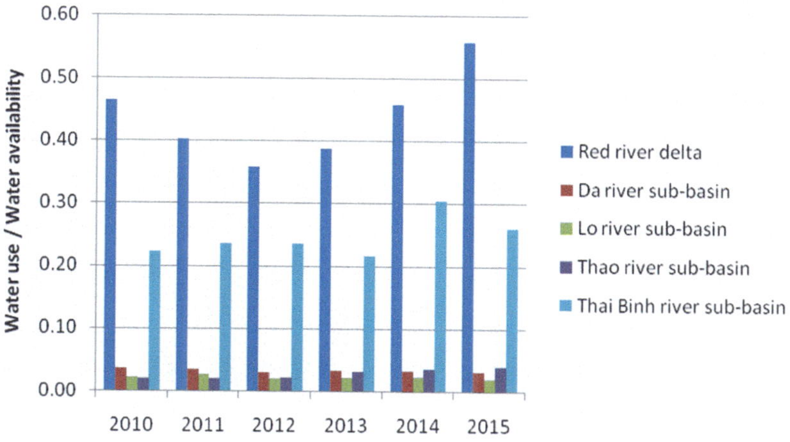

developing activities, the water demand in the whole basin has increased from 15 billion m^3 in 2010 to 20.9 billion m^3 in 2015. Therefore, the ratio of water used per water availability in the whole basin fluctuated from 0.17 to 0.22. In particular, in each sub-basin such as Da, Lo, and Thao River Basins, this ratio is lower than 0.05 (see Fig. 2) with the total water supply in these sub-basins ranging from 0.9 billion to 1.6 billion m^3 per year and water availability ranging from 15 billion to 40 billion m^3 per year. However, this number did not show the water quality of the water supply. This problem will be discussed in the next section.

3.3 Water Productivity

The Red–Thai Binh River Basin has an important role in the economic development of the North area of Vietnam in particular and of the whole country in general. In this basin, the main economic sector, which provides the basis for sustainable development, is agriculture. The water supply for agricultural activities accounts for 70% of the total water supply. However, due to the reduction of agricultural land because of industrial activities, urbanization, services, and the instability of plant structure as well as cultivated areas, yields, and productions have fluctuated. In 2015, the average yield of paddy rice ranged from 4.6 to 6.1 tons per ha.

Based on the data collected from 2010 to 2015, in general, the total water demand for agriculture and agricultural revenue was slightly increased. The ratio of agricultural revenue per water supply in 2010 was 1.95 USD/m^3 and increased to 2.61 USD/m^3 in 2015. Although the planted area was reduced, this result is mostly thanks to the improvement of plant variety, production method, and water supply as well as the improvement of irrigation system. Figure 3 shows that the ratio in Thai Binh River Basin was the lowest compared to other sub-basins. However, considering the revenue from agriculture, Thai Binh sub-basin is ranked second after Red River Delta with the total revenue of 3.5 billion USD/year in 2015. Partially, in the dry season, Thai Binh River sub-basin is usually affected by salty intrusion in the downstream area. This has caused difficulties for water supply activities within the downstream area; the agricultural productivities were inevitability affected.

The industrial development strategy of Vietnam has indicated that the extension of industrial zones together with more appropriate industrial distribution throughout the territory has been promoting the effectiveness of industrial zones and industrial clusters, accelerating the development of scientific and technological industries with high value, joining the global production network and so on. Thus, from 2010 to 2015, the development of industrial parks contributed significantly to industrial development and

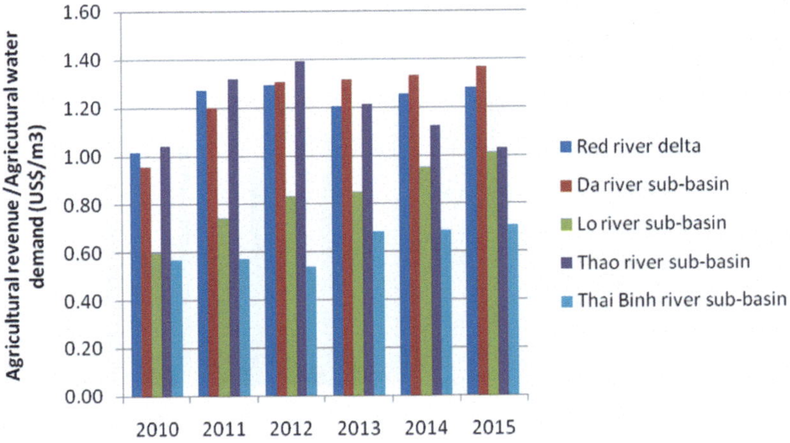

Fig. 3 The ratio of agricultural revenue per water demand in sub-basins of Red–Thai Binh River Basin

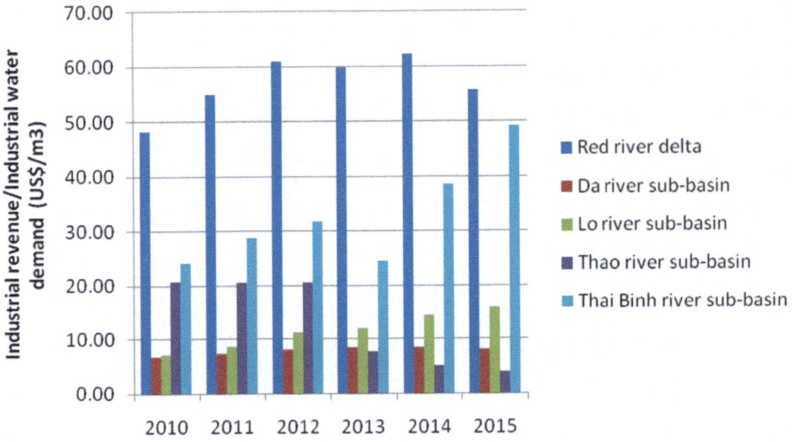

Fig. 4 The ratio of industrial revenue per water demand in sub-basins of Red–Thai Binh River Basin

economic restructuring. The formation and development of industrial zones have created more jobs for the workers, enhanced rapid technology transfer, produced many goods for domestic consumption, and exported products with high competitiveness and attracted considerable sources of investment for industrial development in particular and socio-economic development in general.

Many provinces and cities have been very successful in the development of industrial zones. Until now, the total number of key industrial zones in the Red–Thai Binh River Basin reached 64, in which 15 industrial parks located in Bac Ninh Provinces, 14 located in Hanoi, 11 located in Hai Duong, and others mostly located in the delta area of Red–Thai Binh River.

Even the water supply for industrial activities increased twice in the period of 2010–2015, accounting for 7–10% of the total supply; in overall, the industrial revenue was raised 2.5 times in 2015 compared to that in 2010 (see Fig. 4). However, due to the difference in inflation in this period, the ratio of industrial revenue per water supply fluctuated from 35 to 44 USD per m^3.

3.4 Water-Related Disasters

Flooding is the biggest disaster in Vietnam, especially in the North area where injuries and losses of life often reach high level. According to statistics in the last 40 years, the major cause of flooding in the Red–Thai Binh River Basin is a

Table 3 The elevation of low-laying area where usual submerged

No.	Province	Low-laying area (ha)	Elevation (m)	An usual submerged district
1	Vinh Phuc	8857	<+6.5	Vinh Tuong, Yen Lac, Me Linh
2	Bac Ninh	28,328	<+2.5	Thuan Thanh, Que Vo, Tien Du, Tu Son, Yen Phong, Luong Tai, Gia Binh,
3	Ha Noi	4000	<+5.0	Thanh Tri, Tu Liem, Gia Lam, Dong Anh, Soc Son,
		28,757	<+2.5	Ung Hoa, Phu Xuyen, My Duc, Chuong My
4	Hung Yen	9500	<+2.0	An Thi, Kim Dong, Tien Lu, Khoai Chau, Yen My,
5	Hai Duong	33,815	<+1.5	Thanh Ha, Tu Ky, Ninh Giang,
6	Hai Phong	10,857	<+0.9	Vinh Bao, Tien Lang, An Lao, Kien Thuy,
7	Thai Binh	20,034	<+1.0	Quynh Phu, Kien Xuong, Vu Thu, Hung Ha
8	Ha Nam	5250	≤+1.5	Binh Luc, Thanh Liem, Ly Nhan
9	Nam Dinh	16,275	<+0.8	Y Yen, Vu Ban, Hai Hau, Nghia Hung, Giao Thuy
10	Ninh Binh	16,231	<+0.8	Nho Quan, Gia Vien, Yen Khanh, Kim Son
	Total	181,904		

combination of weather patterns causing heavy rainfall and flood. Typical combinations are (1) tropical convergence zone with cyclones combined with high pressure in the Pacific; (2) tropical convergence cyclones combined with cold air; (3) tropical depression associated with the operation of high pressure in the Pacific; and (4) the hot low convergence zone associated with high pressure in the Pacific.

The seven largest floods occurred in the Red–Thai Binh River Basin in 1968, 1969, 1970, 1971, 1983, 1986, 1996, and the peak flooding level reached 12.0 m, which is higher than the Alarm Water level III 0.5 m (Hanoi hydrological-station). A flood occurred in August in 1971 that made the Red River dyke broken, affecting 100,000 people. The flood in 1971 was listed as one of the biggest floods in twentieth century.

The depth, duration, and area of inundation depend on the total rainfall and drainage conditions of the low-lying regions. Table 3 shows some regions in the Red–Thai Binh River Basin which are usually to be submerged if heavy rainfall occurs. However, the type of land used in these areas mostly is cultivated land.

The flooding in the Red–Thai Binh River is characterized by large volume, wide area, strong intensity, and duration (see in Table 4). It does not only hinder economic activities in tangible term but also causes difficulties to the local people in both tangible and intangible terms. Based on the results of flied, survey and data

Table 4 The flooding area corresponding to the level of total rainfall

Province	The submerged corresponding to the level of total rainfall (ha)		
	100–200 mm	200–300 mm	>300 mm
Vinh Phuc	7360	10,500	19,500
Bac Ninh	7102	14,290	28,328
Ha Noi	22,076	38,694	57,539
Hung Yen	7620	15,800	25,700
Hai Duong	10,990	39,600	47,000
Hai Phong	10,857	13,550	>13,550
Thai Binh	26,712	43,810	64,800
Ha Nam	9950	12,292	13,133
Nam Dinh	26,941	57,635	64,526
Ninh Binh	12,088	21,705	37,478
Total	141,696	267,876	371,554

Fig. 5 The total flood loss (million USD) due to floods in sub-basins in recent years

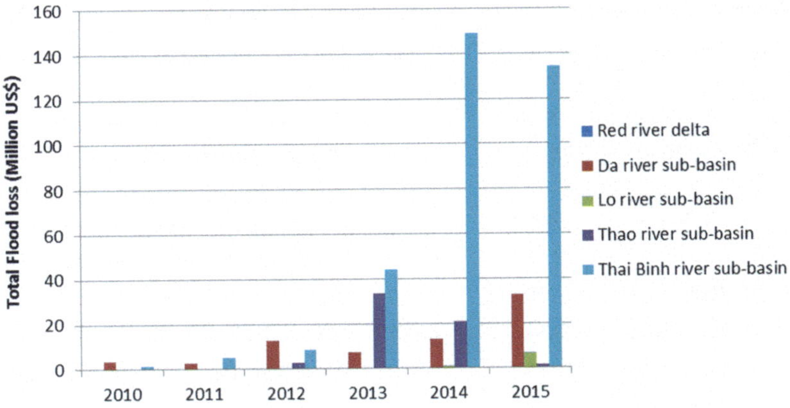

Fig. 6 The number of affected people due to floods in sub-basins in recent years

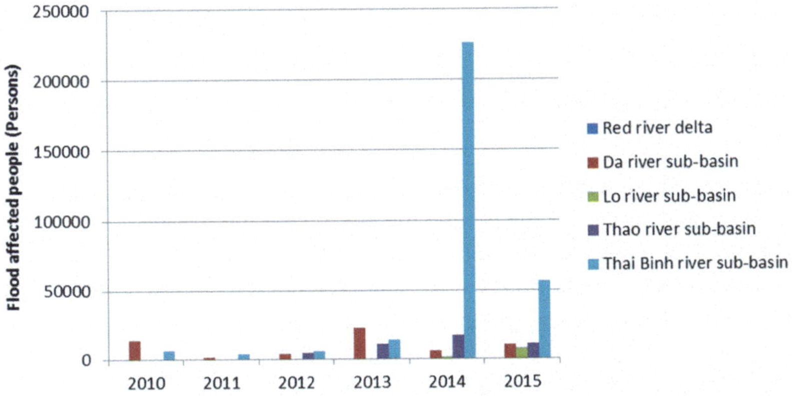

collection, heavy rainfall, and big flood did not occur much in the period of 2010–2015, and then the total losses or damages caused by floods were not high except for the Thai Binh River Basin. In both 2014 and 2015, the Thai Binh sub-basin had been directly affected by the tropical storms and typhoons such as Rammasun and Kalmaegi occurred in 2014, Kujira and Mujigae in 2015, all of which caused big losses and damages in this sub-basin as can be seen in Figs. 5 and 6. Although not affected by heavy rainfalls and big floods, the Red River Basin has faced the problems of drought, which has been increasing in recent years.

Drought is also a natural disaster with having a negative effect on humans. Although there is no fixed definition of this term, the main cause of drought is attributed to the lack of precipitation for a long period, usually several months or longer. The continuous occurrence of drought in the Red River Basin in recent years is mainly due to the following reasons: (1) a progression of more complicated annual weather. The rainy season started later and finished earlier; the annual precipitation in the basin has reduced compared to the average of many other years together with the increase of evaporation. (2) The growth of population has fostered the development of the economic sectors and the complexity of water uses, while the improvement and development of irrigation system is slow and a lacking of regulation system. The irrigation systems were mostly built many years ago that have become deteriorated and did not meet the water demand in the current period. (3) The regulations of water use in some reservoirs such as Hoa Binh and Thac Ba in an early stage of the dry season were not suitable for water demand in the downstream. (4) The forecasting, especially long-term forecasting is limited, causes more difficulties for the assessment and planning. (5) The intake water of irrigation systems located

Fig. 7 The ratio of irrigated area per arable area in sub-basins in 2010

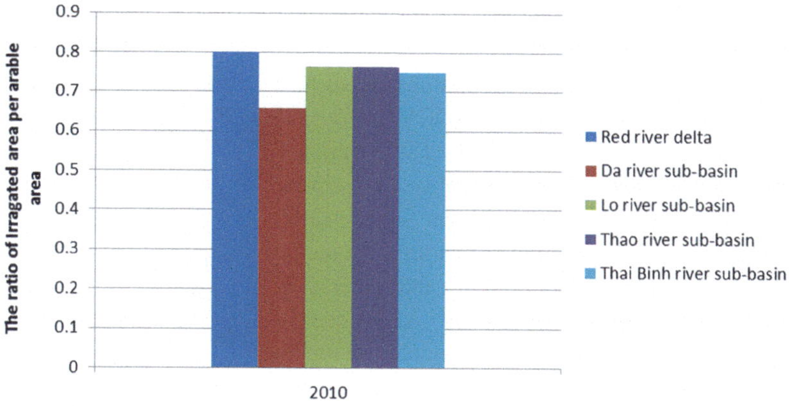

in China territory, upstream of Thao River, led to the low-flow to Vietnam in the dry season.

According to the survey results, in recent years, in the mountainous provinces drought occurred but not severed, scattered, and only in some specific areas. Although in the dry season water for the Red–Thai Binh River Delta is regulated from four major reservoirs such as Son La, Hoa Binh, Thac Ba, and Tuyen Quang, the phenomenon of the lack of water occurs every year. According to the review of the Institute of Water Resources Planning of Vietnam, annually about 60,000–120,000 ha (accounting for 10–20%) of winter-spring paddy-rice area suffer from drought or have difficulty in taking water. It can be mentioned that drought occurred in dry season 1999, 2003, 2004, 2006, and 2009.

The rate of irrigated area over irrigable area in Red River Delta is largest among five sub-basins as shown in Fig. 7 and accounted for 80%. Based on the statistical data in 2014, the total irrigated area in the basin accounts for 69% of the total cultivated area. The irrigated areas which are unsteady and/or without irrigation works cover 273,701 ha (of which the area without irrigation works is 25,843 ha). Some irrigation systems are degraded, and canal systems were filled up.

3.5 Water Health

The pollution, degradation, and depletion of water resources continue to increase, while the control mechanism of polluting sources and deforestation is ineffective. The pressure and challenges in protecting water resources in the Red–Thai Binh River Basins, especially in the Hai Phong, Quang Ninh, Thai Binh, Nam Dinh, and Ninh Binh Provinces, and water pollution continue to increase and become more common. The main reasons include the increase of industrial zones leading to an increase in wastewater, the overuse of fertilizers and chemicals in agricultural, forestry, and fishery production. Moreover, the widespread construction and operation of hydroelectric power plants in the upstream area lead to the loss of self-cleaning ability of rivers; unplanned tourism development, unprofessional operation, and dramatic increase in the amount of wastewater and solid wastes entering the natural environment resulting in severe consequences for the aquatic environment.

On the other hand, the slow progress of the implementation of pollution treatment activities causes high level of pollution in the Red–Thai Binh River. The monitoring and control of waste sources as well as the enforcement of the law on water environment protection from businesses and people are low. The pollution of the Red–Thai Binh River Basin causes serious damage to land and agricultural production, especially in the situation of rise of sea level and extreme weather conditions caused by climate change. Meanwhile, about 80% of water used in agriculture currently comes from the surface water of these river systems. Therefore, the strengthening of water quality management of these river systems is urgently needed and plays an important role in the sustainable development of the Northern provinces of Vietnam.

The national technical regulations on surface water quality of Vietnam (QCVN 08-MT: 2015/BTNMT as can be seen in Table 5) determine four levels of water quality as follows: (1) A1—Used for domestic water supply (after application of conventional treatment), aquatic animal and plant conservation, and other purposes such as A2, B1, and B2. (2) A2—Used for domestic water supply but suitable treatment technology or use purposes such as B1 and B2. (3) B1—For irrigation, irrigation or other uses with similar water quality requirements or uses such as B2; and (4) B2—Waterway and other purposes with low water quality requirements.

If considering water quality at B1 level, most of the water in monitoring positions within Da River Basin (Fig. 8), some positions along Nhue River (Fig. 9), and Duong River (Fig. 10) of the Red River Delta has higher quality than the permissible level. All of these water resources are near the industrial parks and trade villages with limited control and treatment of wastewater. It is requested that the management and improvement of treatment system as well as the production methods should be changed in order to meet the requirements of water quality.

Table 5 The threshold values of some indicators—QCVN 08-MT: 2015/BTNMT

Indicators (mg/l)	Threshold values			
	A		B	
	A_1	A_2	B_1	B_2
pH	6–8.5	6–8.5	5.5–9	5.5–9
BOD_5 (20 °C)	4	6	15	25
COD	10	15	30	50
DO	≥ 6	≥ 5	≥ 4	≥ 2
TSS	20	30	50	100

There are three main rivers in Lo River Basin such as Gam, Pho Day, and Lo Rivers with average value of water quality are computed and indicated in Table 6. These values are in thresholds of The national technical regulations on surface water quality of Vietnam (QCVN 08-MT:2015/BTNMT).

In the Thao River Basin, analytical samples are located in Ngoi Duong in Cam Duong ward and Ngoi Dum in Kim Tan ward, Lao Cai city. The results showed that in the dry season the pH

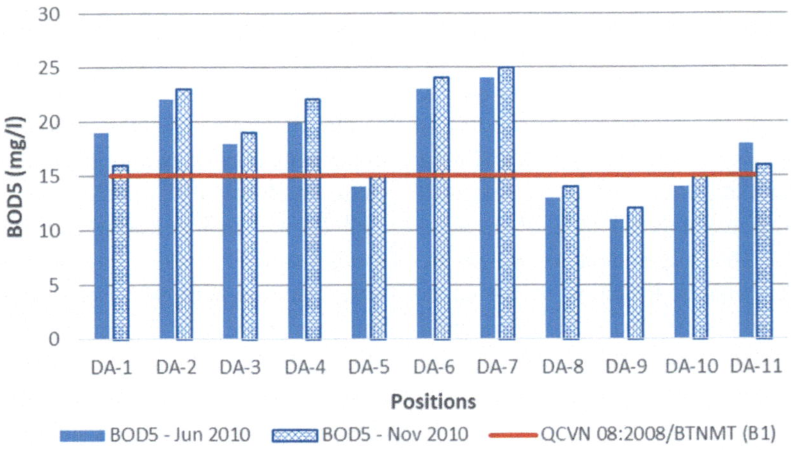

Fig. 8 The BOD_5 (mg/l) in some positions along Da River Basin in 2010. *Notes* DA-1: Cau Bang stream—Cao Phong. DA-2: Đat Tra lake—Cao Phong. DA-3: A river near Chuong three-way crossroads—Da Bac. DA-4: Nanh stream—Da Bac. DA-5: Da River—end of Ky Sơn District. DA-6: Stream in Mat, Dan Ha—Ky Son. DA-7: Cham Mat bridge—Hoa Binh city. DA-8: Hoa Binh bridge—Hoa Binh city. DA-9: Da River—beginning of Ky Son District. DA-10: Da River. DA-11: Hoa Binh reservoir

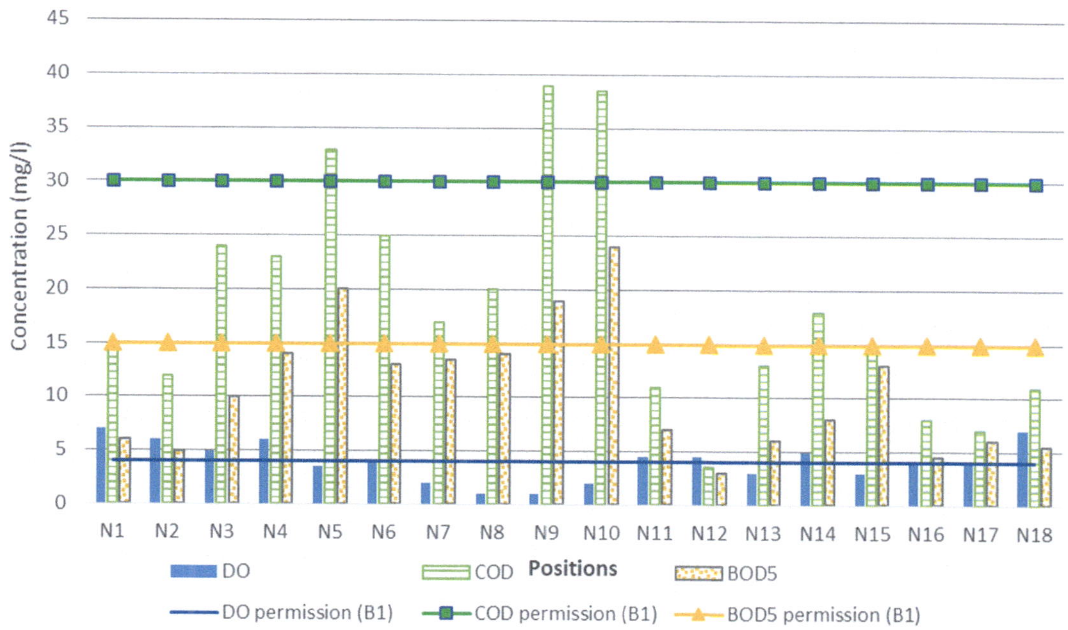

Fig. 9 The DO, COD, and BOD_5 (mg/l) in some positions along Nhue River, Red River Delta in 2011

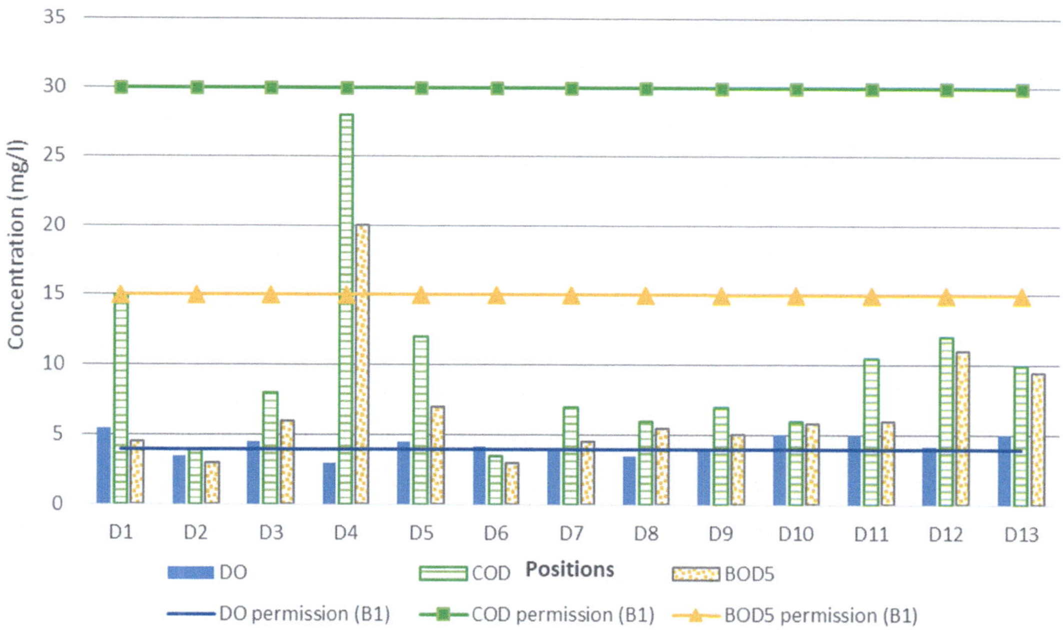

Fig. 10 The DO, COD, and BOD_5 (mg/l) in some positions along Duong River, Red River Delta in 2011

ranged from 6.2 to 6.5 mg/l, TSS in range of 2–3 mg/l, BOD_5 in range of 3–4.2 mg/l, COD in range of 6.4–9.3 mg/l; in the wet season the pH ranging from 6.4 to 6.5 mg/l, TSS in range of 4–6 mg/l, BOD_5 in range of 7–7.6 mg/l, and COD in range of 15–15.7 mg/l.

In the Red River Delta, the water quality samples are collected in Luoc, Tra Ly, and Hoa Rivers in 2011 with times of dry season (February), beginning of flood season (May), middle of flood season (August), and end of flood season (November). The data are drawn in Figs. 11 and 12, and all values indicated that the BOD_5 in observed rivers is accepted for B1 level which is for irrigation or other uses with similar water quality requirements or uses such as B2 (for waterway and other purposes with low water quality requirements).

3.6 Water Governances

In general, water resource management has been increasingly strengthened and attained significant progress in the structure of the water management sector from central level to local level. In particular, in 2014, the Inter-Ministerial Circular No. 50/2014/TTLT-BTNMT-BNV was issued to instruct the functions, tasks, powers, and organizational structure of the Natural Resources and Environment Department of People's committees of provinces and cities under central authority, Natural Resources and Environment Division under the People's Committees of districts, towns, and provincial cities. At the same time, the management institutions of water resources have also been constantly improved and strengthened to meet the management requirements in the new situation. Many legal documents on water resources have been promulgated, creating a legal corridor for the implementation of management function on water resources throughout the country. The organizational arrangements are also focused. The training and human resource development on water resources management is always paid attention, respect, and consistency at all levels.

The integrated water management has become a consistent view of Vietnam and has been reflected throughout the National Water Resources Strategy in 2006. In recently, Water Resource Law No. 17/2012/QH13 is the highest legal document on water resources, which have been promulgated.

The Law on Water Resources in 2012 replacing this document published in 1998 fundamentally overcomes the limitations of this document, which has been reviewed and summarized from the 13-year practical implementation. The objective of management aims at not only the quality and quantity of water but also at the management of the riverbank, as well as at the establishment of instruments and economic measures in water resource management. At the same time, following the Law on Water Resources in 1998, some contents were clearly defined and specified. In addition, many new regulations are added to meet the general requirements for water resource management in the current period and in the coming time.

In the coming time, Vietnamese Government will focus on the following tasks in order to

Table 6 The values of some water quality indicators in Tuyen Quang Province, Lo river basin

Indicators (mg/l)	Gam river		Pho Day river		Lo river	
	2006	2007	2006	2007	2006	2007
pH	7.39	7.25	7.51	7.2	7.4	7.2
DO	7	–	7.8	5.5	7.5	5.5
BOD_5	6.5	5.5	5.5	5.5	6.5	6.5
COD	15.7	10.5	8.1	10.1	13.9	13.1
TSS	–	5	–	11		62.5

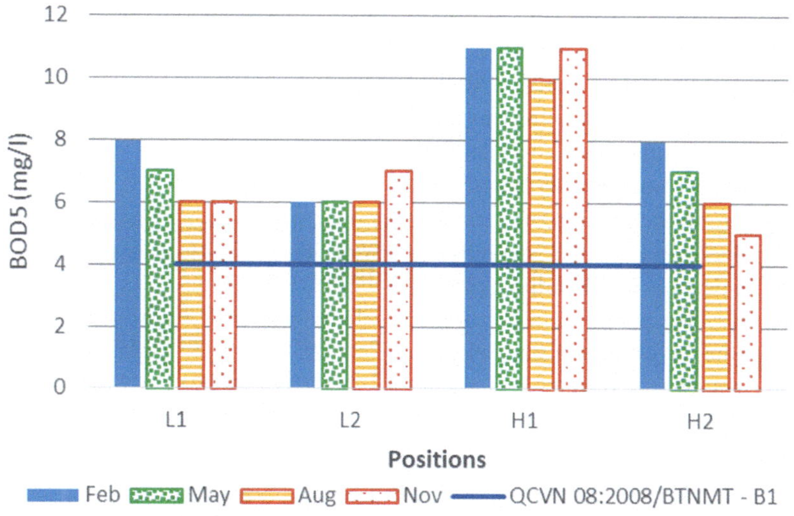

Fig. 11 The BOD_5 in some positions along Luoc and Hoa Rivers (Red River Delta) in 2011. *Notes* L1—Nam Lang sluice, Tan Tien ward—Luoc River. L2—Distributary of Luoc—Hoa River, An Khe ward. H1—Nghin bridge, Hoa River. H2—Thuy Truong ward, Hoa River

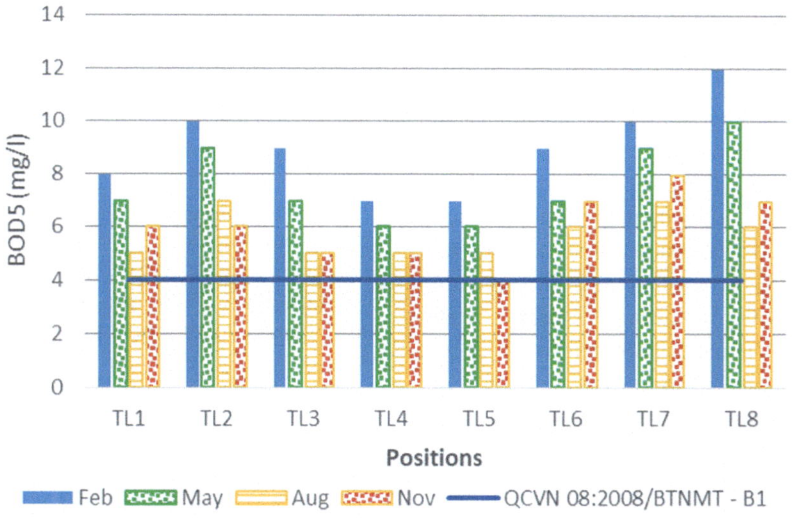

Fig. 12 The BOD_5 (mg/l) in some positions along Tra Ly River (Red River Delta) in 2011. *Notes* TL1—Tra Ly River, at the beginning of Thai Binh Province. TL2—Vuc sluice, Dong Phu ward. TL3—Hoa Binh bridge, Thai Binh city. TL4—At the intake water of Thai Binh irrigation limited company. TL5—Bo bridge, Thai Binh city. TL6—Vu Dong ward, Thai Binh city. TL7—Duc Duong sluice, Tra Giang ward. TL8—Tra Ly bridge

achieve successful and effective integrated water resource management: (1) Focusing on implementing the Law on Water Resources in 2012. (2) Focusing on improving the efficiency of reservoir operation under the inter-reservoir operating procedures; Supervising the exploitation and use of water resources, discharge of wastewater; Monitoring, evaluating, and forecasting the situation of degradation, exhaustion, and saltwater intrusion, especially transboundary river basin; Identifying the minimum flow in some major and important river basins. (3) Promoting and dissemination of the Law on water resources, education, and communication to raise awareness and responsibility, especially in water resource protection, pollution prevention, degradation, and depletion of water sources; At the same time, intensifying the inspection, examination, and handling of administrative violations in the domain of water resources. (4) Improving the quality, speeding up the process of appraisal of licensing dossiers, and at the same time intensifying the inspection and handling of the implementation of the permit's regulations. (5) Strengthening the investigation and assessment of water resources, developing water resource observation stations, creating the national information system and database on water resources; Step up the planning work, first of all carrying out the master plan on basic investigation of water resources. (6) Assessment of the impacts of climate change and sea level rise on water resources in the river delta, and proposing solutions to reduce the impact of climate change on water resources; Developing and adjusting plans and development plans and make them suitable with the fluctuations of water resources. (7) Strengthening the apparatus and the capacity of water resource management at all levels; Establishing the river basin management organizations and implementation of coordination and monitoring missions in some important river basins. (8) Promoting and strengthening cooperation with countries sharing water resources with Vietnam and other countries and international organizations in the field of water resources. (9) Implementation of the "National Action Plan for improving the effectiveness of management, protection and use of water resources".

In the Red–Thai Binh River Basin, the Red–Thai Binh River Basin Organization (RRBO) is a non-profit agency under the Ministry of Agriculture and Rural Development of Vietnam. RRBO was established on April 9, 2001, according to the Decision No. 39/2001 by the Minister of Agriculture and Rural Development of Vietnam. The main tasks of RRBO include (1) Preparing, submitting, and monitoring the implementation of the Red–Thai Binh River Basin plan to ensure consistent management of the river basin plan with the administrative boundary. (2) Coordinating with relevant Ministerial, sectorial, and local agencies in baseline investigation, inventory, and assessment of water resources of the Red–Thai Binh and in preparing, submitting, and monitoring the implementation of river basin plans for tributaries of the Red–Thai Binh system. (3) Proposing a resolution of disputes on water resources in the Red–Thai Binh River Basin.

Table 7 The summary of questionnaire for the assessment of water governance and adaptation

Question	Percentage of responds	
	Yes	No
Governance index		
1. Does the basin have an Integrated Water Resources Management (IWRM) plan?	0.79	0.21
2 Does the basin have a River/Sub-river Basin Organization (RBO)?	0.11	0.89
3. Is there a provision to incentivize water conservation and/or water source protection?	0.26	0.74
4. Is public opinion sought when developing water-related plans?	0.15	0.85
5. Is there a mechanism to monitor pollution offences?	0.23	0.77
Adaptability index		
1. Is there a centralized database of water-related information?	0.74	0.26
2. Is there an Early warning system in place?	0.11	0.89
3. Are future drivers of change (e.g., climate change) taken in consideration when developing long-term DWR master plans?	0.23	0.77
4. Is there a mechanism for staff to upgrade knowledge?	0.60	0.40
5. Is there a flexibility to change water allocation quota for different users, whenever there is need to do so?	0.21	0.79

In order to have better understanding the level of awareness and knowledge of local managers as well as the remaining issues in implementing river basin management, a questionnaire is designed as shown in Table 7. This survey is implemented in Department of Natural

Resources and Environment, Department of Agriculture and Rural Development of Provinces within the Red–Thai Binh River Basin. There are 45 responds but most of them do not have any additional comments or any suggestion.

Together with the questionnaire distributed to some departments of provinces such as Department of Agriculture and Rural development, Department of Natural resources and environment, District People's Committees, the results help for better understanding of water security in Red River Basin.

4 Conclusions

This paper presented the framework for water security assessment in the basin scale and applied for the Red–Thai Binh River Basin. The framework includes seven parameters, five dimensions including water availability, water productivity, water-related disaster, watershed health, and water governance that were established based on the extensive literature review and by consulting and interacting with local water resource managers, decision-makers, and other experts through workshops. The results indicate that the Red River has a moderate level of water security over the analysis period of 6 years from 2010 to 2015. The water availability is very different in various parts of basin; however, the water supply in this period met the demand for social-economic development. The average ratio of water uses per water availability fluctuates in sub-basins such as in Da, and Lo, Thao Rivers ranging from 0.02 to 0.04 while those in Red River Delta and Thai Binh River Delta are 0.44 and 0.25, respectively. The average ratio of industrial revenue and agricultural revenue per water supply in overall is 40 USD/m^3 and 2.37 USD/m^3, respectively. Because the rainfall patterns in the basin change in both space and time, the occurrences of catastrophic floods on the Red River are markedly divergent. Flooding in the delta has severe impact on socio-economic activities. Besides, in dry season, there is also the lack of water supply in cultivated areas in addition to strong tides and sea level rise leading to increased salt intrusion in the downstream. The water quality is still a problem in the numbers of river parts; the BOD_5 is higher 2–3 times compared to the safe level. This requires the integrated water resource management in Red–Thai Binh River Basin to improve the control, monitoring, and management capacity. We suggest to develop a new master plan for the management of water resources in the Red River Delta based on the concept of Integrated Water Resources Management and considering spatial and temporal varieties at water availability.

Acknowledgements The authors would like to thank the Asia-Pacific Network for Global Change Research (APN) for financial support to conduct this research. The authors also would like to thank departments belonging to Ministry of Natural Resources and Environment (MONRE), Ministry of Agricultural and Rural Development (MARD); Provincial Department of Agriculture and Rural Development (DARD), Provincial Department of Natural Resources and Environment (DONRE), Provincial People's Committee (PPC) for responding the questionnaire and support for data collection. Especially, many thanks to CUOMO FOUNDATION for financial supporting to authors to present this research at the Conference of Water Security and Climate Change in Cologne in September 2017.

References

AECOM Asia Co. Ltd. and GFD (2012) Project TA-7629 VIE: capacity building for river basin water resources planning in Vietnam. Available via DIALOG. http://gfd.com.vn/news.html. Assessed 24 July 2017

Asian Development Bank (2016) Addressing water security in the People's Republic of China: The 13th Five-Year Plan (2016–2020) and beyond. Asian Development Bank, Mandaluyong City, Philippines

Bakker K, Allen D (2015) Canadian water security assessment framework: tools for assessing water security and improving watershed governance. Canadian Water Network

Grey D, Sadoff CW (2007) Sink or swim? Water security for growth and development. Wat Policy 9(6):545–571

Siegfried D, Anil M, Christophe C, Gordon Y (2014) Hydrological sciences & water security: past, present and future. In: Proceedings of the 11th Kovacs Colloquium, UNESCO Publication, Paris, France, 16–17 June 2014. It is available in the Website: https://unesdoc.unesco.org/ark:/48223/pf0000228086, assessed on 30/07/2017

The Asian Water Development Outlook (AWDO) (2013) Measuring water security in Asia and Pacific. ISBN 978-92-9092-989-5
The World Bank (2016) water security as a top global risk - in search of more comprehensive approaches. It is available in the website: https://www.worldbank.org/, assessed on 15/07/2017
Tuan LA (2011) Water security under threat. Retrieved August 2011, from The Saigon Times. It is available in the website: http://english.thesaigontimes.vn/Home/business/other/18550/, assessed on 24 July 2017
Yong J (2015) China's water security: current status, emerging challenges and future prospects. Environ Sci Policy 54:106

Contextualizing Farmers' Perceptions of Agricultural Wastewater Reuse in Areas of Water Shortage and Contamination: An In-Depth Case Study of Wadi Al-Far'a Watershed—Palestine

Suha Al-Madbouh, Thomas Claßen, and Claudia Hornberg

Abstract

What shapes farmers perception of agricultural wastewater reuse (WWR) in Palestine under circumstances of water shortage, poor sanitation, and political uncertainty? The current study answers this question by applying the 'cultural approach' to situate farmers' perceptions of agricultural WWR within broader contextual frameworks and offer interpretations beyond individual attitudes provided solely by the psychological and psychometric approaches. It deploys qualitative case study research design and combines six qualitative and quantitative data collections. The study argues that a complex series of interrelated individual and contextual factors shape farmers' perceptions of agricultural WWR. While farmers' economic, agricultural, health, environmental, and political concerns related to agricultural WWR significantly influence their perceptions of using RW, farmers' psychological, social, and religious concerns were not significant. Contrary to what has been perceived by stakeholders in the Palestinian water and wastewater sector, the researchers conclude that the issue of social acceptance of agricultural WWR is not an obstacle to implementing agricultural WWR projects in Palestine. Institutional conservatism, which resists community participation, and weak institutional structure could be what initially contributed to farmers' opposition to past WWR projects. The researchers contribute both conceptually and methodologically to the existing knowledge. They offer a conceptual model for case study research examining farmers' perceptions of agricultural WWR and invent the concept of Case-Driven Significant Factor (CDSF) which promotes a new type of individual and contextual factors derived from the particularities of the case itself.

Keywords

Agricultural wastewater reuse · Farmers' perceptions · Social acceptance · Cultural approach · Case-specific factors · The West Bank-Palestine

S. Al-Madbouh (✉)
Institute for Technology and Resources Management in the Tropics and Subtropics (ITT)/TH-Köln, 50679 Cologne, Germany
e-mail: Suha.al-madbouh@th-koeln.de

T. Claßen
StadtLandschaft & Gesundheit, 33501 Bielefeld, Germany
e-mail: Thomas.Classen@lzg.nrw.de

C. Hornberg
Faculty of Health Sciences, Bielefeld University, 33501 Bielefeld, Germany
e-mail: claudia.hornberg@uni-bielefeld.de

L. Ribbe et al. (eds.), *Towards Water Secure Societies*,
https://doi.org/10.1007/978-3-030-50653-7_3

1 Introduction and Current State of Research

Over the last few decades, the growing problems of water shortage, loss of quality, and lack of alternative water sources became main motives for farmers to use wastewater for irrigation (Scott et al. 2004). Recently, millions of small-scale or commercial-scale farmers in urban and peri-urban areas of developing countries, including the MENA region, have resorted to untreated wastewater or wastewater-polluted water sources to irrigate high-value edible crops for urban markets (Qadir et al. 2010; Scott et al. 2004). The increasing dependence of the livelihood of farmers on these resources became the strongest socioeconomic factor pushing for wastewater reuse (WWR) practice (Scott et al. 2004).

Despite the global trend of prioritizing the use of treated wastewater, also called reclaimed water (RW), for agricultural irrigation in water management master plans, implementing agricultural WWR projects still receives rejection amongst the public and farming communities (Chenini 2011). Hence, the issue of social acceptance of WWR has been viewed as a major "obstacle"; yet, it is the key factor for the success of any WWR project regardless of economic, environmental, and engineering considerations (Beck 2009; Baumann 1983; Po et al. 2003; Wester et al. 2015).

The growing problem of water shortage is one of the main challenges the Palestinian agricultural sector is facing (Jayyousi 2003; Mizyed 2013; The World Bank 2009). Groundwater abstracted by the Palestinians had declined dramatically over the last decade due to the lowered water table, which is attributed mainly to Israeli over-extraction of groundwater (The World Bank 2009).[1] Furthermore, limited Palestinian access to surface and groundwater due to restraints imposed by the Israeli occupation creates a considerable gap between the actual water needs and the actual supply. These restraints include restricting the construction of new wells and limiting annual abstraction rates from existing wells (MA'AN and JVPC 2010; PLO Negotiation Affairs Department 2011; The World Bank 2004). Hence, the problem of water shortage, besides a lack of wastewater collection and disposal systems in most of the Palestinian rural areas that result in inappropriate disposal of untreated wastewater (Mizyed 2013), necessitates the implementation of wastewater treatment and reuse projects in Palestine. The use of RW, therefore, has a high priority in the Palestinian Water Policy (Abdo 2008).

However, opposition of Palestinian farming communities to agricultural WWR projects has been perceived as an "obstacle" to the implementation of these projects in the West Bank and the Gaza Strip, and leads to failure of these projects (Abdo 2008; Abu-Madi et al. 2008). For instance, in the West Bank, the community of Deir Debwan refused irrigating with RW sourced from the nearby al-Bireh wastewater treatment plant despite the high technical performance of the plant (Abu-Madi et al. 2008). In the Gaza Strip, farmers' opposition to agricultural WWR resulted in failure of WWR projects implemented in Gaza City and Jabalia village in the 1980s (Abdo 2008).

Yet, several Palestinian surveys showed high acceptance levels of agricultural WWR amongst Palestinian farmers and communities in the West Bank and the Gaza Strip (Abu-Madi et al. 2008; Al-Kharouf et al. 2008; Arafat 2012; Nassar et al. 2010; Qattawi 2008; Tubail et al. 2004). Even surveys conducted in Deir Debwan revealed high acceptance levels of using RW (87%, n = 320, by Abu-Madi et al. (2003) and 70%, n = 100, by Qattawi (2008)) despite the opposition of this community to agricultural WWR in practice.

Accordingly, there is a paradox between high acceptance levels of agricultural WWR amongst Palestinian communities and farmers, revealed by local surveys, and opposition of these communities to agricultural WWR projects when implemented on the ground. This paradox, however, is not an exception to findings of international surveys on social acceptance of

[1]The Israeli over-extraction by 389 MCM/per year from the West Bank Aquifer System is 80% more than the agreed Oslo II (in 1995) allocation of 483 MCM/per year (The World Bank 2009).

WWR. Different studies on public acceptance of potable and non-potable reuse schemes conducted in the USA, Australia, the UK, and China have shed light on the issue of "salience" to explain this contradiction. They argue that the mere approval of an idea (i.e. a general WWR option) implies neither approval of nor support for its implementation (i.e. a salient WWR option) (Chen et al. 2015; Marks 2004; Russell and Hampton 2006). This is because attitudes change and respondents become less supportive when WWR option becomes more concrete and tangible to them (i.e. more salient) (Bruvold 1988; Hartley 2003; Po et al. 2003).

Being unexceptional, the Palestinian paradox is still puzzling in the case where local WWR projects would contribute to alleviating the prevailing circumstances of water shortage, poor environmental and sanitation conditions, and threat of confiscation of Palestinian (uncultivated) agricultural lands by the Israeli occupation. This study, therefore, addresses the question of "What shapes farmers' perception of agricultural WWR when proposing a salient reuse option under circumstances of water shortage, poor environmental and sanitation conditions, and political uncertainty?". It focuses on examining the independent variable of "farmers' concern of consumer's perception of buying crops irrigated with RW". Thereby, it tests the hypothesis claiming the factor of 'farmers' worry about possible consumers' boycott of crops irrigated with RW' as a prominent factor inhibiting farmers' perceptions of WWR (Abu-Madi et al. 2003; Carr et al. 2011; Nassar et al. 2010; Po and Nancarrow 2004; Schaer-Barbosa et al. 2014).

The current research question is not only viewed as a perceptional and cognitive question but also a contextual socio-cultural one. It draws on the assumption that a complex and interwoven series of individual and contextual factors shapes farmers' perceptions of agricultural WWR, and the interactions and reciprocal relationships between these factors play a major role in shaping these perceptions. Hence, this study employs the theoretical framework of the "cultural approach" according to which farmers' perceptions of WWR are part of a broader pattern of social, cultural, institutional, and political contexts (Baumann 1983; Hartley 2003; Ormerod and Scott 2012; Russell and Hampton 2006; Russell and Lux 2009; Saad et al. 2017; Stenekes et al. 2006). It thus bridges the paradigmatic gaps of the psychological and psychometric approaches that exclusively employ the "yuck factor" (i.e. the psychological barrier of disgust) and "perceived risk", respectively, to explain public attitudes towards WWR (Beck 2009; Bruvold 1985; Ching 2010; Po et al. 2003; Schmidt 2008). While the cultural approach does not oppose the analysis of individuals' attitudes, it sees limited value in explaining them solely as an individual psychological matter (Russell and Lux 2009).

2 Research Design and Methodology

In order to provide an in-depth and context-based understanding of what shapes farmers' perceptions of agricultural WWR, a qualitative case study research design was used, specifically a within-case study design (i.e. a single case study with embedded units) (Yin 2003). This design allowed for offering rich descriptions and in-depth interpretations of local contexts and settings involved in shaping farmers' perceptions of agricultural WWR. Furthermore, it allowed for uncovering variation in farmers' perceptions and contexts shaping these perceptions between different places and performing comparison. Finally, it provided an opportunity to generate hypotheses by uncovering multitude variables and new causal relationships. On the other hand, the current cross-sectional case study research design had two main inherent limitations, including (1) limited empirical generalization of its findings to other cases, and (2) lack of evidence of causality between the dependent and independent variables.

2.1 Study Area

The Wadi al-Far'a watershed, located in the north-eastern region of the West Bank as part of

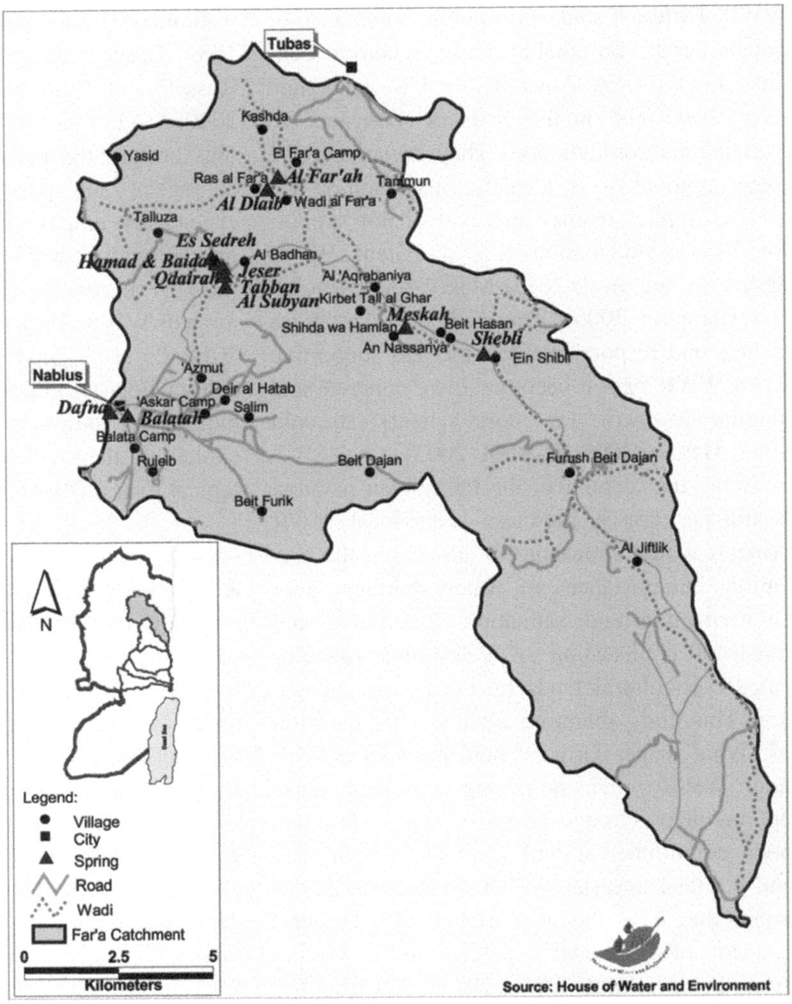

Fig. 1 A close-up geographic map of Wadi al-Far'a watershed (adopted from house of water and environment database)

the Lower Jordan Valley, was the "case" selected for investigation. The two villages of an-Nassariyya and Froush Beit Dajan, located in the central and lower parts of Wadi al-Far'a watershed, respectively, were selected as embedded cases within the watershed for the purpose of cross-village comparison. Wadi al-Far'a watershed is one of the most important catchments and agricultural areas in the West Bank with an area of 331 km^2 that accounts for 6% of the West Bank area (EQA 2006). It includes 20 communities within its borders, out of which 10 communities (around 20,000 inhabitants), including an-Nassariyya and Froush Beit Dajan (1585 and 1191 inhabitants, respectively) (EQA 2006; PCBS 2009), are concentrated around the mainstream in the watershed called al-Far'a Stream (EQA 2006), as shown in Fig. 1.

This stream receives water mainly from "al-Far'a Spring" and constitutes (in addition to irrigation wells) a main water supply system in the watershed (EQA 2006). A main source of pollution in the watershed is the discharge of untreated industrial and domestic wastewater, coming mainly from the nearby city of Nablus, into spring water in the watershed (EQA 2006; Thawabe 2006). The contaminated spring water is then used intensively by farmers for irrigation without any mitigation measures. This takes place particularly in the downstream areas of an-Nassariyya and Froush Beit Dajan, where water shortage is severe. This problem poses great health risks through direct exposure to the contaminated water or consuming crops irrigated with this water (EQA 2006; Thawabe 2006).

The case of the Wadi al-Far'a watershed was selected for two reasons: First, its *typicality* or *representativeness* to a broader set of cases in the Lower Jordan Valley and the West Bank and, second, the issue of *necessity*. The latter is reflected in (1) the ecological and agricultural significance of the watershed in the Palestinian context, and (2) the threatening circumstances under which farming communities in the watershed live and practice their agricultural activities including agricultural water shortage, environmental contamination, and political uncertainty. The Wadi al-Far'a watershed runs through "Area B" and "Area C" under partial (B) or complete (C) Israeli jurisdiction (EQA 2006; MA'AN and JVPC 2010).[2] In these areas, the Israeli policy and practices aim at uprooting Palestinian residents through confiscation of their agricultural lands, restriction on expansion of rural communities, and forced eviction or displacement (ARIJ 2006; PLO Negotiation Affairs Department 2011). An-Nassariyya and Froush Beit Dajan were selected based on variations in the above three contexts across the two villages.[3]

2.2 Data Collection

In 2008, 2009, and 2012, three field visits, with a total duration of 8 months, were conducted to collect data from the study area. A combination of six qualitative and quantitative data collection methods was used in each of the two cases of an-Nassariyya and Froush Beit Dajan, including (1) in-depth interviews, (2) semi-structured questionnaires, (3) direct structured observations, (4) direct semi-structured observations, (5) informal conversational interviews, and (6) analysis of primary sources and review of secondary data. A total of 26 interviews were conducted in Arabic language. These interviews included 16 in-depth open-ended interviews conducted with farmers (8 in an-Nassariyya and 8 in Froush Beit Dajan, including 4 women) and 10 informal conversational interviews conducted with gatekeepers and key informants (6 from an-Nassariyya and 4 from Froush Beit Dajan).

Farmers' interview protocol inquired mainly about farmers' concerns associated with agricultural WWR, including their health and environmental, economic, agricultural, marketing, political, social and cultural, religious, and psychological concerns. Farmers were also asked about prior knowledge and experience associated with agricultural WWR and their trust in Palestinian authorities and stakeholders involved in implementing future WWR projects. Since the current research question inquired about farmers' perceptions of "a salient WWR option", the participants were made aware that the study findings will inform the decision-making process whether agricultural WWR is a socially acceptable intervention in the Lower Jordan Valley (including the Wadi al-Far'a watershed) or not, for a real possibility of implementing future WWR projects in the area at later time. Each farmer interview lasted for 8½ hours on the average and was conducted within 3–4 visits in duration between 7 and 10 days.

Questionnaires were administered during farmers' interviews, and each consisted of 38 questions that were highly related to the content of the interview protocol. Using questionnaires intended at enhancing the validity of the

[2]According to the Oslo II Agreement (1995), most of the watershed area is still located in "Area B" or "Area C", particularly the central and lower parts of the watershed including an-Nassariyya and Froush Beit Dajan (EQA 2006; MA'AN and JVPC 2010).

[3]Researchers' observations in the Wadi al-Far'a watershed and their tracing of the three contexts of irrigation water shortage and contamination and political uncertainty in the watershed disclosed marked variations in these contexts across the central and lower parts of the watershed where An-Nassariyya and Froush Beit Dajan are located, respectively. That is, the problem of water shortage and degree of political uncertainty increases as we move from the upper to the central and lower parts of the watershed, while contamination of spring water with wastewater decreases as we move from the central to the lower parts (EQA 2006; MA'AN and JVPC 2010; Thawabe 2006).

qualitative evidence rather than providing statistical evidence. As for conversational interviews, they helped the researchers learn about social settings and local contexts of an-Nassariyya and Froush Beit Dajan, which added valuable depth to data collected in farmers' interviews. Each conversation lasted for 2½–3 h.

In addition to the interviews, 11 structured observations of farmers' agricultural activities, practices, and health and psychological behaviours associated with the contaminated spring water were carried out. Semi-structured observations were conducted to observe human activities associated with the land and water resources in the watershed in addition to ecological, health, and political conditions under which farmers live. A combination of four non-probability sampling techniques was used to identify the study participants, including a list of potential participants provided by the Palestinian Hydrology Group (a local non-governmental organization that has been working intensively with farmers in the watershed area), snowball sampling, purposive sampling, and theoretical sampling.

2.3 Data Analysis

"Framework Analysis" method was used to analyse the case study data. This method shares many of the features of "thematic analysis" that is based on identifying, coding, and categorizing key themes emerging from the data (Green and Thorogood 2004; Lacey and Luff 2009). Yet, the former is a more sophisticated method that moves beyond simply coding and categorizing the data to identifying and explaining relationships, and thus it provides thicker descriptions and richer interpretations (Green and Thorogood 2004). The researchers, additionally, extended the analysis process and applied a combination of qualitative (Framework Analysis) and quantitative analytic approaches.

Framework analysis consists of five main stages including familiarization, identifying a thematic framework, indexing, charting, and mapping and interpretation (Lacey and Luff 2009). In the familiarization stage, the researchers checked verbatim the handwritten transcripts of farmers' interviews against the digitally recorded interviews. They then read and re-read the transcripts to become familiar with the data. Afterwards, cross-referencing of interview data with field notes of observations and informal conversations was performed, and participants' names were anonymized and coded. Thematic analysis was then undertaken to identify a thematic framework (i.e. the initial coding scheme) based on a combination of issues emerging from the data and prior issues identified in the literature. This scheme was developed and refined using constant comparison as long as new ideas were detected in the data. In the indexing stage, the scheme served as the skeleton of the coding framework that was systematically applied to the whole data set through identifying pieces of text that exemplified the initial codes. Initially and newly emerged codes were clustered into 14 core categories based on linkages found between the codes. Each core category included codes (subcategories) related to the same theme.

After indexing, the coded data was displayed in charts using Excel spreadsheets. This charting process visualized the whole data set to allow comparison across and within interviews. It also allowed for within-village and cross-village analysis as well as gender-based and income-based comparisons. In a further step, the researchers identified four to six core categories of concerns influencing perception of WWR for each of the interviewed farmers based on farmer's high level of emphasis on these concerns. The categories were arranged in positions from "1" to "6". A category in position "1" was the one most emphasized by the farmer while that in position "5" and "6" was the one least emphasized by him/her or not identified as the main core category, respectively. Sometimes, a position was occupied by two categories or no category. Relative significance level was assessed for each of the main core categories by calculating the weighted average (WA) of positions of this category across all participants. Based on the WA, the significance level of the category was determined using a 4-point scale of significance including "highly significant" categories (WA = 1–3.99), "significant" categories (WA = 4–4.99), "slightly significant" categories (WA = 5–5.99), and "not significant" categories (WA = 6) (see Table 1, Sect. 3.2). Weighted averages were also calculated for different clusters of farmers to detect differences in significance levels of influencing factors across an-Nassariyya and Froush Beit Dajan, between male and female farmers, and between low-income and medium- to high-income farmers. In some cases, the significance level of a category was determined based on a combination of quantitative and qualitative assessments (i.e. WA combined with the researchers' assessment of farmers' emphasis on the category in comparison to their emphasis on other categories).

In the last analysis stage of mapping and interpretation, explanatory accounts were emphasized by searching for and explaining associations and patterns within the data. Maps and summary tables were used to visually display and search for these associations and patterns, develop and test interpretations, and draw conclusions. Conclusions were cross-checked and verified by revisiting and re-examining the data.

3 Results and Discussion: Offering a Conceptual Model for Hypothesis Generation

3.1 Farmers' Perceptions of Agricultural WWR: Cross-Village and Gender-Based Comparisons

In-depth interviews with farmers in An-Nassariyya and Froush Beit Dajan revealed positive perceptions and a high acceptance level of the salient agricultural WWR option proposed for the Wadi al-Far'a watershed area. It was found that 50% of interviewed farmers strongly accepted using RW for irrigation and 31.3% accepted using RW, while 12.5% accepted with hesitation and 6.2% refused using RW. Farmers' perceptions were not based on simple linear factors, but a wide range of complex and interrelated individual and contextual factors influencing each other.

In particular, contextual factors specific to the case of the Wadi al-Far'a watershed (i.e. contextual case-specific factors) played a highly significant role in shaping farmers' positive perceptions of agricultural WWR. These factors included irrigation water shortage, contamination of the spring water used for irrigation with WW, and political uncertainty represented in confiscation of agricultural lands and control over water resources by the Israeli occupation. Under these case-specific conditions, farmers living in the Wadi al-Far'a watershed area perceived agricultural WWR application as highly advantageous in (1) alleviating irrigation water shortage; (2) preventing contamination of spring water and adverse health, environmental, and economic impacts resulted from using the contaminated water for irrigation; (3) achieving economic benefits and financial savings for farmers; (4) reducing, or possibly ending, farmers' conflicts over water; and (5) protecting agricultural lands from confiscation and defending farmers' identity. Two statements in point from interviewed farmers: "*…Well, we are telling you that here we are hunting for water, in the sense of the word, and if we have reclaimed water, this will be a big achievement.*" (MAM7FAN q55), and "*….a person thinks to cultivate his land in any way, so what do you think if there is a source of reclaimed and safe water!*" (TH16FAN q183).

The findings on farmers' perceived advantages of WWR enhancing their perceptions of this application agree with findings of Palestinian, American, and Australian studies on social acceptance of potable and non-potable WWR including agricultural WWR. These studies revealed that social acceptance of WWR was likely to be higher, when water conservation and protection of the public health and the environment were clear benefits of WWR (Hartley 2003). They also indicated that individual concerns of costs and the benefits achieved by using RW played a significant role in their acceptance of WWR (Bruvold 1985; Po et al. 2003), particularly for non-potable WWR, again including agricultural WWR (Hurlimann

Table 1 Weighted averages and significance levels of core categories of factors shaping farmers' perceptions of agricultural wastewater reuse

Core category and related factors	Number of farmers emphasized on concerns	Frequency of a core category in positions from 1 to 5					Number of farmers did not emphasize on concerns	Weighted average (WA)	Significance
		1	2	3	4	5	6		
Economic concerns – Price of RW – Economic profits and financial savings achieved by irrigating with RW – Cost and efficiency of RW conveyance and distribution systems – Allocation mechanisms of RW amongst farmers	13	4	3	1	3	1	3	3.25	Highly significant
Concerns of water shortage – Farmers' concerns of adverse agricultural, economic, health, environmental, and social impacts of water shortage	10	5	3	0	1	1	6	3.5	Highly significant
Agricultural concerns – Agronomic quality of RW – Palestinian WWR guidelines of restricted irrigation	14	1	2	5	5	1	2	3.56	Highly significant
Health and environmental concerns – Potential health risks of agricultural WWR – Availability and enforcement of Palestinian WWR standards and guidelines – Environmental benefits of agricultural WWR	12	3	3	1	3	2	4	3.63	Highly significant
Knowledge and experience associated with agricultural WWR	11	1	3	3	0	4	5	4.13	Significant
Farmers' suffering from contaminated spring water	7	3	1	3	0	0	9	4.25	Significant
Political concerns – Practices of the Israeli occupation – Farmers' national affiliation	5	1	2	0	2	0	11	4.94	Significant
Loyalty to agricultural work and lands	5	1	1	1	1	1	11	5.06	Slightly significant
Issue of trust – Farmers' trust in the Palestinian authorities and stakeholders involved in implementing future WWR projects in the Wadi al-Far'a watershed – Farmers' trust in technology-based treatment of wastewater	3	0	1	1	1	0	13	5.44	Slightly significant
Psychological concerns	3	0	1	1	0	1	13	5.5	

(continued)

Table 1 (continued)

Core category and related factors	Number of farmers emphasized on concerns	Frequency of a core category in positions from 1 to 5					Number of farmers did not emphasize on concerns	Weighted average (WA)	Significance
		1	2	3	4	5	6		
– Feeling disgust from RW – Source of wastewater – Eating crops of RW cultivated by other farmers									Slightly significant
Social and cultural concerns – Farmers' concerns of the social point of view on agricultural WWR – Farmers' concerns of opinions of social figures and other farmers on agricultural WWR	1	1	0	0	0	0	15	5.69	Not significant[a]
Marketing concerns – Price of crops irrigated with RW – Farmers' concerns of consumers' perceptions of buying crops irrigated with RW – Farmers' concerns of adverse impacts of agricultural WWR application on internal and external markets of local agricultural crops	1	0	0	1	0	0	15	5.81	Slightly significant
Farmers' concerns about involving them in the decision-making process – Farmers' participation in this study as part of the decision-making process – Farmers' concerns of participating in formulating agricultural WWR standards	0	0	0	0	0	0	16	6	Slightly significant[b]
Religious concerns – Farmers' concerns of the religious point of view on agricultural WWR – Farmers' concerns of opinions of religious figures on agricultural WWR	0	0	0	0	0	0	16	6	Not significant

[a]Scale of significance: highly significant (WA = 1–3.99), significant (WA = 4–4.99), slightly significant (WA = 5–5.99), not significant (WA = 6)

[b]The category "social and "cultural concerns", WA between 5 and 5.99, was classified as "not significant" since farmers' emphasis on it was less than their emphasis on categories classified as "slightly significant". In contrast, the category of "farmers' concerns of involvement in the decision-making process" with (WA = 6) was classified as "slightly significant" category since farmers' emphasis on it was more than their emphasis on other categories classified as "not significant"

et al. 2005; Marks et al. 2003; Mizyed 2013). Furthermore, case-specific contexts of water scarcity and pollution with sewage discharges were key drivers for farming communities to accept agricultural WWR (Keremane and McKay 2006; Mooney and Stenekes 2008; Po et al. 2003; Schaer-Barbosa et al. 2014).

Despite farmers' perceived advantages of WWR, positive perceptions and acceptance of agricultural WWR were contingent upon meeting their expectations from future agricultural WWR projects about (1) providing them with RW with prices considerably cheaper than those of fresh water purchased from local agricultural wells, (2) securing a consistent and reliable supply of RW, (3) ensuring compliance of RW quality with agricultural and health standards of WWR, (4) applying clear and fair allocation mechanisms of RW amongst farmers, and (5) assuring farmers that using RW would not be at the expense of Palestinian rights to fresh water and farmers' rights to fresh water from their own agricultural wells.

Farmers who were hesitant about, or refused, using RW were sceptical about RW quality and potential adverse health effects and agricultural impacts associated with agricultural WWR application. None of them had prior knowledge or experience with agricultural WWR but was willing to run individual trail (or observe those of other farmers) before making a final decision. If these trails did not impose adverse health and agricultural impacts and achieved economic benefits, hesitant and opponent farmers would be encouraged to use RW. In agreement with these findings, Hurlimann (2007) and Russell and Lux (2006) revealed negative association between individuals' lack of knowledge about WWR and the perceived risks of WWR.

Cross-village and gender-based analyses revealed significant differences in farmers' perceptions of agricultural WWR, while income-based differences were not significant. Cross-village analysis revealed that farmers' acceptance level of agricultural WWR was higher in An-Nassariyya than in Froush Beit Dajan (100% accepted, 62.5% strongly accepted). This finding was mainly attributed to cross-village variation in degree of contamination of spring water used by farmers. Structured and semi-structured observations revealed that the quality of contaminated spring water in An-Nassariyya was more deteriorated than in Froush Beit Dajan. Hence, farmers in An-Nassariyya perceived quality of the contaminated water as closer to untreated wastewater while farmers in Froush Beit Dajan perceived it as closer to fresh water. This situation made the proposed option of irrigating with "treated wastewater" more familiar to farmers from An-Nassariyya, closer to their minds, and more compatible with their exigent need for an irrigation source (like RW) of "expectedly" a better quality than the deteriorated quality of spring water. An interviewed farmer from An-Nassariyya said: "*Using reclaimed water for irrigation is 'normal' as we use now untreated wastewater*" (YAK6FAN q44). On the other hand, farmers from Froush Beit Dajan suspected that quality of RW might not be better than that of spring water they used. The female farmer, SAS15FFBD, from Froush Beit Dajan said: "*I have clean [spring] water. What would force me to irrigate with reclaimed water while I do not know about its quality and if it is arable or not!*" (q49).

Gender-based comparison revealed that female farmers were more ambivalent about, and less supportive of, agricultural WWR than were male farmers (50% and 8.3%, respectively). This finding agrees with trends in the literature which shows that males tend to be predisposed more positively towards WWR (Dolnicar and Schäfer 2009; Marks 2004). Interviews revealed that (1) women's lack of knowledge and/or experience with agricultural WWR, (2) women's higher perceived health and agricultural risks of WWR, and (3) their concerns about the social point of view on agricultural WWR and criticism by community or family members tended to predispose them against WWR. A woman farmer from An-Nassariyya stated: "*I cannot use reclaimed water without the village*" (NS3FAN q 189). Also, FAJ13FFBD from Froush Beit Dajan expressed:

> Reclaimed water is socially unacceptable due to its source. People will think about the source of this water, and they will not be certain about its reclamation… . I will be concerned about their

acceptance, and it will make me hesitant about using reclaimed water. (q 162)

3.2 Understanding Farmers' Perceptions of a Salient Agricultural WWR Option

Farmers' perceptions are determined by a wide range of contextual and individual factors. The complexity of the interactions between these factors makes it hard to explain them separately. In fact, these interactions could be what initially shaped farmers' perceptions of agricultural WWR. In the case of the Wadi al-Far'a watershed, the overall interaction between the influencing factors tended to enhance farmers' perceptions of agricultural WWR while considering irrigation water shortage and contamination and political uncertainty. Furthermore, it was found that under these circumstances, four categories of factors (concerns) were "highly significant" in shaping farmers' perceptions of the salient agricultural WWR option proposed in the watershed. These categories included "farmers' economic concerns" followed by "farmers' concerns of water shortage", "farmers' agricultural concerns", and "farmers' health and environmental concerns", as indicated in Table 1. Additionally, three categories of factors had "significant" influence on shaping farmers' perceptions including "farmers' knowledge and experience associated with agricultural WWR" followed by "farmers' suffering from the contaminated spring water used for irrigation" and "farmers' political concerns and national affiliation". Section 3.3 explains findings on these categories.

Besides the independent variables being diverse and interrelated, their significance levels varied across place due to variations in contexts and varied more distinctly in accordance with variations in contexts than variations in farmers' socioeconomic characteristics (mainly gender and income). The researchers used this finding for inventing the concept of "Case-Driven Significant Factor (CDSF)" which offers a novel contribution to research on social acceptance of potable or non-potable WWR applications. This concept was developed for a particular kind of individual or contextual factor that has the following three CDSF-characteristics:

- It is specific to the case being investigated.
- It has significant influence on shaping perceptions of WWR.
- Variation in this factor between locations within the case under investigation results in a marked variation in the significance level of its influence on perceptions of WWR across these locations. The degree of variation in the factor determines the degree of variation in its significance level.

The researchers identified three CDSFs for the case of the Wadi al-Far'a watershed, conceptualized as CDSFs–Wadi al-Far'a watershed, including irrigation water shortage, contamination of irrigation water, and political uncertainly. These factors were case-specific, were amongst the most significant and significant factors shaping farmers' perceptions, and had marked variation in their significance levels across An-Nassariyya and Froush Beit Dajan, as indicated in Table 2. Section 3.3 will elaborate on CDSFs–Wadi al-Far'a watershed.

3.3 Model of Understanding Farmers' Perceptions of a Salient Agricultural WWR Option Proposed in Areas of Water Shortage and Contamination in Conflict Zones

Based on the study findings, the researchers offer a conceptual model for hypothesis formation and middle-range theory building that can be tested and further extended to broader cases of comparable conditions for generalization. The model, as shown in Fig. 2, brings together the study arguments about individual and contextual factors (and the main linkages between these

Table 2 Significance levels of core categories of factors shaping farmers' perceptions of agricultural wastewater reuse, including CDSFs–WFW[a] (case-driven significant factors–Wadi al-Far'a watershed and cross-village comparison in significance levels of these categories)

Level of significance	Categories of factors shaping farmers' perceptions of using reclaimed water for irrigation (16 farmers)	Categories of factors shaping perceptions of farmers from an-Nassariyya (8 farmers)	Categories of factors shaping perceptions of farmers from Froush Beit Dajan (8 farmers)
"Highly significant" categories	1 Economic concerns 2 Concerns of water shortage (CDSF-WFW[a]) 3 Agricultural concerns 4 Health and environmental concerns	1 Economic concerns 2 Suffering from contaminate spring water, Health, and environmental concerns 3 Agricultural concerns	1 Concerns of water shortage 2 Economic concerns 3 Agricultural concerns
"Significant" categories	5 Knowledge and experience associated with agricultural WWR 6 Suffering from contaminated spring water (CDSF-WFW) 7 Political concerns (CDSF-WFW) and national affiliation	4 Knowledge and experience associated with agricultural WWR 5 Concerns of water shortage, Loyalty to agricultural work and lands	4 Health and environmental concerns 5 Knowledge and experience associated with agricultural WWR 6 Political concerns and national affiliation
"Slightly significant" categories	8 Loyalty to agricultural work and lands 9 Issue of trust 10 Psychological concerns 11 Concerns of marketing crops of RW/consumers' perceptions of buying crops of RW 12 Involving farmers in the decision-making process	6 Political concerns and national affiliation 7 Issue of trust, Psychological concerns 8 Concerns of marketing crops of RW/consumers' perceptions of buying crops of RW 9 Involving farmers in the decision-making process	7 Suffering from contaminated spring water 8 Issue of trust, Loyalty to agricultural work and lands 9 Psychological concerns 10 Concerns of marketing crops of RW/consumers' perceptions of buying crops of RW 11 Involving farmers in the decision-making process
"Not significant" categories	13 Social and cultural concerns 14 Religious concerns	10 Social and cultural concerns 11 Religious concerns	12 Social and cultural concerns 13 Religious concerns

factors) shaping farmers' perceptions of a salient agricultural WWR option proposed in areas of agricultural water shortage and contamination in conflict zones, such as the Wadi al-Far'a watershed.

The model suggests that economic concerns are the most significant concerns shaping perceptions of farmers in the Wadi al-Far'a watershed towards agricultural WWR. Farmers in the watershed were highly concerned about providing them with RW at an affordable price that would be considerably cheaper than prices of fresh water purchased from local agricultural wells. They were willing to pay 0.07–0.46 USD/m^3 of RW (avg. 0.19 USD/m^3), which is about 43.8% of the average price (0.42 USD/m^3) of water purchased from agricultural wells in the watershed. Additionally, they were concerned about increasing their economic profits and financial savings when using RW through extending their agricultural lands, diversifying their crops, increasing crop yields, alleviating or preventing economic losses resulted from irrigating with the contaminated water, reducing agricultural water bill, and reducing purchase of fertilizers.

Moreover, farmers were concerned about the efficiency of conveyance and distribution systems of RW and reliability of RW supply. They expressed worries about possible economic losses resulting from intermittency of RW supply due to inefficient conveyance systems or frequent technical problems in wastewater treatment plants. Reliability of RW supply was an important issue for farmers as it would enable them

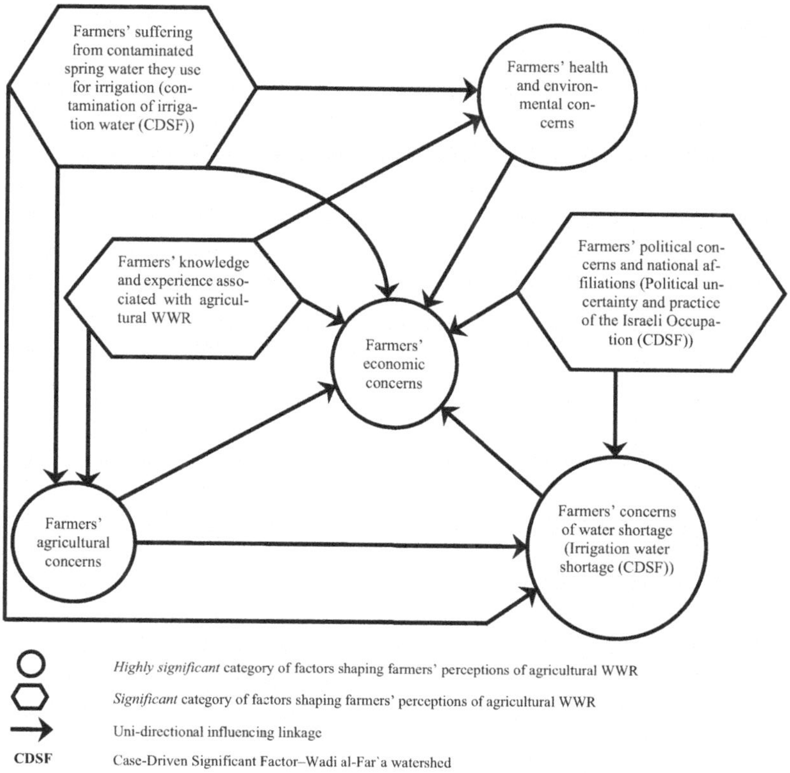

Fig. 2 Conceptual model of understanding farmers' perceptions of a salient agricultural wastewater reuse option proposed in the Wadi al-Far'a watershed

taking right decisions on kinds and quantities of crops they cultivate, size of the cultivated area, and time and period of cultivation, the case that would prevent economic losses. Finally, farmers were highly concerned about fairness of allocation mechanisms of RW amongst them. They expressed that the use of RW should not be confined to high-income or large-scale farmers.

Besides economic concerns, the model suggests that farmers' concerns related to the CDSF of "irrigation water shortage" have highly significant influence on enhancing farmers' perceptions of WWR. Farmers thought that using RW would contribute to alleviating the problems resulting from water shortage, such as restricted cropping patterns, increased areas of uncultivated lands, farmers' conflicts over water, farmers' abstention from agricultural work and coercion to leave their lands, drop in water levels in agricultural wells, and increase in prices of well water. In Froush Beit Dajan, where the problem of irrigation water shortage was intense, influence of this CDSF on farmers' perceptions dominated influences of other factors, including economic ones, contrary to the case of An-Nassariyya, where irrigation water shortage was less intense.

Farmers' agricultural concerns also have a highly significant influence on shaping their perceptions of using RW. These concerns are related to the (1) agronomic quality of RW and (2) the Palestinian WWR guidelines of restricted irrigation that prohibit irrigating vegetables with RW and prohibit irrigating fruit trees with RW 3 weeks before harvesting. Farmers were concerned about specific features of quality of RW including salinity, precipitations, microorganisms, and chemical compounds and their adverse impacts on soil, lands, and water resources. While farmers showed interest in benefiting from the fertilizers RW contains, they expressed worries about providing them with a poor agronomic quality of RW that would turn RW from a beneficial source into a nuisance.

As for the Palestinian WWR guidelines of restricted irrigation, implementing these guidelines would have a dual influence on shaping farmers' perceptions of WWR. On the one hand, they would enhance perceptions of farmers, whose concerns of water shortage dominated perceiving the guidelines as obstacles to their use of RW. They would also enhance perceptions of farmers who thought that quality of RW would be more suitable for irrigating fruit trees and fodder crops than for irrigating vegetables. On the other hand, the guidelines would inhibit perceptions of farmers who perceived restricted irrigation as implication about poor quality of RW, who were worried that other farmers might not comply with the guidelines, and who cultivated exclusively vegetables thinking that vegetables are the most profitable crops. While findings from local studies in the West Bank and the Gaza Strip revealed farmers' preference for restricted irrigation over unrestricted irrigation (Abu Shaban et al. 2006; Al-Kharouf et al. 2008), this preference was not found in this study.

Concerning farmers' health and environmental concerns, the model indicates that these concerns play a highly significant role in shaping farmers' perceptions of WWR. Most interviewed farmers expressed reservations about potential health risks of agricultural WWR and were highly concerned about the compliance of quality of RW with health standards of WWR. Hence, farmers were concerned about availability and enforcement of Palestinian WWR standards, guidelines, and monitoring plans and viewed these standards and plans as safety valves for protecting the public health and preventing misuse of RW. "Perceived health risks of agricultural WWR" being a major determinant of farmers' perceptions of WWR confirmed findings of previous relevant Palestinian studies (Abu Shaban et al 2006; Isaed et al. 2008). Furthermore, data revealed that farmers' obligation to wear protective equipment (if needed) when irrigating with RW would not hinder their acceptance of using the water if the equipment provided flexibility in movement. Besides health concerns, farmers were highly concerned about the environmental benefits of agricultural WWR application (mainly alleviating pollution of water resources and the environment) as they were discontented with the deteriorated environmental conditions resulting from wastewater discharge into al-Far'a Stream.

The interviews also revealed that farmers' sufficient prior knowledge and successful prior experiences associated with agricultural WWR application significantly enhanced their perceptions of using RW. In contrast, lack of sufficient knowledge (about potential adverse impacts of WWR, kinds of crops that can be irrigated with RW, and advantages and disadvantages of agricultural WWR) as well as lack of successful experience with WWR inhibited farmers' perceptions. Research on social responses to WWR suggested inconsistent correlation between individuals' prior knowledge and their acceptance of using RW (Marks 2004; Marks et al. 2006). Yet, a positive correlation between the two variables was reported more frequently (Baumann 1983; Dolnicar et al. 2011; Ghanem 2012; Hurlimann and McKay 2004; Roseth 2008). The study findings also confirm the positive correlation revealed in the literature between individuals' prior experience with agricultural WWR and their acceptance of using RW. While positive experiences enhanced acceptance of WWR (Marks 2003), negative experiences inhibited perceptions of WWR (Carr et al. 2011; Qadir et al. 2010).

Besides the CDSF of "irrigation water shortage" explained earlier, the model suggests a second CDSF, which is "contamination of irrigation water". Interviews revealed that farmers' coercion to irrigate with the contaminated spring water (due to water shortage) and their suffering from its adverse impacts on their personal and families' health, their agricultural lands and crops, and their agricultural income significantly enhanced farmers' perceptions of using RW. Farmers thought that treating wastewater and using it for irrigation would prevent contamination of spring water and its adverse impacts. Despite the enhancing influence of the above CDSF, farmers' suffering from the contaminated water made them more vigilant about the quality of RW and its potential adverse impacts to the extent that they would refuse agricultural WWR in case of incompliance with WWR standards and resulted in adverse effects. In the case of An-Nassariyya, where quality of spring water was highly deteriorated, influence of the CDSF of "irrigation water contamination" on farmers' perceptions was stronger than any dominating influences of other factors except economic ones.

A third CDSF suggested by the model is "political uncertainty due to the Israeli confiscation of agricultural lands and control of water resources". Farmers' fears of this significantly enhanced their perceptions of using RW. Farmers, particularly those in Froush Beit Dajan, believed that providing them with the supplementary irrigation source of RW would be one of the main strategies of protecting their 'uncultivated' lands from confiscation and defending their identity. The farmers also stressed their concerns of using RW at the expense of Palestinian rights to fresh water, given the undefined recognition of Palestinian water rights in the West Bank by the Israeli government according to Article 40 of the Oslo II Agreement in 1995 (Pacheco 2001). They expressed that they would refuse agricultural WWR if using RW would be at the expense of Palestinian rights to fresh water and farmers' rights to fresh water from their own agricultural wells. Besides political concerns, farmers' national affiliation enhanced their perceptions of WWR. Farmers' believed that agricultural WWR projects would contribute to sustainability of Palestinian farming communities and development of the country through encouraging Palestinian labour force working in Israel to return and work in their lands that were left uncultivated.

Interestingly, farmers' marketing concerns, including "farmers' concerns of consumers' perceptions of buying crops irrigated with RW", are excluded from the model, as they have limited influence on shaping farmers' perceptions of agricultural WWR (see Table 1). This finding eliminates the hypothesis claiming that farmers' negative views on consumers' perceptions of buying crops irrigated with RW might dampen their perceptions of using RW. Farmers in the Wadi al-Far'a watershed did not see a negative consumer perception towards crops irrigated with RW, and thus they did not have worries about possible consumers' boycott of these crops. In fact, these worries were mitigated under the local context of lacking grading systems for agricultural crops in the Palestinian market that does not allow consumers to differentiate between crops irrigated with RW and other crops in the market.

Other factors excluded from the model were those related to the "issue of trust". The study evidence on the limited influence of these factors on shaping farmers' perceptions of agricultural WWR refuted evidence from Australia and the USA on "trust" being a fundamental factor shaping acceptance of potable and non-potable WWR, including agricultural WWR (Marks 2004; Mooney and Stenekes 2008). In the case of al-Far'a watershed, the conditions of water shortage, contamination, and political uncertainty threatened farmers' livelihood, and thus did not sensitize their perceptions to the issue of trust as much as to concerns of sustaining their agricultural work and income through WWR. Hence, concerns of economic sustainability (i.e. economic factors) overrode all other concerns influencing farmers' perceptions. This interpretation explains why farmers, such as AM4FAN and KH10FFBD, who expressed distrust in Palestinian authorities and stakeholders involved in implementing future WWR projects in the watershed, still accepted using RW. Despite these findings, encouraging the dialogue and building trust between Palestinian authorities and stakeholders in the water and wastewater sector, on the one hand, and farmers, on the other hand, is still vital. This is because the "issue of trust" might become a significant factor shaping farmers' perceptions when WWR projects were implemented on the ground and farmers could then develop opinions about performance of stakeholders responsible for these projects. Additionally, building farmers' trust in responsible stakeholders would be one strategy to decrease their perception of health risks caused by agricultural WWR given the study evidence on "distrust" contributing to perceived health risks.

3.4 Explaining the Palestinian Paradox Between the Hypothetical Support of Agricultural WWR and Opposition to Agricultural WWR Projects

Acceptance amongst farmers of the salient agricultural WWR option proposed in the Wadi al-Far'a watershed was found to be higher (93.8%) than farmers' acceptance levels of general agricultural WWR options revealed by other Palestinian studies in the West Bank (87% by Abu-Madi et al. (2008), 83.1% by Al-Kharouf et al. (2008), and 70% by Qattawi (2008)) and the Gaza Strip (88.9% by Nassar et al. (2010) and 86.1% by Tubail et al. (2004)). Mizyed's (2013) survey (n = 180) also revealed higher acceptance level of farmers (90%) towards salient agricultural WWR options proposed in different areas in the West Bank. This comparison suggests that findings on the issue of "salience" in the Palestinian case contradict with the argument in the international literature that social support declines when WWR option becomes more salient (Bruvold 1988; Hartley 2003; Po et al. 2003). Accordingly, salient-related explanations fall short of explaining the paradox in the Palestinian case between the hypothetical support for the idea of agricultural WWR and opposition to proposed or implemented agricultural WWR projects.

Using the cultural approach to contextualize farmers' perceptions of agricultural WWR allowed the provision of an explanation for the above paradox other than the salient-related one. It showed that farmers' positive perceptions of agricultural WWR were highly influenced by contextual factors specific to the case of the Wadi al-Far'a watershed. Furthermore, these perceptions were contingent upon meeting farmers' expectations from future agricultural WWR projects about different economic, health, agricultural, and political aspects associated with agricultural WWR application (see Sect. 3.1). These findings imply that farmers' opposition to past Palestinian agricultural WWR projects could be attributed to shortcomings of the projects themselves. That is, the projects did not meet farmers' expectations as they were implemented or proposed without adequate involvement of farmers in the decision-making process. They lacked to understand farmers' concerns associated with these projects and local contexts influencing these concerns. This explanation raises questions about the institutional dimension of the "problem" of farmers' opposition to WWR projects. It could be that the institutional culture of the Palestinian water and wastewater sector, which hinders adequate community involvement in planning and decision-making processes, has considerably contributed to farmers' opposition to these projects. For instance, in the case of Deir Debwan (see Sect. 1), A. Jarrar, in a personal interview conducted on 17 October 2012, disclosed that inappropriateness of the proposed agricultural WWR option to the local context and deep (but usually unarticulated) social values of this community resulted in social opposition to WWR. This indicates that the community of Deir Debwan was not fully involved in the decision-making process associated with planning and implementing al-Bireh wastewater treatment plant. It can be argued that social opposition to agricultural WWR project is always possible in any one case, including the case of the Wadi al-Far'a watershed, if the project is proposed or implemented without being tailored to the local contexts and does not meet farmers' expectations.

Noteworthy, the study finding on the limited influence of "farmers' concerns of involving them in the decision-making process" on farmers' perceptions (see Table 1) did not imply that these concerns were not influential in shaping farmers' perceptions. Rather, this finding was attributed to the local and national contexts under which farmers lived. In fact, farmers acknowledged their participation in this study as 'part' of involving them in the decision-making process and the influential role their participation played in shaping their perceptions of WWR. However, farmers did not emphasize the issue of "participation" in their interviews for two reasons. First, they were overwhelmed with the severity of the problems of irrigation water shortage and contamination they faced and their exigent need for a supplementary clean source of irrigation water. Second, the institutional culture and structure of the Palestinian water and wastewater sector, being based on the traditional paradigm of centralized water management, shrank farmers' expectations about getting them involved in the decision-making process.

3.5 Contributions and Limitations

This study contributes both, conceptually and methodologically to the existing knowledge. First, the model developed by the study offers a conceptual framework for case study research examining farmers' perceptions of agricultural WWR. It shows that the multi-layered contextual analysis performed based on the cultural approach improves knowledge of what shapes farmers' perceptions of agricultural WWR and adds more comprehensive perspectives to research focusing on social responses to WWR. Second, the solid investigative methods used in this study offer a new means of analysing perceptions of WWR examining multiple individual and contextual variables and their collective influence on farmers' perceptions of agricultural WWR. It is, therefore, an improvement of the methodological and analytical frameworks often found in the relevant literature.

Third, the study model adds a new perspective to the existing knowledge on the significant role of a particular kind of individual and contextual factors that shape responses to WWR, conceptualized as Case-Driven Significant Factors (CDSFs). The concept of CDSF offers a novel contribution to methodological approaches of research on social responses to potable or non-potable WWR applications. It attempts to provide a road map to case study research for identifying case-specific factors influencing responses to WWR. That is, the typology of CDSFs of a particular case can be used to identify case-specific factors influencing perceptions of WWR in different locations within this particular case. This typology can also be used and extended to other cases of comparable conditions and contexts. For instance, the typology of CDSFs–Wadi al-Far'a watershed created by the researchers can be borrowed and used for the identification of case-specific factors shaping farmers' perceptions of agricultural WWR in other Palestinian or international cases where farmers live under comparable conditions and contexts. Creating a national Palestinian typology of CDSFs–Palestine, in turn, allows the identification of case-specific factors influencing perceptions of WWR of different local communities. To take this a step further, national CDSFs typologies of different regional countries, such as Palestine, Jordan, and Lebanon, for instance, could be used to create a regional typology of CDSFs, which, in turn, could be extended to a typology of CDSFs for the Middle East.

The concept of CDSF, therefore, attempts to offer a collective typology of case-specific factors that can be applied to comparable contexts and may be further extended. In other words, it paves the way to a predictive model of case-specific factors influencing perceptions of WWR that respects the specificity and contextuality of a particular case. In the field of public health, the concept of CDSF promotes new types of compositional and contextual factors that are derived from the particularity of the case itself rather than generalized models of factors. Additionally, it contributes to further sensitizing socio-ecological models of human behaviours to be tailored to a particular case, community, or group of people. Hence, it calls for sensitizing decision-making processes, community participation initiatives, and intervention strategies in a way that gives attention to the particularity of each case or needs of each group, which promotes concepts of environmental justice and health and social equalities.

Despite the above study contributions, limitations still exist. First, this research intends to provide an understanding of the way perceptions of agricultural WWR are shaped rather than statistical inference. This is because adopting case study research design limits the empirical generalization of the study findings to other cases. The independent variables and the linkages amongst them may vary in other case studies in accordance with contexts and groups of people under investigation. To enhance generalizability, however, the researchers endeavoured to demonstrate representativeness of the case of the Wadi al-Far'a watershed to broader cases of comparable conditions and contexts in the Lower Jordan Valley and the West Bank. Nevertheless, the uniqueness of the Palestinian political context under which the study participants live, reflected mainly in the Israeli occupation, is the main reason limiting generalizability of the existing case study to communities other than the Palestinian ones.

Besides the above limitation in generalizability, findings and conclusions of this study cannot and should not be generalized to case studies on perceptions of WWR applications other than agricultural WWR. Perceptions and concerns may differ from those revealed in this study when inquiring reuse applications involving the degree of contact with RW closer than that involved in agricultural irrigation. Second, this study provides no causality link between the dependent and independent variables due to the cross-sectional nature of the collected data which limits ability to draw casual inferences. Furthermore, the relationships amongst social science factors are complex and interwoven due to different reciprocal interactions between them. These interactions produce many chicken-and-egg relationships (Hartley 2003) and do not allow inferences to be made about causality. Nevertheless, it was attempted to uncover preliminary causal relationships between the dependent and independent variables. Third, researchers' limited control over the case study settings in addition to subjectivity of participants' recollection (i.e. reflections and interpretations of events) are further inherent limitations of the current research design.

4 Conclusions and Recommendations

Three main arguments are made in this study. First, farmers' economic, agricultural, health, environmental, and political concerns regarding agricultural WWR significantly influence their perceptions of salient agricultural WWR options proposed in areas of water shortage and contamination in conflict zones, such as the Wadi al-Far'a watershed. On the other hand, farmers' psychological, social, and religious concerns related to WWR are not significant. Second, farmers' concerns about consumers' perceptions of buying crops irrigated with RW only had a limited influence on shaping farmers' perceptions of agricultural WWR. We can thus reject the hypothesis that farmers' views on consumers' perceptions of buying crops of RW constitute the major factor shaping their perceptions of RW utilization. Third, contrary to claims by the Palestinian authorities and stakeholders in the water and wastewater sector (Abdo 2008), the issue of social acceptance of agricultural WWR is not an obstacle to implementing agricultural WWR schemes in Palestine.

The contradiction between social acceptance revealed by this study and social opposition perceived by the Palestinian stakeholders raises questions about the institutional dimension of the issue of farmers' opposition to past agricultural WWR projects. It could be that stakeholders' misperceptions, institutional conservatism that resists institutionalizing community participation initiatives, and weak Palestinian institutional structure and organization have initially contributed to social opposition to local WWR projects, and hence failure of these projects. Baumann (1983) argues likewise in his review of relevant American studies and concludes: "public acceptance of waste-water reuse is not a formidable obstacle to its adoption, but that a greater constraint is the pessimistic perception by planners of consumer attitudes" (p. 79).

Backed by evidence from this study, it is recommended to implement agricultural WWR projects in the Wadi al-Far'a watershed. These projects would produce identifiable economic, agricultural, health, environmental, social, and political benefits that would contribute to alleviating poverty and sustaining livelihood of farming communities in the watershed. At the national level of the West Bank, providing farmers with RW as a supplementary, reliable, and potentially lower cost source of irrigation water would enhance agricultural water supplies and irrigated agriculture, especially in dry seasons. This, in turn, would improve food security given that irrigated agriculture produces 38% of agricultural production in the West Bank (Mizyed 2013). Agricultural WWR projects would also enhance groundwater recharge through infiltration and contribute to alleviating the water crisis in the West Bank and saving fresh water for domestic and potable uses. Additionally, they would close the loop between water supply and wastewater disposal, which would promote sustainability of water resource management, prevent pollution of water resources and the environment, and protect the public health. Finally, agricultural WWR would become one of the main strategies of defending uncultivated agricultural lands from confiscation and protecting the identity of Palestinian farming communities.

To successfully implement Palestinian agricultural WWR projects, following support is required: (1) technical and economic feasibility and environmental impact assessment studies, (2) clear and effective political and institutional frameworks, and (3) coherent and enforceable legal framework for agricultural WWR. Not less important, farmers should be involved genuinely in the decision-making process during the early stages of planning WWR projects as well as in developing national WWR standards, guidelines, and policies. Ormerod and Scott (2012), who employed the cultural approach to explain public perceptions of potable WWR in Arizona, concluded that "finding innovative ways for the citizenry to truly participate in water policy decisions during the design and planning phases could engender confidence in the public engagement process" (p. 17). Finally, encouraging the dialogue and communication between Palestinian authorities and stakeholders in the water and wastewater sector, on the one hand, and the farming communities, on the other hand, is necessary to develop mutual understanding of challenges in terms of national agricultural WWR projects and create confidence amongst these groups.

In addition to policy implications, this research offers three main propositions for future research based on its findings and arguments. First, there is a need to recognize and encourage the new trend of case study research in using the cultural approach on a case-by-case basis to explain social responses to WWR and offer interpretations beyond socio-demographics and individual's attitudes solely provided by psychological and psychometric approaches. Second, the "institutional approach" should be used to examine the influence of the Palestinian institutional and political contexts (including institutional conservatism and administrative fragmentation) on farmers' opposition to local WWR projects and recommend institutional changes critical to success of these projects. The institutional approach focuses on analysing ways in which WWR options are defined, negotiated, and coordinated by different stakeholders within the existing institutional context (Beck 2009; Baumann 1983). Third, the study conceptual model needs to be tested and further extended to broader cases of comparable conditions to enhance generalization. Fourth, this study calls for extending the typology of CDSFs–Wadi al-Far'a watershed to a national typology of CDSFs–Palestine. The offered model serves as a platform and a starting point for this extension.

Acknowledgements We wish to thank the following organizations and academic institutions for their generous funding of the conduct of this research: the German Federal Ministry for Education and Research (BMBF) through the project Sustainable Management of Available Water Resources with Innovative Technologies (SMART), Fazit Stiftung, and Bielefeld University/ Rektorat Stipendien. We also thank the Palestinian Hydrology Group, director Dr. Abdulrahman Tamimi, for their advice and help with data collection. We would also like to thank Dr. Ingo Heinz, Dr. Ghada al-Madbouh, and Dr. Stephen Gasteyer for advice and criticism.

References

Abdo K (2008) The Palestinian experiences of reusing treated wastewater in agriculture. In: Birzeit water drops, proceedings of the first symposium on wastewater reclamation and reuse for water demand

management in Palestine, vol 6. Birzeit University, Birzeit, Palestine, pp 11–21
Abu Shaban A, Doppler W, Wolff H-P (2006) Determinants of farmers' acceptance of treated wastewater in irrigated agriculture in the northern Gaza strip. In: Deutscher T (ed) 2006 conference on international agricultural research for development. Bonn, Germany
Abu-Madi M, Braadbaart O, Al-Sa'ed R et al (2003) Willingness of farmers to pay for reclaimed wastewater in Jordan and Tunisia. Water Sci Technol Water Supply 3:115–122
Abu-Madi M, Mimi Z, Abu-Rmeileh N (2008) Public perceptions and knowledge towards wastewater reuse in agriculture in Deir Debwan. In: Birzeit water drops, proceedings of the first symposium on wastewater reclamation and reuse for water demand management in Palestine, vol 6. Birzeit University, Birzeit, Palestine, pp 91–100
Al-Kharouf S, Al-Khatib I, Shaheen HQ (2008) Appraisal of social and cultural factors affecting wastewater reuse in the West Bank. Int J Environ Pollut 33:3–14
Applied Research Institute-Jerusalem (ARIJ) (2006) Wadi al Far' a village profile. Report, ARIJ
Arafat W (2012) The role of public awareness towards sustainable use of treated wastewater in agricultural irrigation. Master's thesis, Birzeit University, Birzeit, Palestine
Baumann D (1983) Social acceptance of water reuse. Appl Geogr 3:79–84. https://doi.org/10.1016/0143-6228(83)90007-3
Beck S (2009) Between disgust and trust: state of social science research on the implementation and acceptance of water and wastewater reclamation and re-use (WWT-R). Report, Helmholtz-Zentrum für Umweltforschung (UFZ)
Bruvold WH (1985) Obtaining public support for reuse water. Am Water Work Assoc J 77:72–77
Bruvold WH (1988) Public opinion on water reuse options. Water Pollut Control Fed 60:45–49
Carr G, Potter RB, Nortcliff S (2011) Water reuse for irrigation in Jordan: perceptions of water quality among farmers. Agric Water Manag 98:847–854. https://doi.org/10.1016/j.agwat.2010.12.011
Chen W, Bai Y, Zhang W et al (2015) Perceptions of different stakeholders on reclaimed water reuse: the case of Beijing, China. Sustainability 7:9696–9710. https://doi.org/10.3390/su7079696
Chenini F (2011) Evaluation of the three decades of treated wastewater reuse in Tunisia. In: Barceló D, Petrovic M (eds) The handbook of environmental chemistry, vol 14. Springer, Berlin, pp 215–227. https://doi.org/10.1007/698_2010_65
Ching L (2010) Eliminating "yuck": a simple exposition of media and social change in water reuse policies. Int J Water Resour Dev 26:111–124. https://doi.org/10.1080/07900620903392174
Dolnicar S, Schäfer AI (2009) Desalinated versus recycled water: public perceptions and profiles of the accepters. J Environ Manage 90:888–900. https://doi.org/10.1016/j.jenvman.2008.02.003
Dolnicar S, Hurlimann A, Grün B (2011) What affects public acceptance of recycled and desalinated water? Water Res 45:933–943. https://doi.org/10.1016/j.watres.2010.09.030
Environmental Quality Authority (EQA) (2006) The Far'a and Jerash integrated watershed management project. Baseline study report for Wadi Al-Far'a, EQA
Ghanem M (2012) Gender empowerment and treated wastewater reuse in Ein Qinia Village, Ramallah—Palestine. Report, The Heinrich-Böll-Stiftung Palestine & Jordan
Green J, Thorogood N (2004) Qualitative methods for health research. In: Silverman D (ed) Introducing qualitative methods series, vol 10. SAGE Publications Ltd., London, UK
Hartley TW (2003) Water reuse: understanding public perception and participation. Water Environmental Research Foundation (WERF), Alexandria, VA
Hurlimann AC (2007) Is recycled water use risky? An urban Australian community's perspective. Environmentalist 27:83–94. https://doi.org/10.1007/s10669-007-9019-6
Hurlimann A, McKay J (2004) Attitudes to reclaimed water for domestic use: part 2. Trust Water 31:40–45
Hurlimann A, McKay J, Geursen G (2005) Pricing of drinking water vs recycled: fairness and satisfaction. J Aust Water Assoc 32:50–52
Isaed R, Ghanem M, Abu-Madi M (2008) Willingness to use treated wastewater and to pay for its irrigated products in Dura–Hebron. In: Birzeit water drops, proceedings of the first symposium on wastewater reclamation and reuse for water demand management in Palestine, vol 6. Birzeit University, Birzeit, Palestine, pp. 101–112
Jayyousi A (2003) Water supply and demand development in Palestine: Current status and future prospects. In: Daibes F (ed) Water in Palestine: problems-politics-prospects. Palestinian Academic Society for the Study of International Affairs (PASSIA), Jerusalem, Palestine, pp 57–83
Keremane GB, McKay JM (2006) The role of community participation and partnerships: the Virginia pipeline scheme. Water 29:29–33
Lacey A, Luff D (2009) Qualitative data analysis. The NIHR RDS EM/ YH, Sheffield, England
MA'AN Development Center, Jordan Valley Popular Committees (JVPC) (2010) To exist is to resist—Eye on Jordan Valley. Report, MA'AN Development Center and JVPC
Mark JS, Martin B, Zadoroznyj M (2006) Acceptance of water recycling in Australia: national baseline data. Water 33:151–157
Marks JS (2003) The sociology of "disgust" towards the use of reclaimed water. In: Water recycling Australia 2nd national conference. Unpublished report. Brisbane, Australia
Marks JS (2004) Advancing community acceptance of reclaimed water. Water J Aust Water Assoc 31:46–51
Marks JS, Cromar N, Fallowfield H et al (2003) Community experience and perceptions of water reuse. Water Sci Technol Water Supply 3:9–16

Mizyed NR (2013) Challenges to treated wastewater reuse in arid and semi-arid areas. Environ Sci Policy 25:186–195. https://doi.org/10.1016/j.envsci.2012.10.016
Mooney C, Stenekes N (2008) An analysis of the social aspects of establishing agricultural recycled water schemes—social perspectives on water. Report, Commonwealth of Australia
Nassar A, Tubail K, Moritz A et al (2010) Attitude of farmers towards effluent reuse in the Gaza Strip. J Environ Sci 39:119–134
Ormerod KJ, Scott CA (2012) Drinking wastewater: public trust in potable reuse. Sci Technol Human Values 00:1–23. https://doi.org/10.1177/0162243912444736
Pacheco A (2001) Flouting convention: the Oslo agreements. In: Carey R (ed) The New Intifada: Resisting Israel's Apartheid. R. R, Donnelly and Sons, Chicago, pp 181–208
Palestinian Central Bureau of Statistics (PCBS) (2009) Population, housing and establishment census 2007: main indicators by locality type. Report, PCBS, Ramallah, Palestine
PLO Negotiations Affairs Department (2011) The Jordan Valley: a microcosm of the Israeli occupation. Report, PLO Negotiations Affairs Department, Palestine
Po M, Nancarrow BE (2004) Literature review: consumer perceptions of the use of reclaimed water for horticultural irrigation. Unpublished report, ARCWIS, Australia
Po M, Kaercher JD, Nancarrow BE (2003) Literature review of factors influencing public perceptions of water reuse. Report, CSIRO Land and Water, Australia
Qadir M, Wichelns D, Raschid-Sally L et al (2010) The challenges of wastewater irrigation in developing countries. Agric Water Manag 97:561–568. https://doi.org/10.1016/j.agwat.2008.11.004
Qattawi M (2008) Possibility of social acceptance to reuse recycled sewage water in agriculture in Ramallah and Al-Bireh Governorate: Deir Debwan case study. Unpublished master's thesis, Birzeit University, Palestine
Roseth N (2008) Community views on recycled water—the impact of information. Report. Cooperative Research Centre for Water Quality and Treatment, Adelaide, South Australia
Russell S, Hampton G (2006) Challenges in understanding public responses and providing effective public consultation on water reuse. Desalination 187:215–227. https://doi.org/10.1016/j.desal.2005.04.081
Russell S, Lux C (2006) Water recycling and the community—public responses and consultation strategies. A literature review and discussion (No. 5), Oz-AQUAREC. University of Wollongong NSW, Wollongong NSW, Australia
Russell S, Lux C (2009) Getting over yuck: moving from psychological to cultural and sociotechnical analyses of responses to water recycling. Water Policy 11:21–35. https://doi.org/10.2166/wp.2009.007
Saad D, Byrne D, Drechsel P (2017) Social perspectives on the effective management of wastewater. In: Farooq R, Ahmad Z (eds) Physico-chemical wastewater treatment and resource recovery. InTech, Croatia—European Union, pp 253–267. https://doi.org/10.5772/67803
Schaer-Barbosa M, Santos MEP, Medeiros YDP (2014) Waste water reuse as a mitigating factor to the effects of droughts in the state of Bahia Semi-Arid Viability study. Ambient Soc 17:17–32
Schmidt CW (2008) The yuck factor: when disgust meets discovery. Environ Health Perspect 116:A524–A527. https://doi.org/10.1289/ehp.116-a524
Scott CA, Faruqui NI, Raschid-Sally L (2004) Wastewater use in irrigated agriculture: management challenges in developing countries. In: Scott CA, Raschid-Sally L (eds) Wastewater use in irrigated agriculture: confronting the livelihood and environmental realities. CABI Publishing, UK, pp. 1–10. https://doi.org/10.1079/9780851998237.0001
Stenekes N, Colebatch H, Waite T et al (2006) Risk and governance in water recycling. Sci Technol Human Values 31:107–134. https://doi.org/10.1177/0162243905283636
Thawabe SA (2006) Anthropogenic impact on water resources in the West Bank/Palestine: case from Wadi Fara'a Stream–Nablus area. Environ Forensics 7:241–245. https://doi.org/10.1080/15275920600840628
The World Bank (2004) West Bank and Gaza: wastewater treatment and reuse policy note. Report, The World Bank, Washington DC
The World Bank (2009) West Bank and Gaza: assessment of restrictions on Palestinian water sector development. Report, The World Bank, Washington DC
Tubail KM, Dadah JY, Yassin MM (2004) Present and prospect situation of wastewater and its possible reuse in the Gaza Strip. KA-Abwasser, Abfall 51:866–872
Wester J, Timpano KR, Çek D et al (2015) Psychological and social factors associated with wastewater reuse emotional discomfort. J Environ Psychol 42:16–23
Yin RK (2003) Case study research: design and methods. In: Bickman I, Rog DJ (eds) Applied social research methods series. SAGE Publications Inc, California

On the Importance of Modeling, Optimization, and Control of Anaerobic Degradation Processes for the Treatment of Agricultural Wastewaters: Toward a Biorefinery Challenge

Victor Alcaraz-Gonzalez and Victor Gonzalez-Alvarez

Abstract

The distinctive feature of bioreactors used for wastewater treatment processes (WWTP) is the use of specialized microbial ecosystems as catalysts for the conversion processes. The concept is based on two advantages: its sustainable character and the possible diversification of their products in order to optimize both bioenergy and biomolecule production. These ideas match with a novel concept called "Environmental Biorefinery", in which the main objective is to minimize the pollution rejected to the natural environment while getting as much as possible of high added-value products. One of the most studied processes in this context is anaerobic digestion (AD), whose main product until now is the so-called biogas, which is a mixture of methane (CH_4) and carbon dioxide (CO_2) with trace amounts of other gases such as hydrogen sulfide (H_2S) and hydrogen (H_2). In the past, most of industrial AD WWTP (digesters) were used for reaching minimum standards in the composition of the effluent, and eventually for maximizing biogas production. Nevertheless, the classical concept of anaerobic digester, as conventional wastewater treatment process or methane production process merely can now be overcome with highly innovative outputs. Nowadays, a fundamental research objective in all current variants of AD processes is the search for optimal operating conditions to maximize the productivity of biogas, as well as to optimize their performance as efficient biorefinery using Instrumentation, Control, and Automation approaches (ICA) for wastewater treatment. This paper aims to describe the most important model-based ICA approaches and to analyze their potential for increasing efficiency of AD process. Experimental and simulation results show that these control approaches are a useful, efficient, and low cost alternative that may be implemented at industrial scale to maximize the productivity of biogas.

Keywords

Anaerobic digestion · Wastewater treatment · Process control · Water reuse · Biogas · Biorefinery

1 Introduction

Agro-industrial effluents, which have been considered in the past only as polluting waste, are taking nowadays a particular interest from

V. Alcaraz-Gonzalez (✉) · V. Gonzalez-Alvarez
Departamento de Ingeniería Química, Universidad de Guadalajara, Blvd. Marcelino García Barragán 1421, C.P. 44430 Guadalajara Jalisco, Mexico
e-mail: victor.alcaraz@cucei.udg.mx

L. Ribbe et al. (eds.), *Towards Water Secure Societies*,
https://doi.org/10.1007/978-3-030-50653-7_4

several points of view, e.g., economic, industrial, and even social. Besides the need for their treatment in order to reduce environmental pollution, they are now regarded as potential "Environmental biorefineries" due to its high nutrient content that can be recovered (phosphates, nitrates, etc.), because of its capacity to generate energy (CH_4, H_2), and its potential to reformulate organic matter in new forms of biomass that can be used either as another useful source of organic charcoal or as substrate for biofuels.

There are many different types of agro-industrial effluents treatment processes that include chemical, physical, physicochemical, and biological methods. However, it is well known that biological treatments are the most effective and efficient in terms of costs, time, and performance. Moreover, it is also well known that the only kind of biological process capable of treating directly (without dilution) agro-industrial effluents with chemical oxygen demand (COD) greater than 20 gCOD/l is anaerobic. From an energetic point of view, an advantage of using anaerobic bioreactors for the treatment of organic effluents is the production of biogas, which includes CO_2, CH_4, and BioH_2 (Metcalf and Eddy 2014). Moreover, microbial ecosystems involved in digesters could indeed also be used to produce a whole range of different products of interest. An example of this fact is the so-called two-phase system (Foglia et al. 2010; Liu et al. 2013), where the first acidogenic bioreactor converts the organic matter present in the influent into Volatile Fatty Acids (VFA) and other simple molecules, while a second methanogenic bioreactor specializes in converting the effluent from the first one into CH_4 and CO_2. Among the products of the first bioreactor, it is possible to obtain hydrogen (therefore called bioH_2). In this way, two fuels are generated: bioH_2 and methane, with the advantage that the first one is much more efficient from the energy point of view, and does not produce greenhouse gases after combustion. The main mechanism at the first stage used at far for producing hydrogen is the so-called dark fermentation (DF), in which strict or facultative anaerobic bacteria carry out the degradation of organic matter (carbohydrates) and produce H_2, VFA, and alcohols (Guo et al. 2010; Ntaikou et al. 2010).

Among other biotechnologies involving microbial ecosystems that can utilize the metabolic by-products generated by DF, a recent emerging technology related to Microbial Fuel Cells (MFC) is Microbial Electrolysis Cell (MEC) (Lovley 2006). This relatively new technology is very promising since allowing H_2 production and organic material degradation (Rozendal et al. 2006; Marone et al. 2017). In an MEC, bacteria referred to as exoelectrogens oxidize a substrate under anaerobic conditions and release electrons to the anode (Logan and Regan 2006). Then, a small voltage (0.2–0.8 V) applied externally to the circuit allows H_2 production at the cathode through the reduction of protons. Theoretically, the energy required for this operation is substantially lower—10 times less—than the 1.8–2.0 V used for H_2 production via traditional water electrolysis. Thus, the effluents of DF are suitable to be utilized as influents of an MEC, and then, combining DF with MEC in a cascade two-steps process can result in a potentially complete exploitation of organic waste streams, which maximizes at the same time energy recovery and effluent depollution.

On the other hand, the use of CO_2 from processes such as AD for growing autotroph microalgae has been also widely studied recently for the purification of biogas and biomass production (Cea-Barcia et al. 2014; Meier et al. 2015). Thus, the biogas produced is free of CO_2, while microalgae biomass can be directly exploited as fuel or biofertilizer, or for the production of pigments, anti-oxidants, or biofuels as biodiesel. Besides, VFA from the first phase of AD are suitable to be used as tributaries of bio-electrochemical systems, while the CO_2 produced can be used in the production of microalgae. Thus, seen as a biorefinery, the combination of all these processes can result in a potentially complete exploitation of organic waste streams, which maximizes at the same time energy recovery and effluents depollution.

Nevertheless, from an Instrumentation, Control and Automation (ICA) point of view, the biological complexity associated with such processes results in dynamical systems described by highly nonlinear models, parametric uncertainties, multiple external disturbances, and a lack of appropriate sensors. Considering the limitations on the number and quality of online measurements and mathematical models available, this fact reveals the importance of modeling, and to implement ICA approaches in order to guarantee the best performance of these systems (Bastin and Dochain 1990; Olsson et al. 2014; Jimenez et al. 2015). In this paper, we show some process control approaches that have proved its usefulness in the modeling, monitoring, optimization, and regulation of anaerobic degradation processes (Alcaraz-Gonzalez et al. 2005; Méndez-Acosta et al. 2010, 2011). In particular, three control strategies for the regulation of COD and alkalinities, as well as the estimation of input process, which are relatively easy to implement, are shown (Alcaraz-González et al. 2005, 2013, 2017). In these approaches, the treatment of vinasses by AD is presented as study cases since they represent a serious environmental problem, both global and locally in small cities. However, they can be extrapolated to a large list of similar Waste Water Treatment Processes (WWTP). Approaches showed in this paper are only a representative sample of the potential and usefulness of ICA in WWTP and are based in a relatively simple AD model (Bernard et al. 2001). For a more extensive and recent review, readers may refer to Jimenez et al. (2015) and the references contained therein. In the same way, mathematical developments are not shown here, but readers may refer to the mentioned bibliography. Experimental and simulation results in this paper show that these control approaches are a useful and efficient alternative that may be implemented at industrial scale at a relatively low cost.

2 Modeling

Nowadays, many of the monitoring and control techniques available in the literature belong to those called "model-based". Such techniques require a mathematical model that reproduces, in a suitable way, the dynamics of the key variables of the process in order to be applied. As a consequence, the dynamical modeling of AD processes has been an active research area during the last five decades. In 1968, Andrews introduced the Haldane function to characterize growth inhibition in order to emphasize the process instability due to the accumulation of VFA (Andrews 1968). The modeling of AD processes was later improved by considering three stages: solubilization of organics, acidogenesis, and methanogenesis (Hill and Barth 1974). Afterward, a four-population model with two acidogenesis reactions and two methanization reactions emphasizing the role of hydrogen was proposed (Mosey 1983).

In the later years, other AD models were proposed (Moletta et al. 1986; Kiely et al. 1997; Batstone et al. 1997). However, most of them described in detail particular characteristics of the AD process resulting in complex models which were difficult to use for control purposes (Bastin and Dochain 1990). The latest important AD model was introduced at the beginning of this century for a specialist group of the International Water Association (IWA) proposed a general model for the AD process called ADM1. Such a model includes several bacterial populations and distinct substrates resulting in up to 26 dynamical state variables and 19 biochemical kinetic processes which resulted in a powerful modeling tool for studying and better understanding this complex process, but it is difficult to use for control purposes. Until now, on the knowledge of authors, it has been used only in a few ICA approaches for monitoring process employing asymptotic interval observers (see for instance Montiel-Escobar et al. 2012). On the other hand, most of the proposed models assume that the system behaves as a continuous stirred tank reactor (CSTR). Moreover, in practice one of the main objectives is to increase the contact surface between the organic matter and the microorganisms within the reactor in order to improve the process efficiency. This fact has brought a considerable growth in the use of fixed or fluidized bed reactors in the last decades. In this way, the assumption about CSTR-behavior becomes of a great interest because the different flow regimes that can be induced. Therefore, the proposal of dynamical models aimed for different AD systems, and at the same time suitable for monitoring and control becomes very important as well.

Almost at the same time that ADM1, a model based on mass balances was proposed for the monitoring and control of a fixed-bed AD process (Bernard et al. 2001). This model, named AM2, comes originally from the model developed by Graef and Andrews (1974) and Rozzi (1984), who included alkalinity, bicarbonate, and pH influences within the AD dynamics. Other important features of this model were (a) the structural properties of the model were improved in the seek of simplicity; (b) a second bacterial population was introduced to reproduce better the phase of destabilization due to the excessive accumulation of VFA; (c) gaseous flows with respect to the chemical and biological species into the digester were added; and (d) a simplified kinetic model was proposed. This model was successfully validated and identified using a wide range of operating conditions in a semi-industrial fixed-bed AD process located in the Laboratory of Biotechnology and Environment (LBE-INRA) in Narbonne, France for the treatment of red wine vinasses, but it has been also adapted to describe a wide range of AD systems (e.g., CSTR, UASB, and FBR). This model is considered in the rest of this paper. Considering other biotechnology ecosystems able to be coupled with AD in a biorefinery concept, they can be mentioned some recent research works mainly in microalgae production (Lee et al. 2015) and in MEC systems (Tartakovsky et al. 2011; Pinto et al. 2011; Blanco-Cocom et al. 2013).

3 Process Monitoring

Concerning process monitoring, it is well known that today there exists online sensors more and more accurate providing state information at the industrial scale. However, they are still very expensive, and their maintenance is usually time-consuming, which prevents its acquisition for SME enterprises. Furthermore, biological processes are highly nonlinear time-varying systems in which kinetic parameters are badly or poorly known. To overcome these difficulties, the notion of software sensors has been introduced in the early 80s. In fact, these software sensors simply consist in mathematical model-based algorithms called observers designed to estimate unmeasured states from the available online measurements, and they have demonstrated to be of great help for the automatic control of biological processes (Bastin and Dochain 1990; Gauthier and Kupka 1994; Farza et al. 1998). However, in most of software sensors, the knowledge of all the inputs of the process, including, for example, the substrate input concentration, is needed. In the case of food or pharmaceutical industries, which normally operate in very well-controlled conditions, this information is almost always available. Unfortunately, this is normally not the case in the wastewater treatment field in which process inputs may vary seasonally or even more, to change suddenly in function of a lot of factors like daily use of water resources, agro-industrial operational conditions, and even cultural practices. As a consequence, part of the process inputs vector is considered as unmeasured input disturbances and classical observer schemes cannot be used since the system may become not observable or even not detectable depending on the available output measurements. In this context, two relatively new robust nonlinear observers have found a wide acceptation in biological process, including of course AD. Such robust observers are capable of coping simultaneously with the aforementioned problems while remaining easy to implement with a minimum number of straightforward conditions to verify. The first one is the so-called *asymptotic observer* (Bastin and Dochain 1990; Alcaraz-Gonzalez and Gonzalez-Alvarez 2007), which, even though requiring the knowledge of the process inputs, has the main advantage to allow for the exact cancellation of the nonlinear terms of the systems and this facilitates its design, stability analysis, and implementation. The second one is, in fact, a complete class of observers termed *interval observers*, which allow for reconstruction of a guaranteed interval on the unmeasured states instead of reconstructing their precise numerical values, assuming that only guaranteed lower and upper limits on the inputs are available (Gouzé et al. 2000; Moisan et al. 2009). These two observers as well as the aforementioned AM2 model are used in this paper in the control approaches used as representative example as described in the next section.

4 Process Control and Optimization

Despite the success reported from experiments in which simple (but effective) control strategies have been applied to improve the productivity of biological WTTP, automation is generally not the result of an in-depth analysis of the process using current control theory tools (Bastin and Van Impe 1995; Mailleret et al. 2004). Furthermore, experiments are not always supported by a previous mathematical analysis that ensures good performance in all possible conditions. Moreover, they focus on optimizing operation points (with fixed input variables) and in general they do not consider the dynamical process response to perturbations (e.g., sudden steeps or gradual changes in the input variables). For this reason, it is important to apply generic but innovative control theory approaches, namely, the use of both dynamical mathematical models to numerically simulate the processes and control approaches based on these models in order to avoid instability and to optimize the process. Through the use of suitable control techniques based on this concept, it is possible to numerically simulate the process and to observe phenomena (even discover them) that otherwise could be unperceived, because it is very time-consuming to run dedicated experiments or because it may drastically disturb the experimental system.

In general, it is well known that the implementation of ICA approaches in biotechnological processes is not an easy task due to their always changing dynamics. In fact, they are very often exposed to disturbances that make difficult to reach a stable steady state. In the best cases, they are rather fluctuating around the neighborhood of a desired steady state, but with the risk, always present, to exit from the stable zone or to move to another, possibly not desired, steady-state neighborhood. Furthermore, biological processes in general are known for having highly nonlinear behaviors that make even more difficult the design and implementation of control approaches. This fact reveals the importance to develop control schemes that are robust against model uncertainties, process perturbations, and nonlinearities while using a limited number of online measurements and/or a mathematical model of limited complexity. Among the different control approaches available in the literature, those that optimize some criterion are the most useful ones. They are usually based on a mathematical model of the process and impose constraints on the inputs and outputs of the process to ensure that it operates within the permissible limits (Marsili-Libelli 1989; Méndez-Acosta et al. 2011). Unfortunately, many control approaches often run in open loop, making strong assumptions about the model, such as its linearization around a particular operating point, or assuming ideal conditions in such a way that it is not possible to apply them in a feedback control scheme (Steyer et al. 2006). For example, (Tartakovsky et al. 2011) showed that it is possible to control the production of bioH_2 by manipulating the applied voltage. In other work, Chinellato et al. (Chinellato et al. 2013) showed a technique for regulating the pH in a two-phase semi-continuous reactor by recirculating the digestate. However, in these studies, the methodology for choosing the control inputs is based on experimental evidence, which is relatively good, but in both cases, it is built as a trial-and-error approach rather than a systematically calculated control law. On the other hand, a deep and adequate analysis of the model in combination with other sources of fundamental knowledge may allow designing of relatively simple controllers Van Impe et al. 1992.

4.1 Example 1: SISO Regulation of the COD

The first example showed in this paper is the Single-Input Single-Output (SISO) regulation of COD in a continuous up-flow fixed-bed anaerobic digester for the degradation of red wine vinasses (Alcaraz-Gonzalez et al. 2005). In this case, the objective was to reduce as much as possible the organic pollution expressed as COD. For doing that, an adaptive robust control law was employed resulting in the dilution rate, D (h^{-1}), as control input. Nevertheless, this control law is a function of growth kinetic functions and parameters, as well as online measurements of process inputs and biomass concentrations. All of these were considered uncertain or even unknown. Concerning process inputs uncertainties; this was solved at this time by using available "nominal" values, i.e., information from weekly offline measurements (often monthly) and establishing confidence intervals. However, this issue was more difficult about biomass. As the digester normally cannot be open, it is very difficult to take samples for measuring biomass. Even if some devices exist today for sampling biomass in AD process, their analysis is extremely time-consuming for online monitoring and control purposes. Then, it was necessary using a software sensor for estimating biomass instead of measuring it. However, most of the observers are function of kinetics as well. Then, in this case, the problem caused for the lack of knowledge on kinetics, process inputs, and biomass concentration was solved to a great extent by using a combination of an asymptotic observer (Bastin and Dochain 1990; Alcaraz-Gonzalez and Gonzalez-Alvarez 2007) which is the only that is robust against a totally ignorance on kinetics terms, and it became robust against the process inputs uncertainties by building an interval observer (Kiely et al. 1997; Alcaraz-Gonzalez and Gonzalez-Alvarez 2007). Figure 1 shows the dynamical behavior of the regulated variable, the COD, inside a tolerance interval around the desired set point. In the graphic for D, it can be appreciated two curves: The blue one is the "calculated" control law, while the black one is the actual actuator signal. Fortunately, in this case the actuator followed correctly the controller order, which contributed to the success in this probe. However, it is also well known that faults in sensors and actuators may cause serious malfunction in automated plants. A real study on such cases carried out in this digester was developed in Alcaraz-González et al. (2012), in which an interval observer was also used as Faults Detection and Diagnosis (FDD) tool. Thus, in the current case of Fig. 1, the combination of a good model, an asymptotic interval observer, and the application of these approaches for designing a robust adaptive control law allowed to regulate the COD with a very good accuracy even in the presence of process input perturbations and all the uncertainties mentioned beforehand.

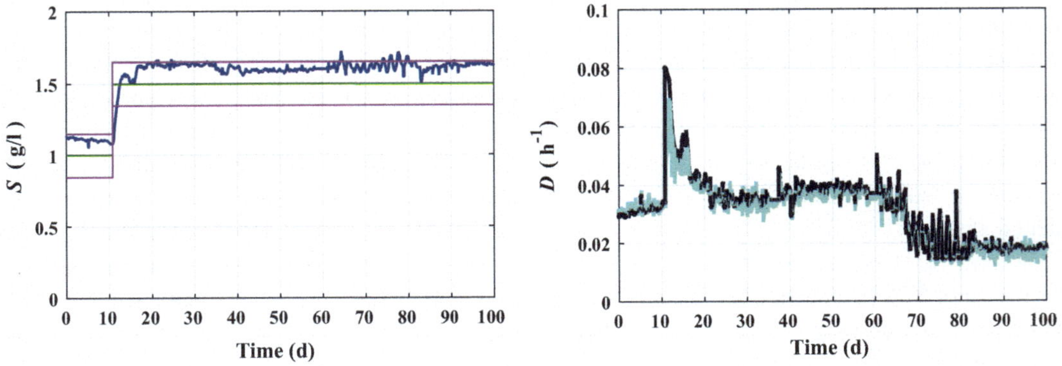

Fig. 1 Robust SISO regulation of the COD, S (g/l): green: set point; blue: measurement; magenta: tolerance interval. The dilution rate D (h^{-1}) is used as control input (black: calculated; blue: measured) (Alcaraz-Gonzalez et al. 2005)

4.2 Example 2: MIMO Indirect Servo-Control of Alkalinities

Other example of successful application of robust adaptive control strategies is presented here as the Multi-Input Multi-Output (MIMO) regulation of alkalinities in AD processes (Alcaraz-González et al. 2013) (see Fig. 2). It is well known that AD suffers from substrate inhibition in the methanogenic phase caused by VFA accumulation, and it is also well known that holding a balanced physicochemical equilibrium inside the digester is the best form to avoid this trouble. In fact, it is well accepted that the key for establishing a healthy physicochemical equilibrium is to hold the total alkalinity TA as TA > 60 mEq/l and the relationship between intermediate alkalinity IA and total alkalinity as IA/TA < 0.3 (Ripley et al. 1986). The problem is that nor IA neither TA are represented as state variables in current models in the literature. Then, the first challenge was to express these variables as true state variables. In the current case presented in this paper, it was done by representing alkalinities as functions of the AM2 state variables VFA (S_2 mmol/l) and the total concentration of strong ions Z (mmol/l). Thus, the original control objectives TA > 60 mEq/l and IA/TA < 0.3 were translated as inequalities for S_2 and Z that were function of the pH. The control inputs in this case were the reactor dilution rate D_1 (d^{-1}) and a special dilution rate D_2 (d^{-1}) for a concentrate NaOH solution in such a way that the total dilution D_T rate was $D_T = D_1 + D_2$. Laws for these control inputs were also designed with the use of interval observers and taking into consideration the same uncertainties that in Example 1. In addition, this approach was derived from the AM2 model that in fact incorporates the physicochemical equilibrium of the system in order to indirectly regulate both total alkalinity and the ratio IA/TA by regulating S_2 and Z, while guaranteeing the so-called operational stability. In fact, this approach is a dynamical optimization rather than a regulation around a set point as it is classically applied in most of the cases. This is because S_2 and Z reference values are represented as inequalities rather than specific set points, which are necessary to hold. In this sense, the so-called regulation problem, like in the former example, becomes a servo-control problem that in this case is a function of the pH. In Fig. 2, in the upper left position there are represented the input pH around 5 and the inside digester pH around 7.5. The dotted line for the digester pH represents a theoretical reference, while the continuous line represents the actual pH. As it can be seen, the pH changes as a consequence of pH and other inputs perturbations, which in turn causes changes in the references for S_2 and Z. In the lower left position in the figure, these references are represented in dashed line, while the actual S_2 and Z are represented in continuous line. As it can be seen, S_2 and Z follow these changing references by the action of the control inputs D_1 (continuous line) and D_2 (dashed line) represented in the upper right position in the figure. Notwithstanding, the more interesting fact is to see how the original control objectives concerning alkalinities follow an exact and constant reference, as it was originally set (lower right position in the figure).

4.3 Example 3: A Virtually Controlled Observer (VCO) for the Estimation of Process Inputs

As it has been mentioned in the former examples, a serious problem in the application of many advanced control approaches in WWTP is the lack of knowledge on kinetics and the process inputs uncertainties. Thus, in this example, it is presented an approach that controls not the actual process but an asymptotic observer for estimating the real inputs in the process (Alcaraz-Gonzalez et al. 2017) (see Fig. 3). Again, this approach was carried out in a continuous up-flow fixed-bed anaerobic digester for the degradation of red wine vinasses. The idea is to design a robust adaptive MIMO control law in which the regulated variables are the estimation errors between the COD measurement (here represented by S_1 (gCOD/l)) and its estimate as well as the measurement of Z and its estimate. However, in this case, the control inputs are not the Dilution rates

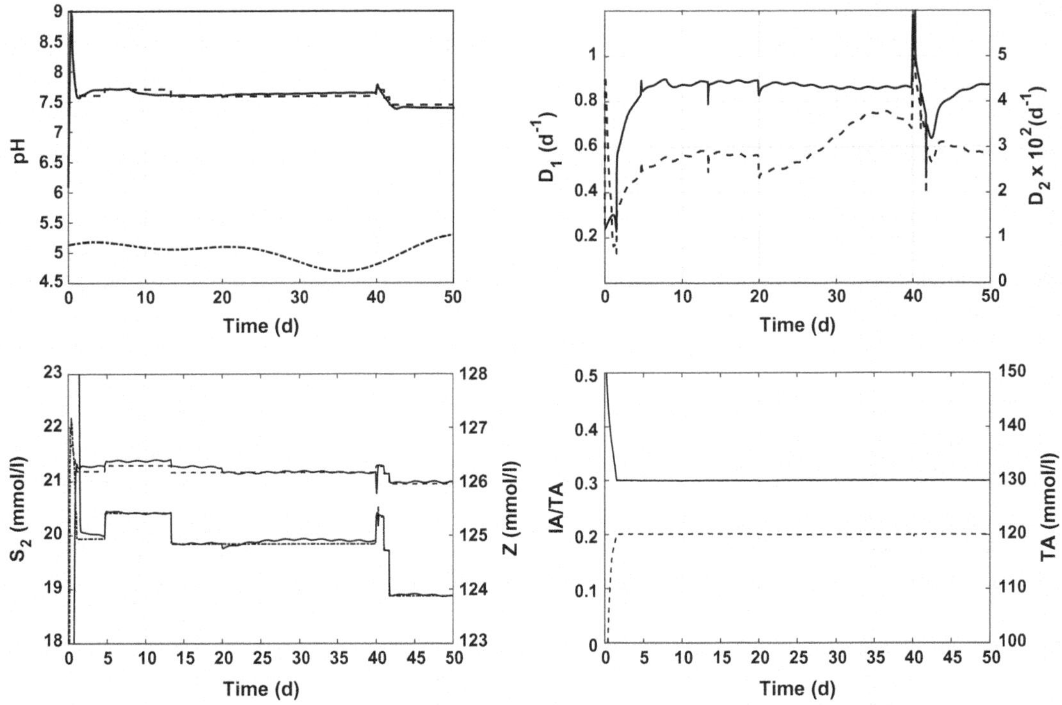

Fig. 2 MIMO robust control of alkalinities (Alcaraz-González et al. 2013)

but the hypothetical process inputs $S^{in}{}_1$ (gCOD/l)) and $S^{in}{}_2$ (mmol/l), which are considered unknown, and they are the variables that are desired to estimate. The set point for the errors is zero in both cases and of course the correct estimation for the process inputs is achieved when the measurements of S_1 and Z match with their respective estimates. In Fig. 3, the model predictions for measured variables are also shown for comparison and discussion purposes. As it can be seen, the estimation is not very accurate when model, measurements, and estimates do not coincide adequately, and it behaves better in this sense when three variables match in a more suitable form. Once again, this fact reveals the importance to have a reliable model for designing these approaches. Nevertheless, even in the presence of process uncertainties, relatively minor model mismatches or measurement faults advanced robust ICA approaches are able to monitor, optimize, and control WWTP in a relatively properly way.

5 Toward the Future: The Biorefinery Challenge

In the previous sections, some literature has been cited in the WWTP field concerning modeling, monitoring, optimization, state variables estimation, and control, all of them belonging to that is named ICA approaches. Moreover, it has been pointed out, mainly in the introduction, that the current tendency in WWTP is not only on the depollution but onto the biorefinery concept in which besides of that, nutrients and energy may be recovered and valorized. However, most of the biotechnologies that are capable to be coupled with WWTP in order to achieve this objective (e.g., dark fermentation, two-stage AD, microalgae production, MEC systems) are still in the stage of understanding and exploration of phenomena of in the best of cases some models already exist. Furthermore, besides modeling, no other ICA approaches have been at least outlined,

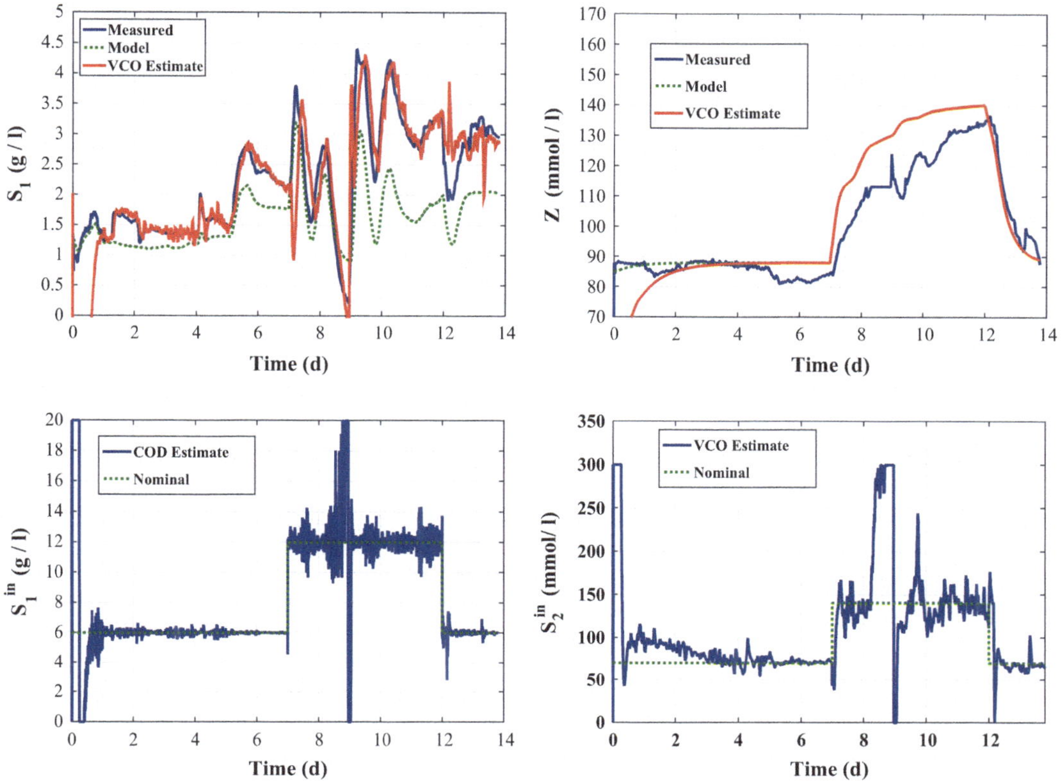

Fig. 3 Process input estimation by using a virtually controlled observer (VCO) (Alcaraz-Gonzalez et al. 2017)

though much of them could be extrapolated directly and relatively easily. Nevertheless, the true challenge is now to make work all these biotechnologies together and to make it efficiently. For achieving this objective, modeling couples of processes or even of a complete biorefinery including more than two processes have to be proposed. Once this task starts to give reliable results, the rest of ICA approaches can start to work also in parallel, for both, in numerical simulations and the real field. Through the three examples shown in this paper, it has been highlighted that ICA approaches are a reliable and efficient way to optimize and control WWTP and that they are necessary when there exists uncertainties, perturbations, and lack of knowledge or information, as it is in the case of WWTP and it will surely in a biorefinery.

6 Conclusions

Designing of feedback control approaches that are based on proper mathematical models and other tools, like software sensors, can significantly improve the performance of the processes. Formal tools of control theory are used to ensure a smooth operation in closed loop. Unlike the usual optimization schemes used in many biotechnological processes, which is carried out around the set points of operation (system's steady states), the use of dynamical models is that it allows also taking into account the system transient dynamical behavior, which add further advantages. The design of controllers can be proven also in simulation, adding realistic conditions such as measurement noise, parametric uncertainty, and the dynamics of actuators and sensors. Methods, like adaptive control or predictive control, that are better adapted to face nonlinear behaviors, have found industrial applications in WWTPs. These control approaches have proven their robustness in the presence of uncertainties, perturbations, high nonlinear behavior, and little online information. On the other hand, AD processes are receiving increased attention as a core process for next-generation wastewater treatment systems. Generically, this process is receiving a high level of international research mainly under the concept of biorefinery, i.e., full recovery of resources, including water, nutrients, and energy. AD is a core component of this new concept, as the key biochemical process for mobilization of energy and nutrients. Indeed, incomplete treatment of wastewater may induce severe process malfunction in the plant, which not only reduces the quality of the treatment, but may also have a major economic impact. Thus, the need for optimally controlled plants will increase due to tightening permit requirements and the need to reduce cost. In order to fulfill the requirements related to water quality and the new tight ecological norms, as well as to reduce costs, the optimal control of WWTPs faces important uncertainties arising from the intrinsic complexity of the plant design.

Acknowledgements This work was partially supported by Projects CONACyT CB-2008-01/101971, FORDECYT/116655. Authors also thank DAAD and the exceed II/Swindon Project for financial support provided.

References

Alcaraz-Gonzalez V, Gonzalez-Alvarez V (2007) Robust nonlinear observers for bioprocesses: application to wastewater treatment. In: Méndez-Acosta HO, Femat R, Gonzalez-Alvarez V (eds) Selected topics in dynamics and control of chemical and biological processes. Springer, Germany

Alcaraz-Gonzalez V, Harmand J, Rapaport A et al (2005) Robust interval-based regulation for anaerobic digestion processes. Wat Sci Technol 52(1–2):449–456

Alcaraz-González V, López-Bañuelos RH, Steyer JPh et al (2012) Interval-based diagnosis of biological systems—a powerful tool for highly uncertain anaerobic digestion processes. Clean Soil Air Water 40(9): 941–949

Alcaraz-González V, Fregoso-Sánchez FA, Méndez-Acosta HO et al (2013) Robust regulation of alkalinity in highly uncertain continuous anaerobic digestion processes. Clean Soil Air Water 41(12):1157–1164

Alcaraz-Gonzalez V, Jauregui-Medina EA, Steyer JPh et al (2017) Simultaneous COD and VFA unmeasured process inputs estimation in actual anaerobic wastewater treatment processes. Control Eng Pract 60: 118–123

Andrews J (1968) A mathematical model for the continuous culture of microorganisms utilizing inhibitory substrates. Biotechnol Bioeng 10:707–723

Bastin G, Dochain D (1990) On-line estimation and adaptive control of bioreactors. Elsevier, Amsterdam

Bastin G, Van Impe JF (1995) Nonlinear and adaptive control in biotechnology: a tutorial. Eur J Control 1:37–53

Batstone DJ, Keller J, Newell RB et al (1997) Model development and full scale validation for anaerobic treatment of protein and fat based wastewater. Wat Sci Technol 36:423–431

Bernard O, Hadj-Sadok Z, Dochain D et al (2001) Dynamical model development and parameter identification of an anaerobic wastewater treatment process. Biotech Bioeng 75:424–438

Blanco-Cocom L, Guerrero-Alvarez A, Dominguez-Maldonado J et al (2013) Mathematical model for a continuous hydrogen production system: stirred fermenter connected to a biocatalyzed electrolysis cell. Biomass Bioenergy 48:90–99

Cea-Barcia G, Buitrón G, Moreno G et al (2014) A cost-effective strategy for the bio-prospecting of mixed microalgae with high carbohydrate content: diversity fluctuations in different growth media. Biores Technol 163:370–373

Chinellato G, Cavinato C, Bolzonella D et al (2013) Biohydrogen production from food waste in batch and semi-continuous conditions: evaluation of a two-phase approach with digestate recirculation for pH control. Int J Hydrogen Energy 38:4351–4360

Farza K, Busawon K, Hammouri H (1998) Simple nonlinear observers for on-line estimation of kinetic rates in bioreactors. Automatica 34(3):301–318

Foglia D, Ljunggren M, Wukovits W et al (2010) Integration studies on a two-stage fermentation process for the production of biohydrogen. J Clean Prod 18(1):S72–S80

Gauthier JP, Kupka I (1994) Observability and observers for nonlinear systems. SIAM J Control Optim 34(4): 975–994

Gouzé JL, Rapaport A, Hadj-Sadok MZ (2000) Interval observers for uncertain biological systems. Ecol Model 133:45–56

Graef S, Andrews J (1974) Mathemathical modelling and control of anaerobic digestion. Wat Res 8:261–289

Guo XM, Trably E, Latrille E et al (2010) Hydrogen production from agricultural waste by dark fermentation: a review. Int J Hydrogen Energy 35:10660–10673

Hill DT, Barth C (1974) A dynamic model for stimulation of animal waste digestion. Wat Pollut Control Assess 10:2129–2143
Jimenez J, Latrille E, Harmand et al (2015) Instrumentation and control of anaerobic digestion processes: a review and some research challenges. Rev Environ Sci Bio/Technol 14(4):615–648
Kiely G, Tayfur G, Dolan C et al (1997) Physical and mathematical modelling of anaerobic digestion of organic wastes. Wat Res 31:534–540
Lee E, Jalalizadeh M, Zhang Q (2015) Growth kinetic models for microalgae cultivation: a review. Algal Res 12:497–512
Liu Z, Zhang C, Lu Y et al (2013) States and challenges for high-value biohythane production from waste biomass by dark fermentation technology. Biores Technol 135:292–303
Logan BE, Regan JM (2006) Elecricity-producing bacterial communities in microbial fuel cells. Trens Microbiol 14:512–518
Lovley DR (2006) Bug juice: harvesting electricity with microorganisms. Nat Rev Microbiol 4:497–508
Mailleret L, Bernard O, Steyer JP (2004) Robust nonlinear adaptive control for bioreactors with unknown kinetics. Automatica 40(8):365–383
Marone A, Ayala O, Trably E et al (2017) Coupling dark fermentation and microbial electrolysis to enhance biohydrogen production from agro-industrial wastewaters and by-products in a bio-refinery framework. Int J Hydrogen Energy 42:1609–1621
Marsili-Libelli S (1989) Modelling, identification and control of the activated sludge process. Adv Biochem Eng Biotechnol 38:90–148
Meier L, Perez R, Azocar L et al (2015) Photosynthetic CO_2 uptake by microalgae: AN attractive tool for biogas upgrading. Biomass Bioenergy 73:102–109
Méndez-Acosta HO, Snell-Castro R, Alcaraz-Gonzalez V et al (2010) Anaerobic treatment of Tequila vinasses in a continuous CSTR-type digester. Biodegradation J 21(3):357–363
Méndez-Acosta HO, Garcia-Sandoval JP, Gonzalez-Alvarez V et al (2011) Regulation of the organic pollution level in anaerobic digesters by using off-line COD measurements. Biores Technol 102(17):7666–7672
Metcalf and Eddy (eds) AECOM (2014) Wastewater engineering: treatment and reuse, 5th edn. McGraw Hill Education, New York
Moisan M, Bernard O, Gouzé JL (2009) Near optimal interval observers bundle for uncertain bioreactors. Automatica 45:291–295
Moletta R, Verrier D, Albagnac G (1986) Dynamic modeling of anaerobic digestion. Wat Res 20:427–434
Montiel-Escobar JL, Alcaraz-González V, Méndez-Acosta HO et al (2012) ADM1-based robust interval observer for anaerobic digestion processes. Clean Soil Air Water 40(9):933–940
Mosey F (1983) Mathematical modeling of the anaerobic digestion process: regulatory mechanisms for the formation of short-chain volatile acids from glucose. Wat Sci Technol 15:209–232
Ntaikou I, Antonopoulou G, Lyberatos G (2010) Biohydrogen production from biomass and wastes via dark fermentation: a review. Waste Biomass Valorization 1(1):21–39
Olsson G, Carlsson B, Comas J et al (2014) Instrumentation, control and automation in wastewater—from London 1973 to Narbonne 2013. Water Sci Technol 69(7):1373–1385
Pinto RP, Srinivasan B, Escapa A et al (2011) Multi-population model of a microbial electrolysis cell. Environ Sci Technol 45(11):5039–5046
Ripley LE, Boyle WC, Converse JC (1986) Improved alkalimetric monitoring for anaerobic digestion of high-strength wastes. J Wat Pollut Control Fed 58(5): 406–411
Rozendal RA, Hamelers HVM, Euverink GJW et al (2006) Principle and perspectives of hydrogen production through biocatalyzed electrolysis. Int J Hydrogen Energy 31:1632–1640
Rozzi A (1984) Modeling and control of anaerobic digestion processes. Trans Inst Meas Control 6:153–159
Steyer JP, Bernard O, Batstone D et al (2006) Lessons learnt from 15 years of ICA in anaerobic digesters. Wat Sci Technol 53(4):25–33
Tartakovsky B, Mehta P, Santoyo G et al (2011) Maximizing hydrogen production in a microbial electrolysis cell by real-time optimization of applied voltage. Int J Hydrogen Energy 36(17):10557–10564
Van Impe JF, Nicolaï BM, Vanrolleghem PA et al (1992) Optimal control of the penicillin G fed-batch fermentation: an analysis of a modified unstructured model. Chem Eng Commun 117:337–353

Evaluation of Wastewater Treatment-via-Dilution for Tala Drain, Nile Delta, Egypt

A. A. Hassan, A. Nada, M. Elshemy, and B. A. Zeidan

Abstract

Agricultural drainage water reuse is well developed in the Nile Delta region in Egypt. The frequently applied treatment practice is the treatment by dilution. This practice consists in mixing fresh water with drainage water. The Rosetta Branch is one of two main Nile Delta branches. It is also the main fresh water resource for the Nile Delta governorates. However, the fresh water quality of this branch is frequently affected by the Tala Drain which transports discharge from agricultural drainage into Rosetta Branch. The wastewater in the Tala Drain is mostly treated by dilution. Hence, this study aims at assessing the impact of this treatment practice on the water quality in the Tala Drain and Rosetta Branch. We defined three water quality indicators to quantify the treatment for Tala Drain and the pollution for Rosetta Branch. This was done based on using seasonal water quality records for one year and two water quality indices. The results indicated that the treatment-via-dilution clearly minimizes the main pollutants in the Tala Drain, such as fecal coliform and raises the dissolved oxygen concentration. For Rosetta Branch, the results showed minor increases in some water quality parameters such as turbidity and nitrate, while the branch water quality status was generally reduced by about 17% according to the used indices. The study concludes that the application of additional drainage water treatment practices is urgently required in the Tala Drain. Furthermore, we recommend to apply this method to assess water quality in other branches and drains in Egypt.

Keywords

Water quality index (WQI) · Treatment · Dilution · Drainage water reuse (DWR) · Rosetta branch

1 Introduction

Egypt depends on agricultural drainage water reuse to fill the wide gap in its water resources (Karajeh et al. 2011). Shaban (2014) referred to the "ever-increasing gap" in the water supply of Egypt and classified drainage water reuse as the

A. A. Hassan (✉)
Faculty of Engineering, Ain Shams University, Cairo, Egypt
e-mail: ahmad9657@yahoo.co.uk

A. Nada · M. Elshemy · B. A. Zeidan
Faculty of Engineering, Tanta University, Tanta, Egypt
e-mail: ali.nada123@yahoo.com

M. Elshemy
e-mail: m.elshemy@f-eng.tanta.edu.eg

B. A. Zeidan
e-mail: b.zeidan@f-eng.tanta.edu.eg

L. Ribbe et al. (eds.), *Towards Water Secure Societies*,
https://doi.org/10.1007/978-3-030-50653-7_5

"most economic and practical option" which Egypt should use to fill this gap. Barnes (2014) reported that agricultural drainage water reuse raised Egypt's available water resources by 20%. Karajeh et al. (2011) reported that the main course of the Nile River, in Egypt, is receiving the effluents of all agricultural drains in the upper part of the country (from Aswan to Cairo). The authors estimated that quantity by about 4 billion cubic meter/year and referred that the reused drainage water in Egypt includes treated and untreated municipal wastewater and industrial effluents (Karajeh et al. 2011). In addition, Barnes (2014) referred to the negative water quality impacts of drainage water on the Egypt's irrigation network.

Nile River is the main source of fresh water for Egypt. Northern to Cairo, the Nile River is divided into two branches, the Rosetta Branch to the west and the Damietta Branch to the east, forming the Nile Delta. Rosetta Branch, which is subjecting to different sources of pollution, is the case study for this work. Five major agricultural drains: ELRahawy, Sabal, El Tahrir, Zaweit El-Bahr, and Tala, directly discharge their drainage water in the Rosetta Branch (Donia 2005). Tala Drain, which is the focus of this study, is considered as the second major source of pollution along the Rosetta Branch after the EL Rahawy Drain (MWRI 2005). It receives agricultural drainage, domestic wastewater, and industrial wastewater. In 2014, the water quality status of this drain was investigated (Mostafa 2014), and the results revealed relatively high levels of BOD and COD; the concentrations of the drain's water quality parameters do not meet the permissible limits of the Egyptian law No. 48/1982.

The Ministry of Water Resources and Irrigation (MWRI) launched a National Water Resources Plan for Egypt (NWRP). The plan describes how Egypt will safeguard its water resources in the future, both with respect to quantity and quality, and how it will use these resources in the best way from an economic and environmental point of view (MWRI 2010). Agricultural drainage water may be reused individually/or after mixing it with natural water in appropriate quantities according to a particular use. Nevertheless, the strategy of operation for drainage water reuse is mainly dependent on the assessment of water quality characteristics.

Water quality index (WQI) is considered as one of the most effective tools which can be used to evaluate the water quality status using a single value indicating the overall quality of water (Abbasi 2012). WQI describes the status of water quality to the public as well as decision- and policy-makers. It is based on the ensemble of the records of physical, chemical, and biological parameters of the water body using specific methods (Simões et al. 2008). Examples for various water quality indices which have been developed are National Sanitation Foundation Water Quality Index (NSFWQI) and Canadian Council of Ministers of the Environment Water Quality Index (CCMEWQI), which are considered as the most frequently used water quality indices for rivers (Elshemy 2010).

The main objective of this study is to assess, how the drainage water treatment-via-dilution for Tala Drain at Kafr El-Zayat city contribute to improve water use efficiency based on field records and water quality index (WQI) approach using three different water quality indicators.

2 Study Area

Figure 1 shows the layout of the study area which covers the outlet of the Tala Drain and the receiving Rosetta Branch at Kafr El-Zayat City. The study area is located at the western border of the Nile Delta. A specific length of the Rosetta Branch (about 2 km) was considered to have a zone before the Tala Drain outlet, and the other zone includes the outlet of the Tala Drain and ends just before the impact of the industrial zone at Kafr El-Zayat City. The first zone starts from at the downstream of Kafr El-Zayat Highway Bridge to the downstream of Kafr El-Zayat Railway Bridge at Kafr El-Zayat City. Selection of the study area limits was intended to avoid the impacts of the industrial wastes from the industrial companies on the water quality conditions of the branch and to have only the impact of Tala Drain effluents.

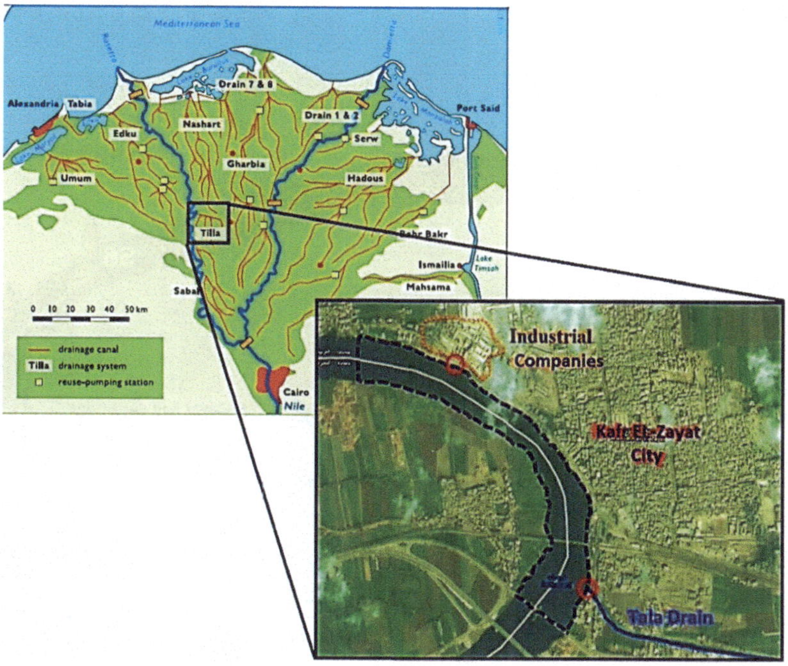

Fig. 1 The layout of the study area (modified after (MWRI 2005))

3 Methodology

To evaluate the treatment-via-dilution practice for Tala Drain by discharging its effluent in Rosetta Branch, some water quality indicators were used (Sect. 3.3). These indicators are based on different seasonal water quality records for Rosetta Branch and Tala Drain and on two different water quality indices. A water quality monitoring program was implemented to seasonally record required water quality parameter measurements (see Sect. 3.1). Two water quality indices were used to evaluate the water quality status of the Rosetta Branch before and after receiving Tala Drain effluents (see Sect. 3.2).

3.1 Data Collection

Six sampling stations (Fig. 2) were considered for this study to monitor the water quality conditions of the Rosetta Branch and Tala Drain at the blending zone. Five stations (S1 & S3–S6) are located in the Rosetta Branch, while one station (S2) is located in the Tala Drain and by its end by about 300 m. S1 and S3 represent the Rosetta Branch conditions before the discharging of Tala Drain, while S4–S6 represent the branch conditions after the drain discharge.

The water quality monitoring program for this work included seasonal measuring of different water quality parameters (13 physical, chemical, and biological parameters) at the considered sampling stations starting from autumn 2014 to summer 2015 (four field trips) according to APHA guidelines (2005). The considered water quality parameters are water temperature (T) [° C], turbidity [NTU], pH, total dissolved solids (TDS) [mg/l], dissolved oxygen (DO) [mg/l], biochemical oxygen demand (BOD) [mg/l], chemical oxygen demand (COD) [mg/l], nitrate (NO3-) [mg/l], phosphate (PO43-) [mg/l], sulfate (SO42-) [mg/l], iron (Fe) [mg/l], manganese (Mn) [mg/l], and fecal coliform (#/100 ml). Some of these records were collected in situ, using the equipment of the Department of Irrigation and Hydraulics Engineering laboratory, Tanta University, Egypt, while the other were analyzed at the Water Supply and Sewerage Company laboratory, Tanta, Egypt.

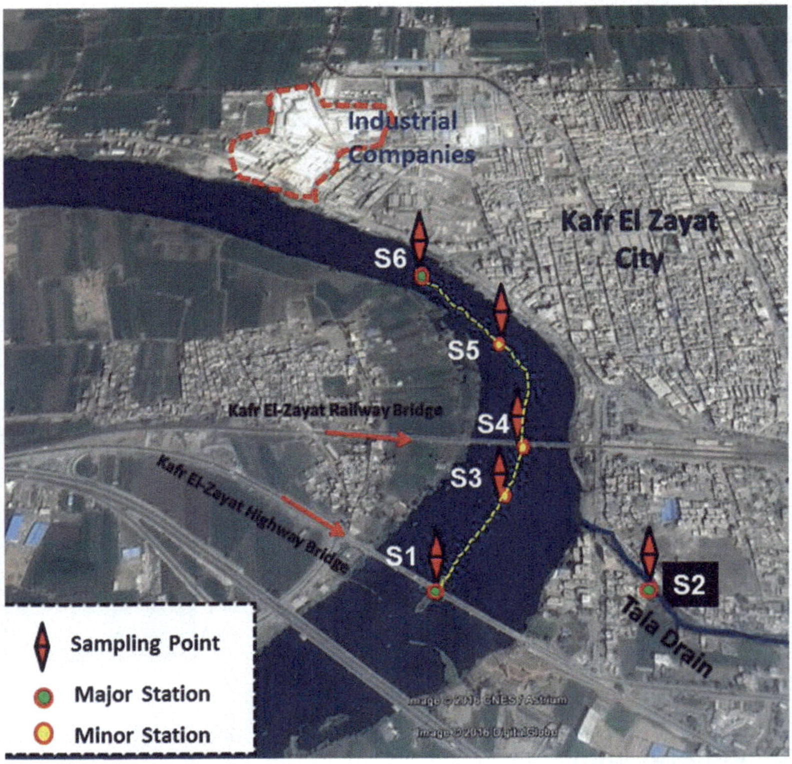

Fig. 2 Layout of the considered sampling stations (modified after Google Earth)

Seasonal and average (2014–2015) records of the considered water quality parameters at selected sampling stations in Rosetta Branch and Tala Drain for the study period are presented in Fig. 3 and Fig. 4, respectively.

3.2 Water Quality Index (WQI)

The Canadian Council of Ministers of the Environment Water Quality Index (CCMEWQI) and the National Sanitation Foundation Water Quality Index (NSFWQI) were used in this study. Both indices are described in the following sections.

3.2.1 CCMEWQI

CCMEWQI was developed in 1997 to provide a simple measure for the water quality records deviation from its local guidelines (CCME 2001). CCMEWQI can be calculated as follows (CCME 2001):

$$\text{CCMEWQI} = 100 - \frac{\sqrt{F_1^2 + F_2^2 + F_3^2}}{1.732} \quad (1)$$

where F_1, F_2, and F_3 represent the scope, frequency, and amplitude, respectively, of the water quality records. These factors are mainly calculated in the context of the applied guidelines. CCMEWQI is described in more detail in its published technical report (CCME 2001). The index categorizes the water quality status of a water body into five categories according to the value of CCMEWQI (see Table 1).

This index was used all over the world and for different cases (Boyacioglu 2010; CCME 2004; Donia and Farag 2010; Ed Quilty 2003; Heuvel 2009; Hurley et al. 2012; Lumb et al. 2006; Mercier et al. 2005; Radwan 2005; Selvam et al. 2014). One of the earliest uses of this index in Egypt was done in 2008, based on the Egyptian guidelines for surface fresh waters, for the River Nile, Egypt (Khan et al. 2008).

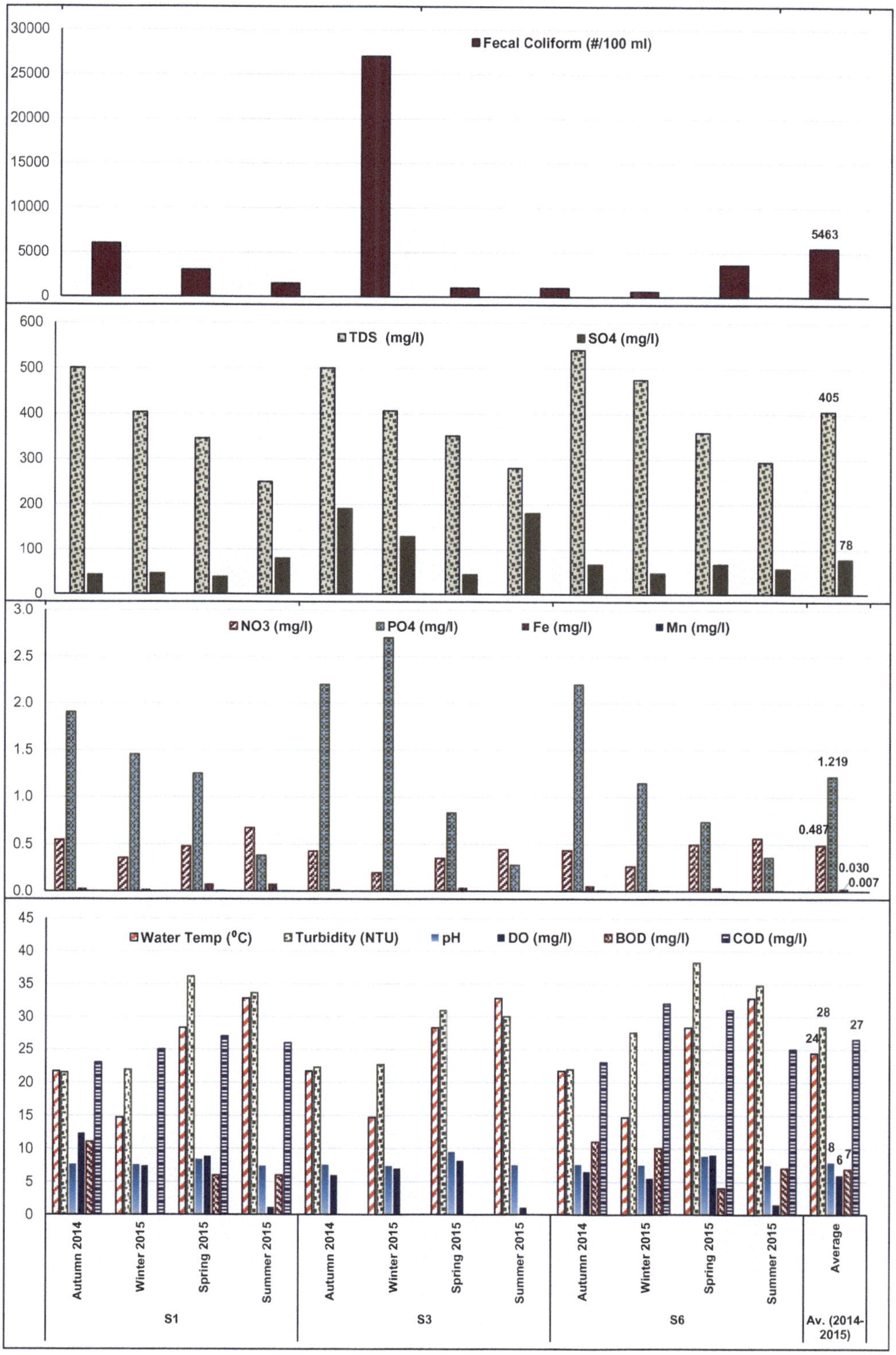

Fig. 3 Seasonal and average water quality records at selected sampling stations in Rosetta Branch (2014–2015)

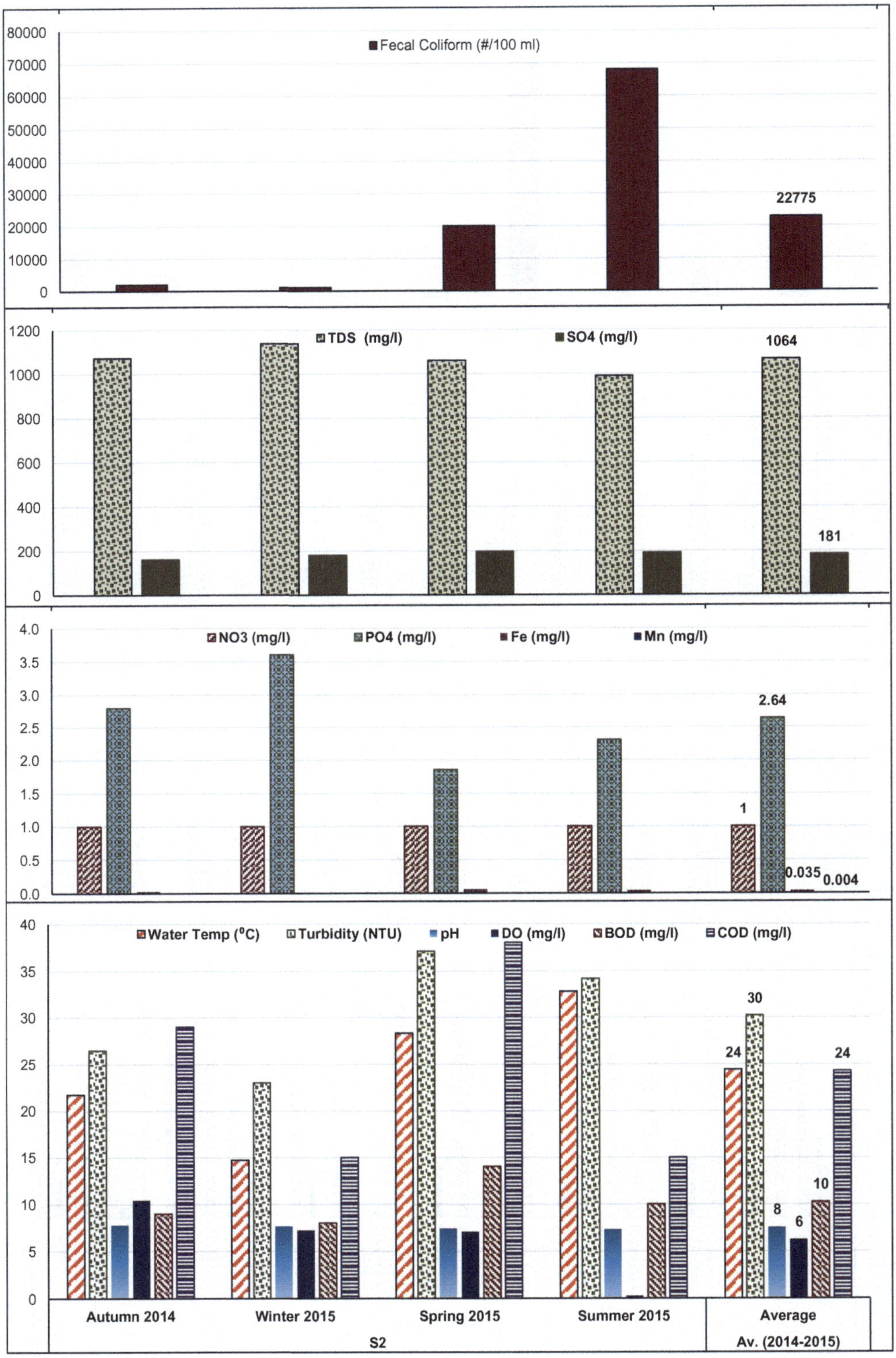

Fig. 4 Seasonal and average water quality records at Tala Drain (2014–2015)

Table 1 CCMEWQI categorization schema

Rank	CCMEWQI score
Excellent	100–95
Good	94–80
Fair	79–65
Marginal	64–45
Poor	44–0

Table 2 Egyptian water quality standards for surface waterways (Law 48/1982—update of 2013)

Water quality parameters	Permissible limits
Turbidity (NTU)	0–100
Water temperature change (degrees C)	+5°C
pH	6.5-8.5
DO (mg/l)	>6
BOD (mg/l)	<6
COD (mg/l)	<10
TDS (mg/l)	<500
No_3 (mg/l)	<2
So_4 (mg/l)	<200
Fe (mg/l)	<0.5
Mn (mg/l)	<0.2

For this study, CCMEWQI was modified according to The Egyptian guidelines for surface fresh waters (Law 48/1982—update of the year 2013, see Table 2). Moreover, the seasonal records of 12 water quality parameters were used for this index; T, turbidity, TDS, pH, DO, BOD, COD, NO_3, SO_4, Fe, Mn, and fecal coliform. The used guideline for fecal coliform, as there is no guideline limit for it in the Egyptian Law 48/1982, was used earlier by Heikal et al. (2007).

3.2.2 NSFWQI

National Sanitation Foundation (NSF) developed NSFWQI in 1970 based on specific selection and assigning different weights for water quality parameters (Brown et al. 1970). Nine water quality parameters were selected to develop NSFWQI; DO, fecal coliform, pH, BOD, temperature change, total phosphate, nitrate, turbidity, and total solids. The water quality status of a waterbody is classified into five categories according to NSFWQI, as shown in Table 3. NSFWQI is calculated as follows (Brown et al. 1970):

Table 3 Water quality classification according to NSFWQI

Rank	NSFWQI score
Excellent	100–91
Good	90–71
Medium	70–51
Bad	50–26
Very bad	25–0

$$\mathrm{NSFWQI} = \frac{\sum_{i=1}^{n}(W_iQ_i)}{\sum_{i=1}^{n} W_i} \tag{2}$$

where

W_i weight factor of the *i*th parameter.
Q_i sub-index quality value for the *i*th parameter.

Parameter weight factors and sub-index quality values are found in Brown et al. (1970). NSFWQI was widely used to assess several studied cases (Abdel Rehim et al. 2002; Chaturvedi and Bassin 2010; Effendi and Wardiatno 2015; Heikal et al. 2007; Ionus 2010; Rajankar et al. 2009; Sharma et al. 2006; Varnosfaderany et al. 2009; Wills and Irvine 1996).

For this study, eight parameters, out of nine, were used; the records of total solids were not available. Moreover, NSFWQI was calculated at the three stations (S3, S4, and S5) based on only six water quality parameters; the records of total solids, fecal coliform, and BOD weren't available.

3.3 Treatment Efficiency

The treatment efficiency of this practice can be evaluated and quantified by using three different indicators as follows:

1. Tala Drain treatment indicator

This indicator quantifies the percentage change of different water quality parameters of the drain after diluting by the Rosetta Branch and can be calculated as follows:

$$\Delta_1 = \frac{C_6 - C_2}{C_2}\% \quad (3)$$

where

- Δ_1 % Change in concentration of the considered water quality parameter.
- C_2 The concentration of the considered water quality parameter at the drain station (S2).
- C_6 The concentration of the considered water quality parameter at the branch station (S6).

2. Rosetta Branch pollution indicator

This indicator quantifies the percentage change of different water quality parameters of the Rosetta Branch after receiving the Tala Drain effluents and can be calculated as follows:

$$\Delta_2 = \frac{C_6 - C_3}{C_3}\% \quad (4)$$

where

- Δ_2 % Change in concentration of the considered water quality parameter.
- C_6 The concentration of the considered water quality parameter at the branch station after receiving the drained water (S6).
- C_3 The concentration of the considered water quality parameter at the branch station just before receiving the drained water (S3). For BOD, COD, and fecal coliform, S1 records of these parameters were used due to the unavailability of their records at S3.

3. Rosetta Branch water quality status indicator

This indicator quantifies the percentage change of water quality status of the Rosetta Branch after receiving the Tala Drain effluents. Two water quality indices were considered for this indicator: NSF and CCME WQIs. This indicator can be calculated as follows:

$$\Delta_3 = \frac{WQI_6 - WQI_3}{WQI_3}\% \quad (5)$$

where

- Δ_3 % Change in concentration of the considered water quality parameter.
- WQI_6 The water quality index of the branch after receiving the drained water (at S6).
- WQI_3 The water quality index of the branch just before receiving the drained water (at S3).

4 Results and Discussion

4.1 Water Quality Records

The collected seasonal water quality records of Rosetta Branch (Fig. 3) were evaluated according to the Egyptian standards (Law 48/1982—update of 2013, Table 2). Due to receiving of industrial wastes through different "agricultural" drains, all COD records were higher than the permissible value (10 mg/l); at S1 the average concentration of COD was about 25.3 mg/l, while at S6 (after receiving Tala Drain effluents too) the average was about 28 mg/l and the maximum was about 32 mg/l in winter 2015. The total average concentration for both stations was 26 mg/l. For BOD, half of the recorded concentrations at S1 and S6 exceeded the permissible limit (6 mg/l), where these records were 11 mg/l (autumn 2014 at S1 and S6), 10 mg/l (winter 2015 at S6), and 7 mg/l (summer 2015 at S6). The total average concentration for both stations was 7 mg/l. All recorded concentrations of DO in summer 2015 were very small, ranging from 1.1 mg/l at S1 to 1.55 mg/l at S6; the permissible DO standard is 6 mg/l. This occurs due to some combined reasons such as the rising in water temperature and the increasing of drains discharges loaded with municipal, industrial, and agricultural wastes. These records explain the "almost annual" fish kill phenomena in Rosetta

Branch which occurs when DO concentrations go down than 2 mg/l (Chapman 1996). Moreover, some records in autumn and winter 2015 slightly decrease than the permissible limit. The total average DO concentration was about 6 mg/l. For pH, almost all records of spring 2015 were higher than the permissible limit (8.5); the maximum was 9.6 at S5, while the total average pH was about 8. TDS records in autumn 2014 (630 mg/l at S5 and 540 mg/l at S6) and in winter 2015 (566 mg/l) at S5 exceeded the permissible limit (500 mg/l); the total average TDS concentration was about 405 mg/l. High concentrations of fecal coliform were also noticed in the study area; the total average concentration was about 5463 cell/100 ml which confirms the presence of wastewater pollution in the branch.

4.2 Rosetta Branch Water Quality Status

Two water quality indices were used to assess the Rosetta Branch water quality status at the study area to quantify the impact of the Tala Drain effluent on the branch water quality status. The spatial distributions of CCMEWQI for Rosetta Branch at different stations are presented in Figs. 5 and 6, while the same distributions using NSFWQI are presented in Figs. 7 and 8.

According to the results of CCMEWQI, which are based on the seasonal records of 12 water quality parameters and the Egyptian standards (Law 48/1982), the average water quality status of the branch is "Good" [83%]. This index shows that the water quality status of the branch in summer 2015 was the worst and ranged from "Marginal" [56%] at S1, which was the minimum CCMEWQI in the branch, to "Fair" [76%] at S5. The average CCMEWQI in summer 2015 was 68% "Fair". A spatial distribution of CCMEWQI in summer 2015 for the study area can be seen in Fig. 6. Moreover, CCMEWQI results showed that S1 had low water quality conditions; the average CCMEWQI at S1 was about 76% "Fair", while the best water quality condition in the study area was at S3; the average was about 91% "Good". The impact of Tala Drain discharge can be clearly noticed at S6 which had the worst water quality conditions in the branch; the average CCMEWQI at S6 was about 75% "Fair".

The results of NSFWQI, which are based on using the records of eight water quality variables, confirmed the findings of CCMEWQI. The average NSFWQI for the branch was 64% which represents a "Medium" water quality status. The summer 2015 experienced the worst seasonal water quality conditions for the branch (58% "Medium"); a spatial distribution for the results

Fig. 5 CCMEWQI along the Rosetta Branch at Kafr El-Zayat City during 2014–2015

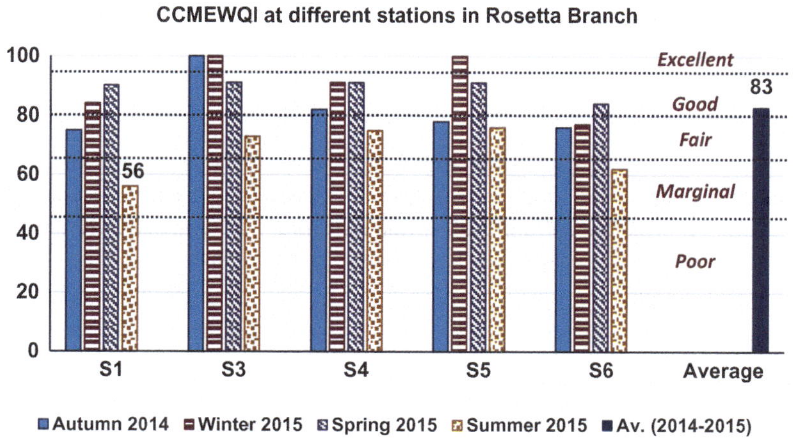

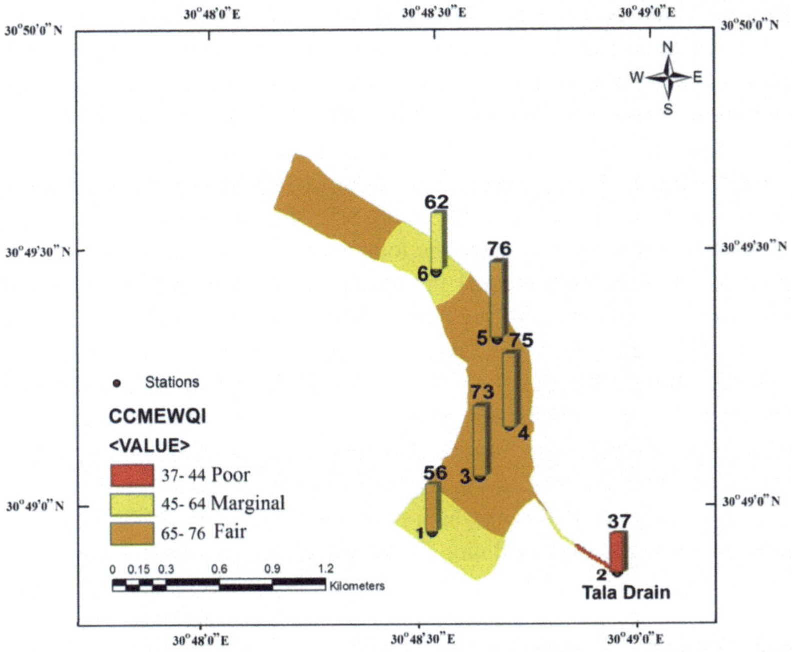

Fig. 6 Spatial distributions of the Rosetta Branch and Tala Drain CCMEWQI (summer 2015)

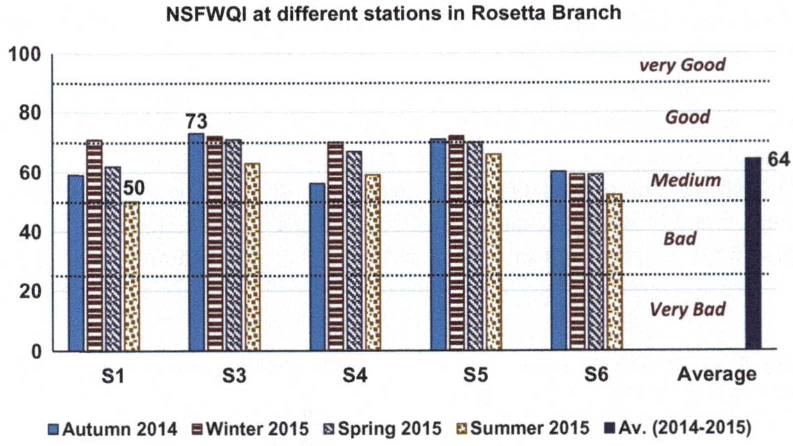

Fig. 7 NSFWQI along the Rosetta Branch at Kafr El-Zayat City during 2014–2015

of summer 2015 can be seen in Fig. 8. S3 and S5 had the best water quality conditions with an average of 70% "Good" for each, while the worst water quality condition in the branch was at S6 (58% "Medium").

4.3 Treatment Efficiency

Treatment efficiency indicators of this process are presented in Figs. (9, 10, 11 and 12) and Table 4 for some selected water quality parameters and two water quality indices (at specific stations in Tala Drain and Rosetta Branch).

The seasonal Tala Drain treatment indicator changes for some selected water quality parameters can be seen in Fig. 9, and the annual averages changes of the indicator are presented in Table 4. Decreases in the TDS [61%], BOD [14%], nutrients (NO_3 & PO_4) [by about 60%], SO_4 [67%], and fecal coliform [63%] average concentrations are clearly noticed. While a slight

Fig. 8 Spatial distributions of the Rosetta Branch and Tala Drain NSFWQI (summer 2015)

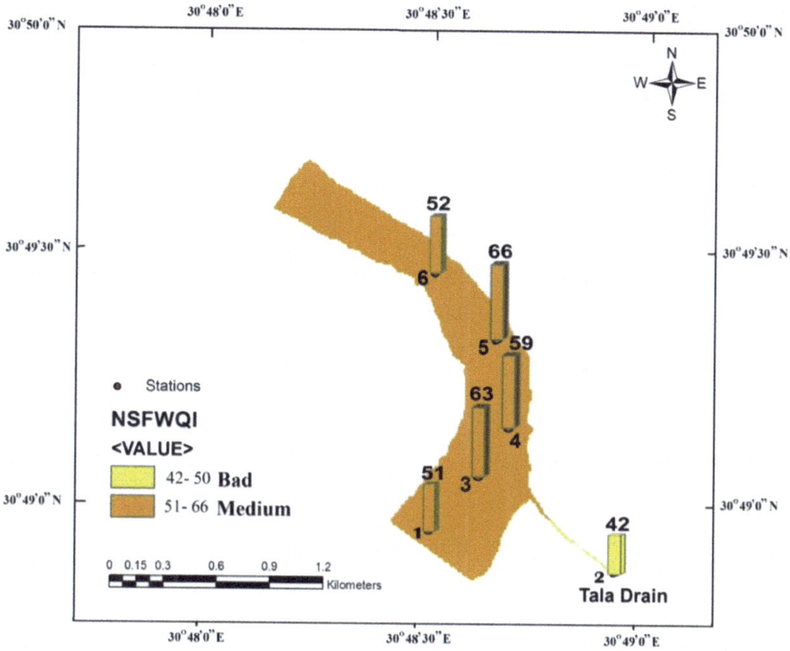

Table 4 A summary of annual average changes of the Tala Drain treatment indicator treatment (Δ_1) and the Rosetta Branch pollution indicator for some selected parameters

Water quality parameters	Tala Drain treatment indicator (Δ_1%)	Rosetta branch pollution indicator (Δ_2%)
	Average change% (2014–2015)	Average change% (2014–2015)
TDS	−61	8
Turbidity	2	15
pH	4	−1
DO	116	11
BOD	−14	−6
COD	35	10
NO_3	−56	26
PO_4	−59	−10
SO_4	-67	−36
Fe	23	75
Mn	127	48
Fecal Coliform	−63	−74

increase in pH average values by 4% and high increases in DO [116%], COD [35%], Fe [23%], and Mn [127%] average concentrations are noticed, these changes due to dilution reveal the pollution sources of both the drain and the receiving branches; Tala Drain mainly receives agricultural drainage water and municipal wastes, while the Rosetta Branch mainly suffers from industrial waste. So, decreases in TDS, BOD, and nutrients concentrations and increases in COD, Fe, and Mn are expected. For DO, the annual averages of both Tala Drain and Rosetta Branch in the study area are the same [6 mg/l]. An increase in DO changes by about 500% in summer 2015 were recorded due to the very small records of DO in Tala Drain [0.26 mg/l] and Rosetta Branch [1.55 mg/l].

Figure 10 shows the seasonal changes of Rosetta Branch pollution indicator, while Table 4 presents the annual average changes of this indicator, and, as can be seen and due to the drain discharging into the branch, increases in the average concentrations of most water quality parameters such as TDS [8%], turbidity [15%],

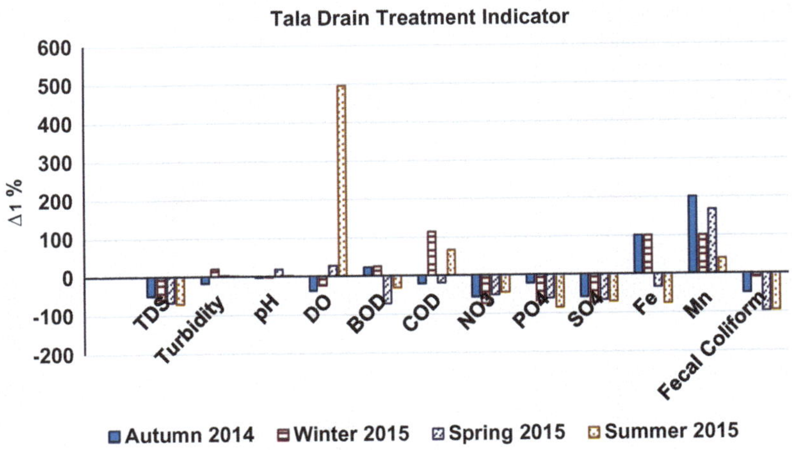

Fig. 9 Seasonal Tala Drain treatment indicator for selected parameters

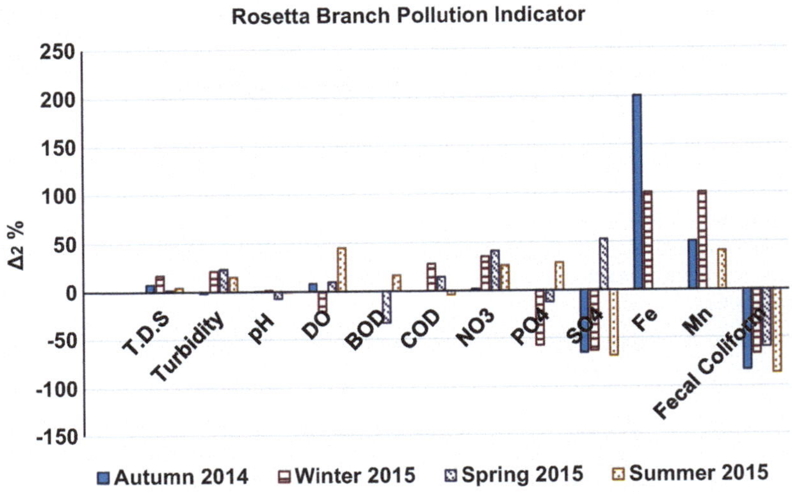

Fig. 10 Seasonal Rosetta Branch pollution indicator for selected parameters

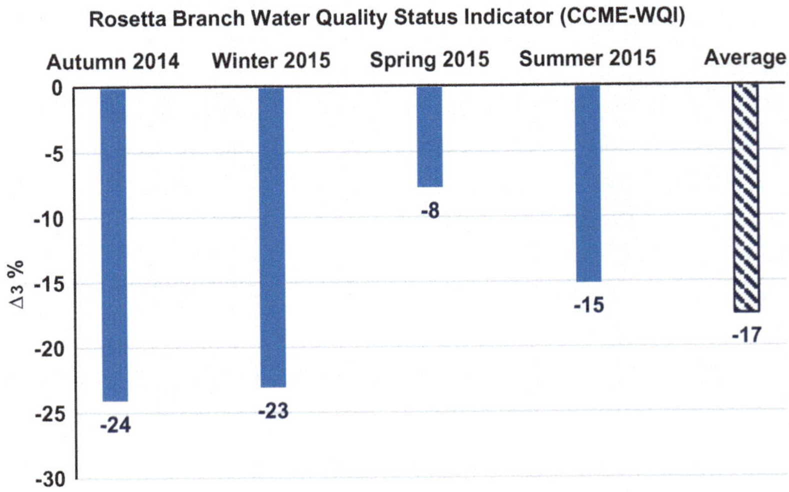

Fig. 11 Seasonal Rosetta branch water quality status indicator (CCME WQI)

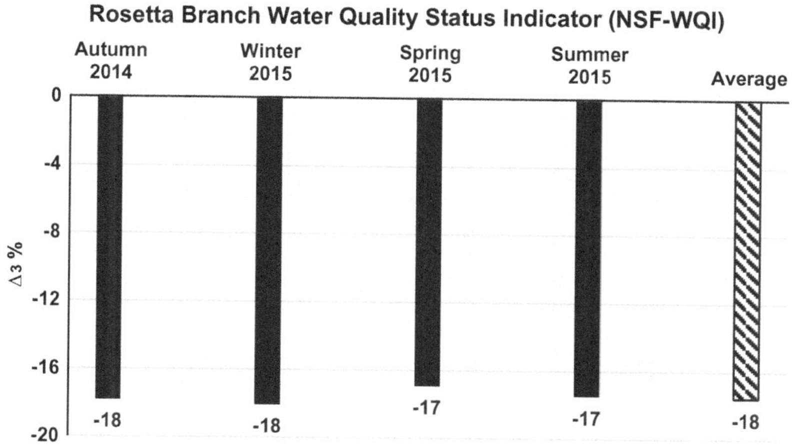

Fig. 12 Seasonal Rosetta branch water quality status indicator (NSF WQI)

NO_3 [26%], Fe [75%], and Mn [48%] and decreases in BOD [6%], PO4 [10%], SO4 [36%], and fecal coliform [74%]. For DO, an increase by about 11% was recorded due to the increase by about 45% in summer 2015 concentrations from 1.07 mg/l (at S3) to 1.55 mg/l (at S6).

To investigate the overall effect of this practice on the Rosetta Branch water quality status, two water quality indices were used: CCMEWQI and NSFWQI. The seasonal Rosetta Branch water quality status indicator changes based on CCMEWQI can be seen in Fig. 11. The figure clearly reveals an average decrease in the water quality status indicator of the branch by about 17%; a severe effect was noticed in autumn 2014 [by about 24%], while the minimum effect was in spring 2015 [by about 8%]. NSFWQI results for the water quality status indicator (Fig. 12.) ensure the same results of CCMEWQI with slight differences due to the considered parameters in each index, and the average decrease in the water quality status indicator of the branch was about 18%.

5 Conclusions

Based on field records for the period from autumn 2014 to summer 2015 and two water quality indices, CCMEWQI and NSFWQI, three different indicators were used to assess the drainage water treatment-via-dilution for Tala Drain in the Rosetta Branch at Kafr El-Zayat city. These indicators are Tala Drain treatment indicator, Rosetta Branch pollution indicator, and Rosetta Branch water quality status indicator. The results of this study revealed that application of the drainage water treatment-via-dilution can reduce the concentrations of most nutrients, TDS and BOD of the drainage agriculture with a negative impact on the overall water quality status of the receiving fresh water channel. For this case study, the negative impact on the branch overall water quality was about 17%, according to CCMEWQI. Considering the degraded water quality status of Rosetta Branch, according to the assessment of the branch water quality status using WQIs, water quality management plans should be applied to enhance the branch water quality status and reduce the effect of Tala Drain drainage water. It is highly recommended to experiment with some low-cost applicable technologies, such as in-stream control structures and treatment systems, and to implement constructed wetland treatment systems on major Egyptian drains. This study should be done for similar cases on the Nile River in Egypt to quantify the drainage water impacts on the Nile River water quality.

References

Abbasi T, Abbasi SA (2012) Water quality indices. Elsevier

Abdel Rehim SS, Abdel Bari MR, Khalil MM, Heikal MT, Salem TA (2002) Impact of heat on the physical and chemical properties of Lake Nasser water. J Environ Sci 5:325–349

APHA (2005) Standard methods for the examination of water and wastewater. American Public Health Association (APHA), Washington, DC, USA

Barnes J (2014) Mixing waters: The reuse of agricultural drainage water in Egypt. Geoforum 57:181–191

Boyacioglu Hl (2010) Utilization of the water quality index method as a classification tool. Environ Monit Assess 167:115–124. https://doi.org/10.1007/s10661-009-1035-1

Brown RM, McClelland NI, Deininger RA, Tozer RG (1970) A water quality index-do we dare

CCME (2001) Canadian Council of Ministers of the Environment, Canadian water quality guidelines for the protection of aquatic life. CCME Water Qual Index 1:34–31

CCME (2004) CCME National water quality index workshop. Water, CCME

Chapman DV (1996) Water quality assessments: a guide to the use of biota, sediments and water in environmental monitoring. World Health Organization

Chaturvedi MK, Bassin JK (2010) Assessing the water quality index of water treatment plant and bore wells, in Delhi, India. Environ Monit Assess 449–453. https://doi.org/10.1007/s10661-009-0848-2

Donia N (2005) Rosetta branch waste load allocation model. In: 9th international water technology conference, IWTC9, Sharm El-Sheikh, Egypt. pp 17–20

Donia NS, Farag HA (2010). A waste load model analysis for el noubariya canal drinking water abstraction, Egypt 2. Description el noubariya canal. In: Fourteenth International Water Technology Conference, IWTC 14, 2010, Cairo, Egypt, 2010. IWTC, pp 675–690

Ed Quilty QE (2003) Water Quality Objectives Attainment Monitoring Fraser River Tributaries from Hope to Kanaka Creek 2002. Environmental Protection. Ministry of Water, Land & Air Protection Lower Mainland Region.

Effendi H, Wardiatno Y (2015) Water quality status of Ciambulawung River, Banten Province, based on pollution index and NSF-WQI. Procedia Environ Sci 24:228–237. https://doi.org/10.1016/j.proenv.2015.03.030

Elshemy MM (2010) Water quality modeling of large reservoirs in semi-arid regions under climate change: Example lake nasser (egypt). PhD, Technische Universität Braunschweig, Germany

Heikal MT, El-Sherbini A, Salem TA (2007) Yousry MM Temporal and spatial variation of water quality parameters along the River Nile. In: NAWQAM Final Conference on Egypt Paradigm in Integrated Water Resources Management, February 17–19. Egypt, Sharm El Sheikh, p 21

Heuvel MRVD (2009) Site specific guidelines for phosphorus in relation to the water quality index calculations for Prince Edward Island, p 35

Hurley T, Sadiq R, Mazumder A (2012) Adaptation and evaluation of the Canadian Council of Ministers of the Environment Water Quality Index (CCME WQI) for use as an effective tool to characterize drinking source water quality. Water Res 46:3544–3552. https://doi.org/10.1016/j.watres.2012.03.061

Ionus O (2010) Water quality index—assessment method of the Motru River Water Quality (Oltenia, Romania) Indicele De Calitate a Apei – Metod Ă De Determinare a Calit Ăţ Ii Apei Râului Motru (Oltenia, România), vol 13, pp 74–83

Karajeh F, Oweis T, Swelam A, El-Gindy A, El-Quosy D, Khalifa H, El-Kholy M, El-Hafez SA (2011) Water and agriculture in Egypt: Technical paper based on the Egypt-Australia-ICARDA Workshop On-farm Water-use Efficiency. International Center for Agricultural Research in the Dry Areas (ICARDA), Egypt

Khan AA, Abdel-Gawad S, Khan HA (2008) Real time water quality monitoring network and water quality indices for River Nile. In: 13th IWRA world water congress, 1-4 September. France, Montpellier, p 15

Lumb A, Halliwell D, Sharma T (2006) Application of CCME Water Quality Index to monitor water quality: a case study of the Mackenzie River Basin, Canada. Environ Monit Assess 113:411–429

Mercier V, Fox D, Khan H, Taylor D, Raymond B, Bond W, Caux P-Y (2005) Application and testing of the water quality index in Atlantic Canada—report summary. Atlantic, CCME

Mostafa M (2014) Modeling of pollutant transport in the nile delta, Egypt. PhD, University of Alabama at Birmingham, Graduate School

MWRI (2005) Water for the future, national water resources plan for Egypt-2017

MWRI (2010) A summary of the Egypt national strategy for the development and management of water resources until 2050

Radwan M (2005) Evaluation of different water quality parameters for the Nile River and the different drains. In: Ninth International Water Technology Conference, Sharm El-Sheikh, Egypt, pp 1293–1303

Rajankar PN, Gulhane SR, Tambekar DH, Ramteke DS, Wate SR, Impact E, Division RA (2009) Water quality assessment of groundwater resources in Nagpur Region (India) based on WQI. E-J Chem 6:905–908

Simões F, Moreira AB, Bisinoti MC, Gimenez SMN, Yabe MJS (2008) Water quality index as a simple indicator of aquaculture effects on aquatic bodies. Ecol Ind 8:476–484

Selvam S, Manimaran G, Sivasubramanian P, Balasubramanian N, Seshunarayana T (2014) GIS-based evaluation of water quality index of groundwater resources around Tuticorin coastal city, South India. Environ Earth Sci 71:2847–2867. https://doi.org/10.1007/s12665-013-2662-y

Shaban M (2014) Drainage water reuse: state of control and process capability evaluation. Water Air Soil Pollut 225(11):2168
Sharma MP, Sharma S, Gael V, Sharma P, Kumar A (2006) Water quality assessment of Behta River using benthic macroinvertebrates. Life Sci J 3
Varnosfaderany MN, Mirghaffary N, Ebrahimi E, Soffianian A (2009) Water quality assessment in an arid region using a water quality index. Water Sci Technol 60:2319–2327
Wills M, Irvine KN (1996) Application of the national sanitation foundation water qualily index in the cazenovia creek, NY, pilot watershed management project. Middle States Geogr 95–104

Upcoming Challenges of Water Reclamation from Unconventional Sources

Frank Rögener

Abstract

Greenhouse gas emissions are having an increasing impact on water cycles. This interconnection is referred to as the carbon–water nexus (CWN). Especially, changing precipitation patterns, decreasing freshwater availability, rising sea levels, and mounting frequency of extreme weather events threaten the social structure, the infrastructure and the environment in many countries. Environmental engineers can help to combat this negative development by, e.g., implementing sophisticated water treatment technologies for water reclamation from industrial, commercial, and residential sources. However, these engineers have to overcome new challenges to do so. In comparison to surface water or groundwater, these still unconventional water sources show a more complex composition and a higher concentration of potentially hazardous substances. This gives rise to additional technical efforts and causes additional costs. Furthermore, when exploiting these unconventional sources, psychological issues of the customers as well as increasing health concerns due to prevalence of pathogens have to be addressed.

Keywords

Water treatment · Water reclamation · Hazardous substances · Unconventional water sources

1 Impact of the Climate Change on Water Cycles

Worldwide, countries are being confronted with increasing problems in terms of providing people with sufficient amounts of clean water. Climate change is not the only reason for this challenge, but it 'represents an additional burden' for natural water cycles (Bates et al. 2008).

Though the prediction of climate change effects is largely based on assumptions, there are foreseeable implications on natural water cycles, such as

- Rising sea levels
- Increasingly extreme weather conditions
- Impacts on biodiversity and whole waterborne ecosystems
- Impacts on the global ocean conveyor belt.

It is expected that in Germany all economic sectors will be affected by the ongoing climate change (Brasseur et al. 2017), although most likely Germany's water supply will not be impaired.

F. Rögener (✉)
TH Köln, Betzdorferstr. 2, 50679 Cologne, Germany
e-mail: frank.roegener@th-koeln.de

L. Ribbe et al. (eds.), *Towards Water Secure Societies*,
https://doi.org/10.1007/978-3-030-50653-7_6

Water supply is crucial for social and economic progress. Thus, concepts for improved water supply should be implemented, especially in developing countries. One pillar of this concept could be water reclamation from unconventional sources, see Table 1. Seawater desalination is already being commonly used (Elimelech and Phillip 2011; Kong 2017) and water reuse in agriculture (Starkl et al. 2015), urban areas, and water resource management are known (Anderson 2003; Levine and Asano 2004). Also, most modern industrial recycling processes are based on water reclamation, e.g., (Schmidt et al. 2007).

2 Basic Principles of Water Reclamation

Basically, there are two options for separation techniques in industrial processes: removal of the solvent—in this case water, or removal of undesired substances in the process solutions, whereby the processed solutions are then fed back, see Fig. 1.

The reclaimed water does not necessarily have to meet drinking water quality criteria; but in many cases, the quality must even be better to meet the requirements of the industrial process. A tool for the systematic design of a water network is the water pinch analysis (Tan et al. 2007), which is based on the determination of the following:

- Water sources (effluent of one specific process step)
- Water sinks (influent to another process step, which requires a certain quality) and
- Key contaminants (crucial for the reliability of the process).

According to (Fischwasser et al. 2003), separation techniques can be divided into regenerators and concentrators. This concept is shown in Fig. 2.

Table 1 Strategies for water supply

Water source	Requirements	Advantages	Disadvantages
A: conventional sources			
Surface water (rivers, lakes)	Constant availability	Generally, high availability	Feed quality may depend on climate and seasonal conditions
Well water	Constant availability	Normally, low efforts for treatment Generally, constant feed quality	Possible drawdown of the groundwater table
Capture and storage of rainfall	Annual rainfall	Independency	Required area Influence on ecosystems Dependence on climate and seasonal conditions
B: Unconventional sources			
Seawater desalination	Coastline	Constant feed quality	Expensive technology Concentrate (brine) discharge required Influence on ecosystems
Water reclamation	Availability of municipal and/or industrial wastewater infrastructure	Drought-resistant source of water Generally, constant feed quality	Several treatment steps required to ensure compliance with all drinking water regulations Expensive technology Concentrate discharge required
Water import	Long-term contracts Piping infrastructure	Lower costs	Dependency

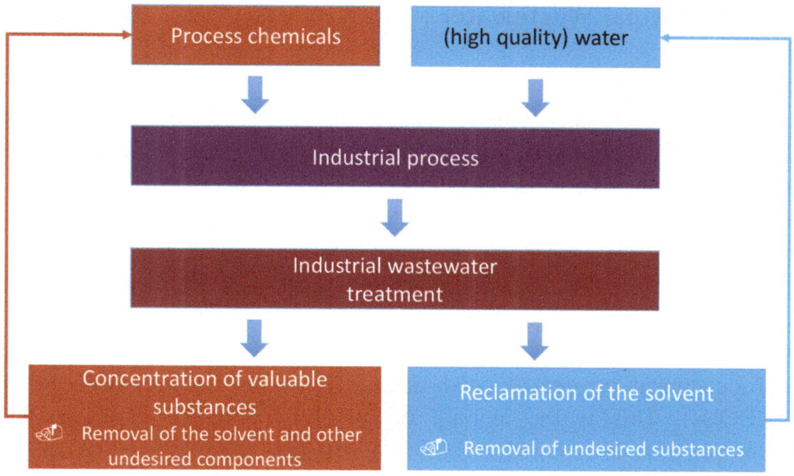

Fig. 1 Options for separation techniques

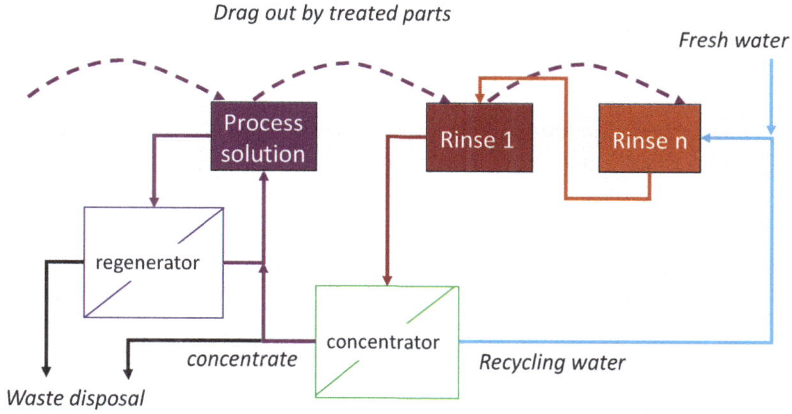

Fig. 2 Process water treatment and water reclamation, after (Fischwasser et al. 2003)

Regenerators restore the initial composition of the desired substance in the process solution, whereas separators separate and concentrate the undesired substances and reclaim the initial solvent (water). The whole concept allows zero liquid discharge (ZLD), (Amosa et al. 2013; Katsoyiannis et al. 2015; Tong and Elimelech 2016), which means that the formation of liquid waste products is completely avoided by technical means.

Basically, the reclamation of water from mixed effluents is complicated and usually uneconomical. Therefore, it is advisable to collect different wastewater streams and to treat them separately. Furthermore, we can conclude that the best wastewater is the one that is not generated. To achieve this aim, organizational and technical measures must be combined.

3 Composition of Municipal and Industrial Wastewaters

To obtain drinking water, all pollutants present need to be removed from the feed water until the levels of drinking water legislation are achieved. The treatment of municipal and industrial wastewaters poses completely different challenges: In comparison to surface water or groundwater, unconventional water sources show a more complex composition and a higher concentration of potentially hazardous substances.

3.1 Municipal Wastewater

Municipal wastewater contains a multitude of organic and inorganic components; 60-70% of them are dissolved. According to legislation, not all components have to be determined individually. The so-called sum parameters describe the characteristics of specific substance groups. Important sum parameters in water technology are

- Conductivity.
- Biochemical oxygen demand within 5 days (BOD_5). This is the amount of dissolved oxygen that is required for the biochemical degradation and refers to easy degradable organic compounds.
- Chemical oxygen demand (COD): This is the amount of dissolved oxygen that is required for chemical degradation of organic substances.
- Total organic carbon (TOC).
- Total organic and inorganic nitrogen compounds (total N).
- Organic nitrogen compounds (organic N).
- Total organic and inorganic phosphorous compounds (total P).
- Organic phosphorous compounds (organic P).

The concentration of pollutants in untreated municipal wastewater is in the milligram range, see Table 2.

In the last few years, further groups of pollutants have come into the focus of water engineers and politicians, namely micropollutants, microplastics, and bacteria that are resistant to multiple antibiotics. Conventional wastewater treatment plants are overstrained with removing these pollutants completely.

Micropollutants are an unspecific group of toxic, persistent, and bioaccumulative organic or mineral substances that are present in extremely low concentrations (nanograms per liter to micrograms per liter). Micropollutants are potentially hazardous for the aquatic environment (an example is the feminization of fish) and human health. These substances are predominantly discharged via municipal wastewater (Umweltbundesamt 2015).

Next to the dissolved micro pollutants, small fragments of plastic debris—so-called microplastics—have been detected in many aquatic environments. According to Thompson (Thompson 2008), there are concerns that microplastics have a negative effect on wildlife and human health, as they can release plasticizers

Table 2 Analysis excerpt of a typical municipal wastewater sample (https://fenix.tecnico.ulisboa.pt/downloadFile/3779571777)

Pollutant	Unit	Concentration level		
		Low	Medium	High
Total solids	mg/L	390	720	1,200
Total dissolved solids		270	500	860
Settable solids		5	10	20
BOD_5		110	190	250
COD		250	430	800
TOC		80	140	260
Total N		20	40	70
Organic N		8	15	25
Total P		4	7	12
Organic P		1	2	4
Total coliform bacteria	CFU/ 100 mL	10^6-10^8	10^7–10^9	10^7–10^{10}
Cryptosporidium		10^{-1}-10^0	10^{-1}–10^1	10^{-1}–10^2
Giardia		10^{-1}-10^1	10^{-1}–10^2	10^{-1}–10^3

CFU: Colony-forming units

Table 3 Typical pollutant parameters in industrial wastewaters (Brito et al. 2007; Rögener et al. 2012; Ammary 2004)

Pollutant	Unit	Beer production	Stainless steel pickling	Olive mills	Pulp and paper
COD	g/L	2.0–6.0	n.d.	97	0.4
BOD	g/L	1.2–3.6	n.d.	31	0.2
Total suspended solids	g/L	0.2–1.0	4.5	n.d.	n.d.
pH	.	4.5–12	<0	n.d.	n.d.
Total N	mg/L	25–80	n.d.	530	13
Total P	mg/L	10–50	n.d.	180	4
Conductivity	mS/cm	n.d.	450	n.d.	n.d.
Fe	g/L	n.d.	30	n.d.	n.d.
Cr	g/L	n.d.	5	n.d.	n.d.
Ni	g/L	n.d.	5	n.d.	n.d.
Fluoride	g/L	n.d.	50	n.d.	n.d.
Nitrate	g/L	n.d.	180	n.d.	n.d.

n.d. not determined

and adsorb hazardous substances, such as persistent organic compounds. Accordingly, water reclamation systems are also affected by the occurrence of these pollutants.

3.2 Industrial Wastewater

Each industrial branch generates a specific combination of pollutants, which need to be treated individually, s. Table 3.

Usually, the concentrations of pollutants in untreated industrial wastewaters are in the gram range. Industrial wastewater treatment plants must be designed specifically for each particular type of effluent. The achievable quality of the reclaimed water depends on feed water quality, legislative demands, and the further internal re-use.

4 Wastewater Treatment

In many cases, different wastewater treatment processes can be applied to achieve the same separation aim. Table 4 depicts common treatment processes and their most effective applications. Basically, separation techniques differ in performance, application range, costs, and handling. In any case, to meet the required water quality, different techniques have to be combined in such a way that the following objectives can be achieved (Ranade and Bhandari 2014):

- Targeted elimination of the respective pollutants
- Efficient removal of pollutants through combination of the best-suited technologies
- Cost-effectiveness of the process.

4.1 State-of-the-Art in Municipal Wastewater Treatment

State-of-the-art effluent treatment plants (ETP) have physical, biological, and chemical treatment steps. Nowadays, a fourth cleaning step for the removal of micropollutants and other contaminants is being discussed.

Mechanical cleaning removes only 20–30% of the pollutants—namely the undissolved floating and suspended material.

Biological cleaning comprises the aerobic and anaerobic microbiological degradation of organic matter.

Chemical cleaning denotes chemical reactions such as oxidation and precipitation.

Table 4 Wastewater treatment processes

Process	Pollutants removed								
	Particles	Turbidity	Microorganisms	Color	Heavy metals	Organic load	Salinity	Nitrates	Micro pollutants
Screening, sieves	x								
Sedimentation	x								
Lamella clarifier	x								
Sand filtration	x	x	x						
Coagulation/ flocculation		x	x	x	x				
Precipitation					x		x		
Activated sludge			x			x		x	
Microfiltration	x	x	x						
Ultrafiltration		x				x			
Nanofiltration				x	x	x	x		
Reverse osmosis					x		x	x	x
Electrodialysis					x		x		
Ion exchange					x		x	x	
Adsorption/ activated carbon				x		x			x
Extraction					x	x	x		
Crystallisation					x		x		
Ozonation			x	x		x			x
UV radiation			x			x			x

The effluent of municipal effluent treatment plants is far from being drinking water, as microorganisms, microplastics, phosphorous compounds, and micropollutants cannot be rejected.

Advanced treatment—also denoted as the fourth treatment step—improves the quality of the cleaned effluent, so that the reuse for multiple purposes is possible.

4.2 Advanced Treatment of Municipal Wastewater

At present, municipal wastewater treatment plants act as sinks for particulate load, oil and fat, and a multitude of easy degradable organic molecules. Investigations have shown that they also act as point source inputs for pollutants (Brun et al. 2006; Talvitie et al. 2015). Especially the removal of microplastics, micropollutants, and multiresistant germs will play an important role in future water treatment technology. Only a targeted combination of various technical approaches will help to meet the different objectives of water pollution control. Furthermore, effective emission reduction has to consider high efficiency of the respective technologies, implementation of the polluter pays principle, and promoting acceptance for water reclamation from unconventional sources.

Advanced treatment technologies include, e.g.:

- Membrane separation (especially reverse osmosis)
- Activated carbon (granular or as powder)
- UV treatment
- Ozonation
- Hypochlorite dosage.

4.3 Treatment of Industrial Wastewater

Zero liquid discharge (ZLD) is an approach for water conservation in industry. It describes a concept of avoiding any liquid waste discharge from a plant. For the successful implementation of ZLD systems, it is crucial that the operating costs are much smaller than the costs of waste disposal.

5 Examples of Water Reclamation from Wastewater

Water reclamation from municipal wastewater is technically feasible; however, it requires a psychological changeover, when the reclaimed water is to be used as drinking water. Thus, for psychological reasons, water reclaimed from wastewater is commonly used in agricultural irrigation, as industrial process water, for groundwater replenishment, or as a seawater intrusion barrier in coastal regions that suffer from sinking groundwater tables (Levine and Asano 2004; Franks et al. 2007).

In a literature study, Po et al. (2003) conclude that the key factor for the successful implementation of water reclamation projects is public involvement to improve the acceptance. This partnership has to already start in the decision-making phase and should be accompanied by public education and outreach programs after the project's conception. Two large-scale projects on water reclamation are briefly described in the following.

5.1 Orange County, California

Orange County Water District is dependent on the supply of water from the Colorado River, as the groundwater is overdrawn, which caused both, lowering of the groundwater table and seawater intrusion from the Pacific Ocean. The river water is transported over a distance of 150 miles. In addition to the river water, another unconventional water source that has been used since 1975 is water reclaimed from wastewater (secondary effluent). In 2007, a supplementary system was implemented: A so-called groundwater replenishment system. It has a capacity of about 265,000 m^3/d and is based on

- Mechanical pretreatment (MF),
- Chemical pretreatment: pH adjustment and dosage of antiscalants), as especially $CaPO_4$ scaling is a big issue,
- Reverse osmosis, and
- UV radiation in combination with hydrogen peroxide dosage for sanitation. The membrane manufacturer claims 85% recovery at a flux of 20.4 l/m^2 h (Franks et al. 2007).

5.2 Singapore's NEWater Strategy

NEWater is high-grade reclaimed water produced from treated municipal effluent (secondary effluent) that is further purified. NEWater is one of Singapore's four water sources. With a capacity of 651,090 m^3 per day, NEWater currently covers around 40% of the urban water demand. By 2060, the share should be increased up to 55% (The Straits Times 2017).

Up to 2017, five NEWater plants with slightly different setups have been implemented (PUB 2017a). The first step of the treatment is microfiltration, which removes turbidity, colloids, and microorganisms (bacteria, viruses, and protozoan cysts). The main treatment step for the generation of purified water is reverse osmosis. For post-treatment both, UV disinfection and pH adjustment, are employed (PUB 2017b), making NEWater ultra-clean and safe to drink.

6 Summary

Water reclamation from unconventional sources has the potential to overcome the reduced availability of freshwater resources. However, only an efficient combination of the different techniques can meet the different objectives of water reclamation. Obviously, this costs money.

Removal of microplastics and micro pollutants will play an important role in future water production. As more and more wastewaters become sources for water supply, indirect dischargers (e.g., medical facilities, metal working, and process industry) must reduce their emissions. Additionally, the polluter pays principle must be installed everywhere.

References

Ammary BY (2004) Nutrients requirements in biological industrial wastewater treatment. Afr J Biotechnol 3 (4):236–238

Amosa MK, Jami MS, Muyibi SA et al (2013) Zero liquid discharge and water conservation through water reclamation & reuse of biotreated palm oil mill effluent: A Rewiev. Int J Acad Res 5(4):169–182

Anderson J (2003) The environmental benefits of water recycling and reuse. Water Sci Technol 3(4):1–10

Bates BC, Kundzewicz ZW, Wu S et al (eds) (2008) Climate Change and Water. Technical Paper of the Intergovernmental Panel on Climate Change, IPCC Secretariat, Geneva, p 210

Brasseur GP, Jacob D, Schuck-Zöller S (eds) (2017) Klimawandel in Deutschland: Entwicklung, Folgen. Risiken und Perspektiven, Springer/Spektrum, Berlin

Brito AG, Peixoto J, Oliveira JM et al (2007) Brewery and winery wastewater treatment: Some focal points of design and operation. In: Oreopoulou V, Russ W (eds) Utilization of by-products and treatment of waste in the food industry, vol 3. Springer, New York, pp 109–131

Brun L, Bernier M, Losier R et al (2006) Pharmaceutically active compounds in atlantic canadian sewage treatment plant effluents and receiving waters, and potential for environmental effects as measured by acute and chronic aquatic toxicity. Environ Toxicol Chem 25(8):2163–2176

Elimelech M, Phillip WA (2011) The future of seawater desalination: energy, technology, and the environment. Science 333(6043):712–717

Fischwasser K, Schwarz R, Süß M (2003) Stoffverlust-minimierte Prozesstechnik – Wann lohnen sich Regeneratoren und Konzentratoren? Chem-Ing-Tech 75(6):781–786

Franks R, Bartels CR, Andes K et al (2007) Implementing Energy Saving RO Technology in Large Scale Wastewater Treatment Plants. In: Abstracts of the IDA World Congress-Maspalomas, Gran Canaria-Spain, 21–26 October 2007

https://fenix.tecnico.ulisboa.pt/downloadFile/3779571777486/IT_Caracterizacao_qualitativa_aguas_residuais_2_pdf. Accessed 28 Aug 2017

Katsoyiannis I, Castellana M, Cartechini F et al (2015) Application of zero liquid discharge water treatment units for wastewater reclamation: possible application in marine ports. In: Stylios C, Floqi T, Marinski J et al (eds) Sustainable development of sea-corridors and

coastal waters - the TEN ECOPORT project in south east europe. Springer, Heidelberg, pp 39–45
Kong L (2017) Prospect of the seawater desalination technology. J Energy Power Eng 11:511–516
Levine AD, Asano K (2004) Recovering sustainable water from wastewater. Environ Sci Technol 38 (11):201A–208A
Po M, Kaercher J, Nancarrow BE (2003) Literature review of factors influencing public perceptions of water reuse. CSIRO Land and Water. Technical report 54/03. https://www.clearwater.asn.au/user-data/research-projects/swf-files/16-laying-the-foundation-for-confident-barrier-free-water-conservation-and-reuse-literature-review.pdf. Accessed 19 Sept 2017
PUB (2017a) Singapore water story. https://www.pub.gov.sg/watersupply/singaporewaterstory. Accessed 28 Aug 2017
PUB (2017b) NEWater Technology. https://www.pub.gov.sg/Documents/NEWater%20Technology.pdf. Accessed 28 Aug 2017
Ranade V, Bhandari V (2014) Industrial Wastewater Treatment, Recycling and Reuse. In: Ranade V, Bhandari V (eds) Industrial Wastewater Treatment, Recycling and Reuse: An Overview). Butterworth-Heinemann, Oxford, pp 23–68
Rögener F, Sartor M, Ban A et al (2012) Metal recovery from spent stainless steel pickling solutions. Resour Conserv Recycl 60:72–77
Schmidt B, Wolters R, Kaplin J et al (2007) Degreasing bath care in the steel and metal working industry. Desalination 211:64–71
Starkl M, Brunner N, Amerasinghe P et al (2015) Stakeholder views, financing and policy implications for reuse of wastewater for irrigation: A case from Hyderabad, India. Water 7:300–328
Tan YL, Manan ZA, Foo DCY (2007) Retrofit of water network with regeneration using water pinch analysis. Process Saf Environ 85(4):305–317
Talvitie J, Heinonen M, Pääkkönen JP et al (2015) Do wastewater treatment plants act as a potential point source of microplastics? Preliminary study in the coastal Gulf of Finland. Baltic Sea. Water Sci Technol 72(9):1495–1504
The Straits Times (2017) Singapore launches fifth Newater plant at Changi, boosting water supply. http://www.straitstimes.com/singapore/singapore-launches-fifth-newater-plant-at-changi-boosting-treated-used-water-supply-by-10. Accessed 28 Aug 2017
Thompson RC (2008) How concerned should we be about microplastics? In: Proceedings of the international research workshop on the occurrence, effects, and fate of microplastic marine debris, Tacoma, 9–11 September 2008
Tong T, Elimelech M (2016) The global rise of zero liquid discharge for wastewater management: drivers, technologies, and future directions. Environ Sci Technol 50:6846–6855
Umweltbundesamt (2015) Measures to reduce micropollutant emissions to water. http://www.umweltbundesamt.de/publikationen/measures-to-reduce-micropollutant-emissions-to. Accessed 28 Aug 2017

How Membrane Bioreactor Technology Can Help to Solve Both, German and Russian Wastewater Problems

Frank Rögener, Sven Theus, Alexander Chusov, and Julia Lednova

Abstract

Municipal wastewater treatment plants focus on the elimination of particles, carbon, nitrogen compounds to comply with local environmental standards. State-of-the-art is a three step procedure based on the combination of physical, chemical, and biological treatment. Investigations have stressed the need for a fourth level of cleaning to remove so-called microplastics and micropollutants, as well as phosphorous compounds, which cannot be eliminated by state-of-the-art technologies. This, together with the higher probability of natural disasters as a consequence of climate change requires considerable adaptation of existing water supply and wastewater systems. The construction of new wastewater treatment plants is not expected in Germany in the near future; the total number of plants is even decreasing. Instead, the focus will be on improving the cleaning performance and retrofitting existing wastewater treatment plants. In contrast to Germany, the existing technology in the Russian Federation is often in poor condition and performance is limited by climate conditions. Accordingly, the construction of new plants with state-of-the-art technology is essential for improving water security. Membrane bioreactor technology (MBR) offers both, the opportunity to reduce the footprint of wastewater cleaning plants and to implement a fourth treatment stage. In particular, MBRs can be operated in combination with upstream UV radiation. Recent research has also shown that the addition of powdered activated carbon (PAC) in submerged MBRs can increase the total cleaning performance. Thus, MBR technology offers major advantages for municipal wastewater treatment in both, Germany and the Russian Federation.

Keywords

Membrane bioreactor · Water treatment · Water quality · Microplastics

F. Rögener (✉)
TH Köln, 50679 Cologne, Germany
e-mail: Frank.roegener@th-koeln.de

S. Theus
DAR - Deutsche, Abwasser-Reinigungs-GmbH, Adolfsallee 27/29, 65185 Wiesbaden, Germany
e-mail: sven.theus@dar.de

A. Chusov · J. Lednova
Peter the Great St. Petersburg Polytechnic University, 195251 St. Petersburg, Russia
e-mail: chusov17@mail.ru

J. Lednova
e-mail: lednovajulia@mail.ru

L. Ribbe et al. (eds.), *Towards Water Secure Societies*,
https://doi.org/10.1007/978-3-030-50653-7_7

1 Impact of Climate Change on Existing Water Problems

1.1 Present Situation

Due to the large size and specific characteristics of the geographical structure of the Russian Federation, climate change effects strongly depend on the respective region in question. General trends in the water sector are depicted in Table 1. It can be stated that all economic sectors both, in Germany and Russia, will be affected by climate change.

Climate change will enhance existing water-related problems; however, the main effects of climate change on the water sector are different in Germany and the Russian Federation. They are strongly dependent on the respective socioeconomic and geographical background. Table 2 depicts the main effects of climate change on the German water sector, including wastewater treatment plants (WWTP).

Table 3 depicts the main effects of climate change on the Russian water sector.

2 Biological Wastewater Treatment

2.1 General Considerations

In biological wastewater treatment plants, organic and inorganic contaminants serve as nutrients for a variety of microorganisms—such as bacteria, protozoa, lower plants as well as multicellular organisms—which degrade the contaminants in specific biochemical reactions. The degradation processes result in the formation of stable simple molecules. The so-called aerobic microorganisms need the supply of oxygen for these degradation processes. In aeration tanks, appropriate technical effort is required to ensure both oxygen supply and sufficient mixing for homogeneous distribution of dissolved and suspended substances. Thus, the microorganisms can be uniformly supplied with nutrients and oxygen (Dieckmann and Metz 1991).

The microbial metabolism is characterized by complex biochemical reactions, which contribute either to the synthesis of cell substance (anabolism) or to their degradation (catabolism). The formation of larger aggregates of the involved microorganisms in biological wastewater treatment plants (flocks) is made possible by natural, so-called extracellular polymeric substances (EPS), which combine cells, cell fragments, and foreign matter. The increase in biomass must be removed from the process as so-called waste activated sludge (secondary sludge) (Köhne 1998). Some parameters that influence the performance of activated sludge flocks are summarized in Table 4.

According to Möbius (2006), operating wastewater treatment plants at a higher sludge age could potentially allow a further reduction of difficult degradable organic substances (persistent COD) which are components of so-called micropollutants). These—until now—cannot be removed in conventional municipal wastewater treatment plants. Furthermore, the removal of plastic particles in the micrometer range (microplastics) is not possible. Both, micropollutants and microparticles end up in the aqueous environment, where they can enter the food chain.

2.2 Membrane Bioreactors

The cleaning performance of conventional activated sludge systems depends on the separation efficiency of undissolved substances in the secondary settling tank. These substances significantly contribute to the COD, BOD5, inorganic nitrogen and phosphorus load of the WWTP effluent.

Membrane bioreactors (MBR) combine biological wastewater treatment in aeration tanks with membrane filtration: Bio-degradation of organic pollutants in MBR is analogous to conventional aeration tanks, but dry matter concentration and sludge age are much higher with 10–20 g/l (Judd and Jefferson 2003) and 17–25 days, respectively. Consequently, sludge loading increases, and this leads to improved

Table 1 Comparison—general trends in the water sector (Brasseur et al. 2017; Climate Change Post Homepage 2017; Henry and Douhovnikoff 2008; Federal Service for Hydrometeorology and Environmental Monitoring (ROSHYDROMET) 2008; Saritas et al. 2015)

Parameter	Germany	Russia
Country size	Relatively small	Largest country of the world
Climate zones	More homogeneous climate conditions (except for the Alps)	Multitude of different climate zones
Temperature increase	About 1.3 °C (between 1881 and 2014)	About 1.3 °C (between 1907 and 2006) Temperature increase in Russia as a whole is larger than the global warming tendency
Policy	Public enterprises in water supply and wastewater disposal Market mechanism of pricing Tendency to remunicipalize water supply and waste disposal enterprises	Public enterprises in water supply and wastewater disposal Due to political reasons, the prices of water supply and wastewater disposal are extremely low The absence of efficient economic mechanisms for the introduction of modern water saving technologies
Water resources	Rich in water resources: per capita, approximately 2,300 m^3/a of water available	Uneven distribution of water resources across the country Water sources and drinking water are highly contaminated by chemical and biological agents In total, an increase of water resources is observed, but in arid regions water resources decrease Increasing threat of water deficits in Russia's eastern water basins caused by rising industrial consumption in neighboring China (transboundary effects) Pollution of water resources by an increasing volume of open wastewater containing micropollutants (e.g., rain water)
Water saving	Widely practiced: Average per capita consumption of water is 123 L (2016)	Not commonly practiced: Average per capita consumption of water is ca. 200 L (with slowly declining tendency)
Impact of climate change	Decrease of the average annual precipitation Decrease of the runoff during summer A high runoff is observed earlier in the year and the lowest runoff later in the year In many regions, heavy winter precipitation has increased; however, precipitation is the form of rain rather than snow Increased potential for extreme weather events with high importance for agricultural and urban hydrological issues Between 1970 and 2014, economic losses caused by climate-related natural hazards amount to 90 billion euros	Increase of the average annual precipitation Decrease in winter precipitation in the northeastern regions of Siberia including the Magadan District, the northern part of Khabarovsk land, and the eastern part of the Chukchi autonomous region Increase in spring precipitation in the western and northeastern regions of Siberia and in the European part of Russia Increased potential for extreme weather events with high importance for agricultural and urban hydrological issues Alterations in the river flow – Changes in the water inflow to reservoirs Decreasing water levels in lakes Modification of the thermal regime of permafrost regions (= permanent ground freezing; permafrost regions comprise 60% of the Russian area) – Influence on drainage leading to increased volume of surface water – Negative effects of frosting and thawing on infrastructure and buildings Desertification observed over the arid lands Boundaries of vegetation zones will shift to the north Tendency toward the degradation of glaciers in the Arctic islands and mountain glaciers Decrease in sea ice observed for the last two decades

Table 2 Main effects of climate change on the German water sector (Brasseur et al. 2017; Dechema 2004)

Effect of climate change	Influence on the water sector	Measures
Changes of rainfall during the year/heavy winter precipitation/heavy rainfall events during summer	Inadequate capacity of WWTP Excessive pollution of the WWTP effluent	Adopting wastewater disposal systems to strongly varying input flows Retrofit of existing WWTP in densely populated areas Intelligent management systems to control water distribution and water use
Decreasing runoff during summer	Inadequate industrial water supply	Optimization of production processes using integrated production techniques

Table 3 Main effects of climate change on the Russian water sector (Climate Change Post Homepage 2017; Dudarev et al. 2013)

Effect	Influence on the water sector	Possible measures
Changes in the water inflow to reservoirs	• Inadequate drinking water supply • Influence on the hydro-power industry	• Water reclamation from unconventional sources • Revision of the operating mode required
Changes in the water inflow to wastewater treatment plants	Excessive pollution of the WWTP effluent	• New construction required • Optimization of regional water use is essential • Change of the mindset required to save the environment
Regions with decreasing water resources	Falling groundwater table	• Alternative and additional sources of water to secure industrial water supply • Preventive maintenance and repair of water supply facilities and networks instead of repair efforts after accidents • Increased replacement rates of water pipelines to reduce water losses

interception of load peaks (Mudrack and Kunst 2010). Membrane filtration instead of sedimentation in a subsequent clarifier separates the activated sludge from the treated sewage. The separation effect of the membranes is worsened by the formation of a surface layer composed of biomass and other organic and inorganic compounds. However, small organic molecules and ions can pass through the membranes.

The advantage of MBR systems is the significantly lower generation of waste activated sludge in comparison with classical systems. This, in return, results in decreased operational costs, due to reduced pump capacity. The operation of MBR with a high concentration of activated sludge, allows 2–3 times smaller dimensions of the activated sludge tank compared to conventional aeration tanks, where the dry matter concentration is between 3 and 5 g/l (Judd and Jefferson 2003).

The MBR process offers significant advantages compared to conventional biological cleaning methods (Möbius 2006; Wett 2005; Deutsche Vereinigung für Wasserwirtschaft et al. 2014):

- Improved biological degradation performance
- Improved effluent quality due to complete rejection of the particulate load,

Table 4 Parameters influencing the biological degradation of pollutants in activated sludge flocks

Influencing parameter	Specification
Microorganisms	Kinds/composition Cell number Growth phase Sludge age
Flock surface	Surface charge Surface tension Hydrophilicity
Wastewater characteristics	Temperature pH Viscosity Nutrient content and composition Oxygen content
Wastewater treatment plant technology	Flow conditions Aeration conditions Volumetric load Sludge age

- Controlled sanitation due to complete rejection of all bacteria and partial rejection of viruses,
- Compact setup due to higher volumetric load and possible elimination of a subsequent clarifier,
- Independence of the biomass concentration from the sedimentation behavior of the activated sludge flocks,
- Independence of the sedimentation properties of the activated sludge,
- No sludge overflow caused by bulking sludge and floating sludge,
- High operational reliability through secure retention of activated sludge and almost solids-free outflow.

Thus, the use of MBR in municipal wastewater treatment is regarded as a key technology, e.g., for the retention of microscopic debris originating from cosmetic products or from the fragmentation of plastic waste (microplastics). Furthermore, the experience of several countries show that drinking water-associated outbreak of diseases often go hand in hand with missing or inadequate wastewater treatment (Gornik et al. 2001; Kramer et al. 2001; Jakubowski et al. 1991; Karanis et al. 2006). This problem could be solved sustainably using MBR technology.

Arriaga et al. (2016) suggest the application of submerged MBR as a polishing step for conventional wastewater treatment plants. In pilot-scale examinations using GE hollow fiber ultrafiltration membranes, they determined that biomass can be adapted to remove selected micropollutants by targeted addition of respective agents during start-up. Both, the removal efficiency for micro-pollutants and the biomass showed long-term stability.

Generally, the disadvantages of MBR technology over conventional biological purification are that

- It requires trained personnel,
- It is more vulnerable to shock loads due to lower tank volumes,
- It has higher investment costs due to the membrane plant and the greater demands for the mechanical pretreatment (Wett 2005) and
- It has higher operating costs, especially due to the energy demand of the aeration systems (Möbius 2006).

However, according to Schaller et al. (2010), the operating costs of MBR strongly depend on the following local conditions:

- Costs of energy
- Costs for sludge treatment and disposal of sewage sludge
- Composition of the sewage sludge
- Size of the wastewater treatment plant
- Local requirements concerning the quality of the water.

Furthermore, the most pressing problems in MBR technology are membrane clogging and fouling (Judd 2016).

3 Comparison of the Municipal Wastewater Treatment Systems in Germany and the Russian Federation

3.1 Wastewater Treatment in the Federal Republic of Germany

In Germany, mixed sewer systems are widely used. They discharge both, domestic and business wastewater and contaminated rainwater (Köhne 1998). In addition, a significant proportion of external water infiltrates into the sewage network (infiltration water) interfering with the smooth operation of the wastewater treatment plants (Statistisches Bundesamt Homepage 2013).

Since 1998 the total number of municipal wastewater treatment plants in Germany has decreased, while an upward trend of the arising amount of wastewater can be determined, see Fig. 1. In 2013, approximately 9,300 municipal wastewater treatment plants were operated in the Federal Republic of Germany, about two-thirds of them have a size that allows the treatment of max. 500,000 m^3/a wastewater. Only 285 systems are designed for the treatment of 6 million and more m^3/a wastewater (Statistisches Bundesamt Homepage 2013).

In 2013, the average percentage of the population connected to the public wastewater sewer system was 96.9%; in seven German states practically the entire population is connected. About two-thirds of the municipal wastewater treatment plants use a biological treatment with additional procedures such as nitrification, denitrification, nitrogen and phosphorus elimination or filtration. Only 22 plants possess a selective elimination of micro-pollutants (Statistisches Bundesamt Homepage 2013).

In Germany, there is an ongoing discussion whether municipal wastewater treatment plants larger than 100,000 inhabitant equivalents should be obliged to install further sewage treatment systems, as this would allow the elimination of micropollutants (2017), microplastics (Hidalgo-Ruz et al. 2012; Talvitie and Heinonen 2014, and multiresistant germs (Ottoson et al. 2006). MBR technology can be an option for these cases.

3.2 Wastewater Treatment in the Russian Federation

In the Russian Federation, the responsibility for the disposal of wastewater is largely in the public sector, State Unitary Enterprises, which operate mostly centralized largescale facilities. For the year 2012, the official statistics showed a total sewage volume of 15.7 billion m^3 (Federal State Statistics Service (Rosstat) 2014).

The vast majority of the wastewater treatment plants in the Russian Federation are in a worn-out state, as they were constructed in the 50–80 s

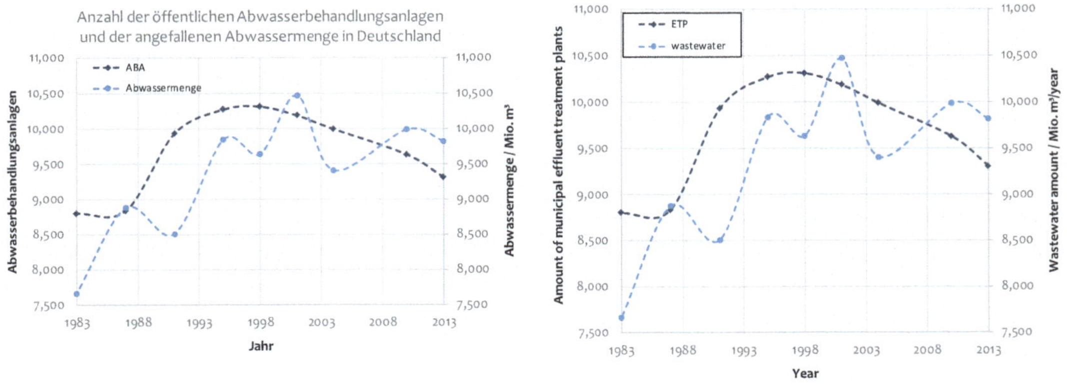

Fig. 1 Number of municipal wastewater treatment plants and arising wastewater volume in Germany, based on (Statistisches Bundesamt Homepage 2013)

of the last century (Gogina and Makisha 2013; Waterworld Homepage 2011).

Unless action is taken, the wear and tear of utilities in several municipalities is expected to increase 2–3% annually. Thus, the cleaning performance often does not comply with the current requirements. This situation further explains why about 17% of the wastewater was discharged without any treatment in 2007 (Waterworld Homepage 2011).

Further challenges for wastewater treatment plants in the Russian Federation arise from the heterogeneous population density and the extreme climatic conditions—long cold periods with temperatures up to −40 °C. Accordingly, particularly in sparsely populated areas, low water temperatures can prevail. Thus, the operation of wastewater treatment plants nearly comes to a standstill as the biological activity is reduced. In addition, power outages must be expected in winter, which can cause a die-off of microorganisms. Hence, more than 80% of the wastewater treatment plants in small towns within the Russian Federation do not reach the required limits (2017).

The costs for design and construction of wastewater treatment plants by domestic producers are estimated with approximately €3 million for a capacity of 10,000 m^3/day, and about €1.5 million for a capacity of 5,000 m^3/day.

4 Technical Aspects of the MBR Technology

4.1 Membranes and Membrane Modules

In MBR technology, the term membrane describes a thin layer of material that rejects biomass and other particulate load from an activated sludge suspension, while the treated water can pass, when a transmembrane pressure across the membrane is applied. The separation effect depends on the pore structure of the membranes, which acts like a sieve: particles larger than the pore size will be rejected, while water, smaller particles and dissolved substances permeate. This is referred to as size exclusion of substances. Technically employed membranes are in the microfiltration and ultrafiltration range with nominal cut-offs between 0.03 and 0.4 μm (Judd 2016).

According to a recent study (Judd 2016), about 45% of all technically applied MBR membranes are based on PVDF, other prevailing polymers are PES, PP, and PE. Ceramic membranes are applied to a minor extent.

Depending on the deployment of the membranes, two general types of MBRs can be distinguished:

- Submerged Membranes: Integration of the membranes directly in the bioreactor
- External Membranes: Installation of the membranes as a separate unit outside the bioreactor.

In submerged systems for municipal wastewater treatment, membranes are mainly applied in two different module designs: plate and frame modules and hollow fiber modules. In plate and frame modules, the plate-shaped membranes are arranged in parallel and joined together to form so called cassettes. A multitude of cassettes are then mounted to form a membrane module. In commercial applications, hollow fiber membranes are the most important type in terms of the installed area (Judd 2016). A large number of hollow fibers is assembled and welded together at one or both ends to form a module. In both module types, the permeate is drained from the outside to the inside of the membrane (so-called outside-in mode).

4.2 Performance

The performance of membrane plants can be described with the following parameters:

Permeate flux P

The permeate flux specifies the hydraulic performance of filtration units. It is defined as the volume flow Q that permeates through a given membrane area. At a given transmembrane pressure, the permeate flux mainly depends on the temperature and the dry matter substance of the wastewater. German technical bulletin DWA M 227 (Deutsche Vereinigung für Wasserwirtschaft et al. 2014) describes that plants in Germany were designed based on a temperature range between 8 and 12 °C and a resulting specific flux between 15 and 30 L/(m^2•h). Judd (2016) states an average flux of 18.5 ± 4.8 L/(m^2•h). Permeate flux directly influences the required membrane area.

Transmembrane pressure p

The transmembrane pressure is the pressure difference between the feed and the permeate phase that creates the driving force for the transport of material through the membrane.

Rejection R

Rejection depends on the size exclusion of substances caused by the porosity (pore size) of the respective membrane and is intensified by the formation of a surface layer of biomass and other organic and inorganic compounds. Membrane bioreactors allow the complete rejection of particulate pollutants including organic, nitrogen, and phosphorus compounds, as well as microorganisms, such as bacteria and large viruses. Thus, a better product quality can be achieved compared to conventional wastewater treatment systems. Ottoson (Talvitie and Heinonen 2014) reports a high separation efficiency concerning the removal of vegetative bacteria, such as E. coli (5 log removal), and enterococci (4.5 log removal).

The removal of phages and spores by the considered MBR was less efficient. DWA (Deutsche Vereinigung für Wasserwirtschaft et al. 2014) reports that the rejection of micropollutants by MBR systems was only slightly better compared to conventional wastewater treatment systems, when operated with comparable sludge loads.

Physicochemical properties strongly influence the removal efficiency of specific trace organic contaminants (2017). Alturki et al. come to the conclusion that especially hydrophilic organic compounds are poorly rejected by MBRs (Nghiem et al. 2009).

Fouling

Fouling refers to the decrease of performance over the operation time due to deposition of particles (accumulation of activated sludge) and growth of biofilms on the membrane surface (surface layer formation), adsorption of macromolecules in the pore structure and deterioration of the membrane material (aging). Depending on the adhesion forces between the foulants and the membrane, physical or chemical cleaning can be applied to restore the performance (Alturki et al. 2010).

Physical cleaning removes only reversible fouling layers. It involves the application of strong shear forces and cyclic reverse flow of permeate through the membrane (backflushing) (Deutsche Vereinigung für Wasserwirtschaft et al. 2014).

For the removal of irreversible fouling layers, chemical cleaning is required. It involves the use of, e.g., oxidizing or chelating chemical agents, which must be selected according to the type of contaminants (Deutsche Vereinigung für Wasserwirtschaft et al.2014).

Crossflow velocity

In addition to the previously described parameters, the performance of a membrane system is also influenced by the overflow speed parallel to the membrane surface (crossflow velocity), as a high crossflow velocity allows the control of the surface layer. Crossflow conditions in submerged membrane systems are achieved by aeration with coarse air bubbles (Judd and Jefferson 2003), which improves both, hydraulic and biological performance. Crossflow conditions in external membrane systems are achieved by pumping the feed over the membrane surface.

4.3 Design

Basically, following aspects have to be carried out for the proper design of retrofitted and newly constructed MBRs:

Table 5 Important aspects for the design of MBR

Aspect	Detailed information
Compilation of process data	**Hydraulic load (rated water volume)** e.g., storm and dry weather flow **Composition of the wastewaters**: temperature, pH value, polluting load (C,N,P compounds), turbidity, etc. **Effluent quality**: minimum requirements (permitted limits)
Required pretreatment	**Safeguard of the membrane modules by** Raking and sifting Sand and fat trap Primary clarification Intermediate screening
Bioreactor	**Biological degradation** Dry matter content Sludge loading Sludge age p-value Oxygen supply **Homogeneous distribution of the sludge in the tanks**
Membrane filtration	**Membrane configuration, tank, permeate collection system** At least two independent filtration lines for small systems or more than two independent filtration lines for large WWTPs (redundancy allows maintenance to be carried out without affecting plant performance; energy efficient operation at fluctuating feed flow) Defined water level above the membrane modules required Defined distance between module and the tank bottom required (air supply) Footprint should be selected twice as large as the cross-sectional area of the module (consideration of internal and external piping)

- Compilation of process data
- Design or check of existing wastewater pretreatment
- Design of the bioreactor
- Design of the membrane part.

More details are shown in Table 5.

Net flux

For continuous operation, it is important to consider that membrane bioreactors are operated cyclically. One cycle consists of the actual filtration phase, switching time, filtration breaks for membrane relaxation, and—if necessary—backflushing. Relaxation is standard practice; only if the filterability of biomass becomes low, relaxation is replaced by backflushing. Typical values for the cycle times are given in Table 6.

In addition, for the determination of the net permeate flux, permeate loss during backflushing and operation breaks due to chemical cleaning have to be taken into account.

Membrane area

The required membrane area is calculated based on the rated water volume divided by the net flux. It is well known that temperature has a significant impact on biological processes and influences, for instance, the rate of biochemical reactions, the microbial composition, and thus the structure of flocs. At the same time, the modified sludge morphology as well as the increasing viscosity of the water at lower temperatures will affect the performance of the membrane filtration (Judd 2004; Chiemchaisri

Table 6 Typical values for the cycle times (Deutsche Vereinigung für Wasserwirtschaft et al. 2014)

Cycle step	Typical duration
Filtration	10 min.
Relaxation	60–90 s
Backflushing	30 s

and Yamamoto 1994). Especially, regions with constant annual temperature are favored for MBR applications. Accordingly, van den Brink (Chiemchaisri and Yamamoto 1994) recommends installing surplus membrane area in countries with seasonal temperature variations to cope with winter conditions.

Module installation

MBRs can be installed either in standing or in suspended position. The installation of MBRs is usually determined using heuristic models. These take into account

- A defined water level above the module,
- A defined ground clearance for air distribution.

Taking the hydraulic system inside and outside the module into account, the installation area should be selected twice as large as the cross-sectional area of the module.

Operating costs

MBR operating costs are largely determined by the energy consumption of the permeate pump and membrane aeration, consumption of cleaning agents, and membrane exchange. The specific power consumption is about 100–200 Wh/m^2 membrane area and 0.5 kWh/m^3 permeate, respectively (van den Brink et al. 2011).

5 Application of Membrane Bioreactors

In Germany, the concept of membrane bioreactors in municipal wastewater treatment has been applied and further developed since 1999. According to (Deutsche Vereinigung für Wasserwirtschaft et al. 2014), 21 large-scale MBR systems with flow capacities between 4 and 1,880 m^3/h were retrofitted in existing wastewater treatment plants in Germany between 1999 and 2011. This corresponds to less than 0.5% of the existing WWTPs in Germany. Together, the German WWTPs retrofitted by MBR technology have an installed membrane area of up to 84,500 m^2. Seven of these MBR plants were built by the GE subsidiary Zenon and seven by Kubota. In 2008, these two companies were also the main equipment suppliers throughout Europe (Bertanza et al. 2017).

Compared with the world's largest MBR for the treatment of municipal wastewater, the German plants are small. In contrast, a recent study (Lesjean and Huisjes 2008) indicates that by the year 2016, throughout the world 28 MBR plants with a peak load design of >100,000 m^3 population equivalents had been installed.

Compared to other countries, the application of MBR technology is not prevailing in Russia. However, apart from international plant constructing companies, smaller domestic enterprises are active here. For example, "Akvametosintes" supplies MBR plants (MBR series "LKOY") with capacities between 10 and 1500 m^3/day (Krzeminski et al. 2017). They are especially used for decentralized water treatment in the suburbs of the capital towns. Ivanova (Miklashevskiy and Muraveva 2014) presents MBR technology using ultrafiltration membranes for the reclamation of high-quality water which is "discharged into water bodies used for drinking water supply, recreation or fish breeding."

Murashev (2016) describes MBR tests performed at the North Aeration Station of "Vodokanal-St. Petersburg." In test trials, hollow fiber membranes made of PVDF were used. In general, results of wastewater treatment used here are comparable with traditional water treatment technologies, see Table 7.

The data show that treatment with MBR:

- allows obtaining a more stable quality of treated water,
- is more effective in removing hardly degradable organic matter (indicated as COD),
- results in a high grade of P removal without using additional reagents.

Despite aeration and backflushing, the measured permeate flow rate gradually decreased from 470 L/h to 200 L/h within 20 days of operation in this WWTP. As recommended by the membrane manufacturer, chemical cleaning was conducted with 2% HCl for 15 min after 1.5 months of operation. After chemical

Table 7 Results of MBR tests performed by "Vodokanal-St. Petersburg" (Ivanova 2016)

Parameter	Untreated water	Treated water
Suspended solids	160–400 mg/L	6.3 mg/L
pH	7.2–7.6	7.2–7.6
COD	280–550 mg O_2/L	21 mg O_2/L
BOD	100–180 mg O_2/L	2.5 mg O_2/L
NH_4–N	16–45 mg N/L	2.9 mg N/L
NO_3–N	0.18–0.29 mg N/L	7.5 mg N/L
PO_4–P	1.4–3.6 mg P/L	0.56 mg P/L

treatment, 70% of performance was restored and the permeate flow rate became 350 L/h.

6 Summary

The higher probability of natural disasters as a consequence of climate change requires considerable adaptation of existing water supply and wastewater systems.

Membrane bioreactors combine activated sludge treatment and membrane separation. The future application of MBR systems will have a different focus in Germany and Russia. While in Germany, retrofitting of existing plants is more likely, the existing municipal effluent treatment plants in Russia require a completely new design. In Russia, also decentralized water treatment plants could be equipped with MBR technology to reclaim high-quality water. In both countries, MBR technology can be applied as a key technology that will help to solve urgent problems connected with the release of microplastics, micropollutants, and multiresistant germs as so-called fourth purification stage. In addition, MBR technology leads to a lower footprint of the wastewater treatment plan.

Furthermore, MBR technology can become a vital step for water reclamation systems in arid regions of Russia. Although large-scale MBR technology is in operation worldwide, research is still required to cope with the specific Russian wastewater conditions.

References

Alturki AA, Tadkaew N, McDonald JA et al (2010) Combining MBR and NF/RO membrane filtration for the removal of trace organics in indirect potable water reuse applications. J Membr Sci 365:206–215

Arriaga S, de Jonge N, Nielsen ML et al (2016) Evaluation of a membrane bioreactor system as post-treatment in waste water treatment for better removal of micropollutants. Water Res 107:37–46

Bertanza G, Canato M, Laera G et al (2017) A comparison between two full-scale MBR and CAS municipal wastewater treatment plants: techno-economic-environmental assessment. Environ Sci Pollut Res 24(21):17383–17393

Brasseur GP, Jacob D, Schuck-Zöller S (eds) (2017) Klimawandel in Deutschland: Entwicklung, Folgen, Risiken und Perspektiven. Springer/Spektrum, Heidelberg

Chiemchaisri C, Yamamoto K (1994) Performance of membrane separation bioreactor at various temperatures for domestic wastewater treatment. J Membr Sci 87(1–2):119–129

Climate Change Post Homepage (2017) https://www.climatechangepost.com/russia/climate-change. Last accessed 8 Oct 2017

Dechema (ed.) (2004) Trends und Perspektiven in der industriellen Wassertechnik. Positionspapier der ProcessNet-Fachgruppe Produktionsintegrierte Wasser- und Abwassertechnik. Frankfurt

Deutsche Vereinigung für Wasserwirtschaft, Abwasser und Abfall e. V. (ed) (2014) Merkblatt DWA-M 227: Membran-Bioreaktor-Verfahren (MBR-Verfahren). Hennef

Dieckmann H, Metz H (1991) Grundlagen und Praxis der Biotechnologie. Gustav Fischer, Stuttgart

Dudarev AA, Dushkina EV, Sladkova YN et al (2013) Food and water security issues in russia II: water security in general population of russian arctic, siberia and far east, 2000–2011. Int J Circumpol Heal 72:1–10

EVD Market survey Russia—Waste and wastewater. http://docplayer.net/37939823-Market-survey-russia-waste-and-waste-water.html. Accessed 10 Oct 2017

Federal Service for Hydrometeorology and Environmental Monitoring (ROSHYDROMET) (2008) Assessment report on climate change and its consequences in Russian Federation—General Summary. Roshydromet, Moscow

Federal State Statistics Service (Rosstat) (2014) Russia 2014 statistical pocketbook. Rosstat, Moscow

Gogina E, Makisha N (2013) Reconstruction of waste water treatment plants in Russia, approaches and solutions. Appl Mech Mater 361–363:628–631

Gornik V, Behringer K, Kölb B et al (2001) Erster Giardiasisausbruch im Zusammenhang mit kontaminiertem Trinkwasser in Deutschland. Bundesgesundheitsblatt Gesundheitsforschung Gesundheitsschutz 44(4):351–357
Henry LA, Douhovnikoff V (2008) Environmental issues in Russia. Annu Rev Env Resour 33:437–460
Hidalgo-Ruz V, Gutow L, Thompson RC et al (2012) Microplastics in the marine environment: a review of the methods used for identification and quantification. Environ Sci Technol 46:3060–3075
Ivanova OV (2016) The experience of introducing advanced technologies based on ultrafiltration membranes. Water Supply Sanit Tech 4:40–42 (in Russian)
Jakubowski W, Sykora JL, Sorber CA et al (1991) Determining Giardiasis prevalence by examination of sewage. Water Sci Technol 24(2):173–178
Judd S (2004) A review of fouling of membrane bioreactors in sewage treatment. Water Sci Technol 49(2):229–235
Judd SJ (2016) The status of industrial and municipal effluent treatment with membrane bioreactor technology. Chem Eng J 305:37–45
Judd S, Jefferson B (2003) Membranes for industrial wastewater recovery and re-use. Elsevier, Oxford
Karanis P, Sotiriadou I, Kartashev V et al (2006) Occurrence of *Giardia* and *Cryptosporidium* in water supplies of Russia and Bulgaria. Environ Res 102 (3):260–271
Köhne M (1998) Analyse und Regelung biologischer Abwasserreinigungsprozesse in Kläranlagen. Automatisierungstechnik 46(5):215–234
Kramer MH, Quade G, Hartemann P et al (2001) Waterborne diseases in Europe—1986–1996. J Am Water Works Assoc 93(1):48–53
Krzeminski P, Leverette L, Malamis S et al (2017) Membrane bioreactors—a review on recent developments in energy reduction, fouling control, novel configurations, LCA and market prospects. J Membr Sci 527:207–227
Lesjean B, Huisjes EH (2008) Survey of the European MBR market: trends and perspectives. Desalination 231:71–81
Miklashevskiy NV, Muraveva TS (2014) Cleaning of wastewater by MBR technology plumbing heating conditioning. Energy Eff 12:34–41 (in Russian)
Möbius CH (2006) Water use and wastewater treatment in paper mills, 1st edn. Books on demand GmbH, Norderstedt
Mudrack K, Kunst S (2010) Biologie der Abwasserreinigung, 5th edn. Springer/Spektrum, Wiesbaden
Murashev SV, Soloveva EV, Shilova NK (2016) The experience of wastewater treatment technology approbation on the basis of a membrane bioreactor. Water Supply Sanit Tech 2:52–57 (in Russian)
Nghiem LD, Tadkaew N, Sivakumar M (2009) Removal of trace organic contaminants by submerged membrane bioreactors. Desalination 236:127–134
Ottoson J, Hansen A, Bjölenius B et al (2006) Removal of viruses, parasitic protozoa and microbial indicators in conventional and membrane processes in a wastewater pilot plant. Water Res 40:1449–1457
Saritas O, Proskuryakova L, Kyzyngasheva E (2015) Water resources—an analysis of trends, weak signals and wildcards with implication for Russia. Working Papers WP BRP 35/STI/ 2015, National Research University Higher School of Economics, Moscow
Schaller J, Drews A, Kraume M (2010) Development of a cost model for membrane bioreactors including sludge handling costs. Desalin Water Treat 8:1–6
Statistisches Bundesamt Homepage (2013) Umwelt: Öffentliche Wasserversorgung und öffentliche Abwasserentsorgung – Strukturdaten zur Wasserwirtschaft. Fachserie 19, Reihe 2.1.3. https://www.destatis.de/DE/Publikationen/Thematisch/UmweltstatistischeErhebungen/Wasserwirtschaft/WasserOeffentlich.Html. Accessed 8 Oct 2017
Talvitie J, Heinonen M (2014) Preliminary study on synthetic microfibers and particles at a municipal waste water treatment plant. HELCOM 2014, BASE project 2012–2014
Umweltbundesamt: Measures to reduce micropollutant emissions to water—Summary. http://www.umweltbundesamt.de/publikationen/measures-to-reduce-micropollutant-emissions-to. Accessed Oct 8 2017
van den Brink P, Satpradit OA, van Bentem A et al (2011) Effect of temperature shocks on membrane fouling in membrane bioreactors. Water Res 45:4491–4500
Waterworld Homepage. http://www.waterworld.com/articles/2011/09/russian-water-and-wastewater-market.html. Accessed 10 Oct 2017
Wett M (2005) Foulingverhalten des Membranbelebungsverfahrens und Auswirkungen auf die Leitungsfähigkeit. Schriftenreihe des Fachgebiets Siedlungswasserwirtschaft der Universität Kassel, vol. 26. Kassel University Press, Kassel

Experiences on Wastewater Treatment and Reuse for Different Types of Industries in Turkey

Eyüp Debik

Abstract

In the face of increasing water scarcity, the industrial wastewater reuse has become an important practice of improving water use efficiency. However, industrial wastewaters have very high pollution potential and industrial wastewater treatment is very challenging. Furthermore, new and emerging pollutants cannot be adequately removed by currently applied state-of-the-art methods. New technologies and a combination of biological and chemical treatment methods are urgently required. Hence, the objective of this chapter is to discuss the most adequate wastewater treatment methods from the studies dealing with the treatment and reuse for the wastewaters, coming from the dominant industries in Turkey: the textile, poultry slaughterhouses and olive processing industries. In the studies, membrane processes as well as physical biological, and conventional treatment methods were applied. The performance of ultrafiltration (UF) and nanofiltration (NF) membrane processes after aerobic and anaerobic pre-treatment for the textile wastewater was researched. The use of centrifugation as a pretreatment step for the UF, NF, and reverse osmosis (RO) membranes was performed for the poultry slaughterhouse wastewater with high particulate COD content. Finally, the effects of the application of UF, NF, and RO membranes after preliminary treatment with centrifugation for olive wastewater with high solids content was analyzed.

Keywords

Industrial wastewater treatment · Industrial wastewater reuse · Pre-treatment · Centrifugation · Biological treatment · Membrane processes

1 Introduction and Objective

In the face of increasing industrial activities and growing numbers of new contaminants industrial wastewater treatment has become very challenging in the last decades. In Turkey, currently applied processes in the treatment of wastewaters coming from main industries such as textile, poultry slaughterhouse, and olive oil production facilities are not sufficient due to growing water volumes and new pollutants. New methods, as well as biological and chemical processes, are urgently required to address these challenges.

Hence, this article has two objectives. The first part presents an overview of the current practice of industrial wastewater reuse and applied state-of-the-art methods for the treatment

E. Debik (✉)
Environmental Engineering Department,
Yildiz Technical University, Istanbul, Turkey
e-mail: debik@yildiz.edu.tr

L. Ribbe et al. (eds.), *Towards Water Secure Societies*,
https://doi.org/10.1007/978-3-030-50653-7_8

of wastewaters coming from main industries in Turkey. The second part describes experiments and results coming from laboratory studies dedicated to testing how different biophysical and biological process influence the treatment of the main pollutants. The performance of ultrafiltration (UF) and nanofiltration (NF) membrane processes after aerobic and anaerobic pretreatment for the textile wastewater was investigated. The use of centrifugation as a pretreatment step for the UF, NF, and reverse osmosis (RO) membranes was investigated for the poultry slaughterhouse wastewater with high particulate COD content. Finally, the effects of the application of UF, NF, and RO membranes after preliminary treatment with centrifugation for olive wastewater with high solids content was analyzed.

2 Wastewater Treatment Methods and System Selection

2.1 Physical Processes

Wastewater flows usually include some solid materials that can be removed by physical methods such as screening, sedimentation, centrifugation, and filtration. These solid materials need to be removed from the water in order to increase the efficiency of the processes to be applied in the following steps and to prevent damage to the equipment to be used. These processes, which are carried out at the beginning of the treatment processes, usually make size-based separation and are often referred to as pretreatment in the treatment flow diagram.

2.2 Biological Treatment

Biological treatment is a widely used method for the treatment of organic wastewater, and it can be applied in two ways as aerobic and anaerobic. If the wastewater has a BOD/COD ratio higher than 0.3 (Na et al. 2005), it is primarily to be evaluated to treat with a biological treatment alternative, then other treatment alternatives can be applied if the desired limits are not reached. Aerobic and anaerobic treatment methods are briefly described below. It should also be noted that both systems have membrane bioreactor applications and are more efficient than traditional systems that have sedimentation tanks. It should be known that the cost of membrane bioreactors (MBR) is slightly higher than that of traditional systems, and this should be taken into account for a choice to be made.

2.2.1 Aerobic Treatment

Aerobic treatment is a process in which the colloidal or dissolved organic substances in the wastewater are converted to end products by microorganisms in an oxygenated environment. In order to operate the process effectively, some of the biomass, sludge, that is formed is recycled to the system to maintain the biomass concentration in the bioreactor, aeration tank, which is provided to be effectively aerated. This commonly used system is known as the "activated sludge system" because the biomass is kept active in the aeration tank. It is necessary to dispose of excess sludge which is formed with end products and removed from the system. The resulting sludge must be disposed of by one of the final disposal methods such as land use, storage, incineration, etc., after a number of treatments such as thickening, aerobic/anaerobic stabilization, dewatering, drying, etc.

2.2.2 Anaerobic Treatment

While anaerobic treatment has been used extensively for the sludge stabilization in the past, it has been widely used for the treatment of industrial wastewater, especially after the 1980s. In this treatment process, microorganisms convert the organic and inorganic substances in the wastewater into end products in an oxygen-free environment. The most important end product is methane, which makes anaerobic treatment advantageous in many cases. Due to methane formation in the system, less sludge is produced than aerobic systems and methane can be used as an energy source. In addition to the use of methane as a source of energy, anaerobic treatment does not require aeration like aerobic treatment, which makes the system even more advantageous. On the other hand, the fact that the organic wastes with high particulate content can be treated effectively by anaerobic methods, according to the type of the process to be used, cause these systems to become prominent, especially in the treatment of organic wastewater with high organic matter content. Taking all these advantages into consideration, if the BOD concentration of any industrial wastewater is above 1000 mg/L, the anaerobic treatment alternative should be considered as a priority over aerobic treatment (Speece 1995). However, it should be kept in mind that anaerobic treatment will often not to meet discharge standards alone and some additional processes will be necessary.

2.3 Membrane Processes

The membranes developed with different structures and techniques perform as a selective barrier under the certain pressure and soluble and particulate matter are separated from the feed stream entering the membrane. In the separation process, the feed solution flows through the membrane surface by the influence of the driving forces to the other side of the membrane and forms a permeate stream, while concentrate stream in which the contaminants are present cannot pass through the membrane. Today, membrane types commonly used in wastewater treatment are microfiltration (MF), ultrafiltration (UF), nanofiltration (NF) and reverse osmosis (RO) membranes. While MF membranes are mostly used as pretreatment in wastewater treatment, clean water is obtained with RO membranes. MF and UF membranes are generally used in MBR systems, and especially with UF membranes, the desired quality for the discharge can be achieved. Membrane processes have some advantages such as low space requirement and operational simplicity, but they have some disadvantages such as high initial investment and operational costs.

2.4 Treatment System Selection

In order to obtain an effective reclaiming, it is first necessary to evaluate the quantity and character of the water used in the process and the wastewater formed from the industrial process with a point-by-point manner. This will allow the selection of the best available technology to be used in the production, the methods that can be applied to reduce the usage of water and other sources, and the selection of the most appropriate treatment process that can be used for wastewater treatment and/or reclaiming. The selected treatment process for wastewater treatment and/or reclaiming requires, in some cases, the separate collection of wastewater streams in the production facility, which will ensure both efficient operation of the treatment process and effective reclaiming. For this reason, careful consideration of the production process, water and wastewater streams will help to decide which wastewater streams should be separated. In addition to this, knowing the amount and quality of water used in the industrial process with a point-by-point manner will help to reduce water usage and determine how much water can be reused.

3 Treatability Study Results of Industrial Wastewaters

The chapters below present the results of laboratory experiments the treatment and/or reclaiming work carried out using wastewater from cotton textile, poultry slaughterhouse, and olive oil production industry, which are important industrial facilities in different countries of the world, are shared. These industries consumes a high amount of water and the pollution of the wastewater is very high.

3.1 Specifications of the Laboratory-Scale Membrane Process

GE Sepa™ CF2 membrane cell supplied from Osmonics®, Inc. was used in the studies (Fig. 1). In the membrane system, permeate flow is

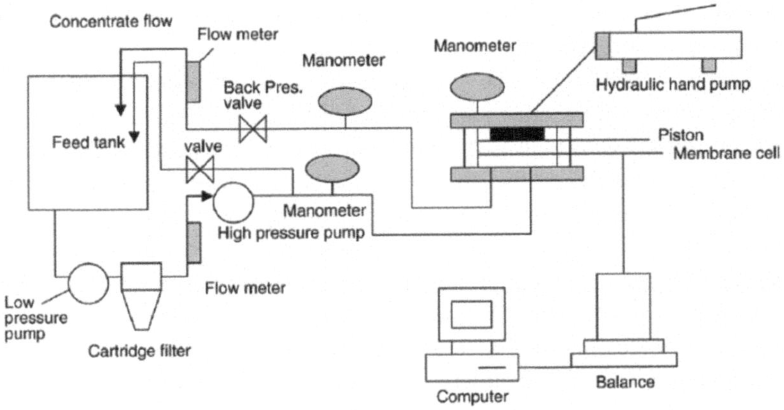

Fig. 1 Schematic diagram of the membrane process (Debik et al. 2010)

collected with a beaker and weighed with a balance while the concentrated flow is separated from the permeate flow and recycled to the feed vessel. In the system, there is a cartridge filter having 10-μm pore size to remove the coarse particles from the wastewater before the membrane cell. All membrane experiments were carried out at 25 °C with a heat exchanger in the feed vessel.

Table 1 Characteristics of the raw (untreated), aerobically, and anaerobically pre-treated textile wastewater

Wastewater	pH	COD, mg/L	Color, Pt–Co
Raw (untreated)	12 ± 1	2300 ± 400	775 ± 175
Aerobically pre-treated	7.1	650	822
Anaerobically pre-treated	7.2 ± 0.5	630 ± 10	300 ± 100

3.2 Cotton Textile Wastewater

In this research, reclaiming studies were carried out with raw wastewater, aerobic SBR (Sequencing Batch Reactor) effluent, and lab-scale SGBR (Static Granular Bed Reactor) anaerobic treatment system effluent (Debik et al. 2010). Wastewater characteristics given to the membrane processes, which were only NF and UF + NF, are shown in Table 1 and the properties of the membranes used in the studies are given in Table 2. Aerobic and anaerobic treatment steps have been proposed as a pretreatment for membrane processes and the results have been evaluated within this scope.

Results from the membrane studies using raw textile wastewater, aerobically, and anaerobically treated textile wastewater were given in Table 3. According to the studies in the literature given in Table 4, the properties of waters can be used for various operations in the textile sector.

Each fiber such as wool, silk, cotton, polyester etc., needs different water qualities to process.

Table 2 Characteristics of the membranes used for the textile wastewater treatment

Membrane type	Material	Membrane property
NP 010	Polyethersulfone	Hydrophilic
UC 010	Cellulose	Not available

Table 3 Results of the treatability studies for textile wastewaters

Wastewater	COD, mg/L	Color, Pt–Co
Raw (untreated)	250–320 (86–89%)	87–131 (83–89%)
Aerobically pre-treated	15–35 (95–98%)	25–31 (96–97%)
Anaerobically pre-treated	95–113 (82–85%)	60–62 (79–80%)

For this reason, it is quite difficult to give a general standard for water quality for the textile industry. According to the treatability results and literature (Erdumlu et al. 2012; Wang et al.

Table 4 Properties of waters can be used for various operations in the textile industry (Debik et al. 2010)

COD, mg/L	Color	Purpose
0–160	0–2 Lovibond	Reuse for reactive and acid dyeing process (Li and Zhao 1999)
30	0.01 (426 nm)	Dyeing of synthetic fibers (Rozzi et al. 1999)
178–218	20–30 Pt–Co	Reuse for reactive dyeing of 100% cotton fabric (Uzal 2007)
40	0.02 (426 nm)	Reuse for low quality wool (Comodo et al. 1993)

Table 5 Characteristics of the raw poultry slaughterhouse wastewater

Parameter	Value
COD, g/L	7.97 ± 0.14
TSS, g/L	2.76 ± 0.70
VSS, g/L	2.41 ± 0.60
Conductivity, mS/cm	2.75 ± 0.10
pH	6.6 ± 0.1

2012), reclaimed water obtained with the UF + NF membrane configuration after aerobic pretreatment has acceptable COD and color. Using this configuration, the treated textile wastewater is reusable for delicate operations of dyeing.

3.3 Poultry Slaughterhouse Wastewater

Poultry slaughterhouse wastewaters were tried to reclaim using different kind of treatment methods in the literature (Del Nery et al. 2016; Malmali et al. 2018), while some of them obtained very good reuse potential (de Nardi et al. 2011; Sardari et al. 2018). In the study, poultry slaughterhouse wastewater obtained from a factory in Sakarya was used in laboratory studies as raw wastewater (Coskun et al. 2016). The properties of the raw wastewater used are given in Table 5. During the studies, the centrifugation process was used as pretreatment and then NF, RO, UF + NF, UF + RO alternatives were evaluated as membrane processes. The properties of the membranes used in the study are given in Table 6.

For the treatment of poultry slaughterhouse wastewater, approximately 80% of COD and more than 90% TSS removal were achieved by centrifugation process. The COD concentration was 1,200 mg/L and 260 mg/L after centrifugation and UF processes, respectively. Thus, while the influent COD concentration was 1,200 mg/L for the UF-only, NF-only, and RO-only processes, it was 260 mg/L for the NF and RO processes after the UF process.

Using the UF process generally affected positively NF and RO performances for all TMPs. When TMP increased from 5 to 10 bars, COD removal efficiencies quickly increased and became more stable in all configurations except the UF + NF configuration. The highest COD removal efficiency was obtained using RO after UF, which was 96.8%, and COD concentration of the effluent was less than 10 mg/L. The conductivity of the treated water after UF + RO process was about 100 μS/cm.

Table 6 Characteristics of the membranes used for the poultry slaughterhouse wastewater treatment

Membrane type	Polymer	MWCO[a], Da	Salt rejection efficiency, %
Ultrafiltration (UF)	Polysulfone	30,000	NA[b]
Nanofiltration (NF)	TF (Thin film)	150–300	98 (for Mg_2SO_4)
Reverse osmosis (RO)	Polyamide	0	99.5 (for NaCl)

[a]MWCO: Molecular weight cut-off
[b]NA: Not available

Table 7 Characteristics of the OMWW

Parameter	Value
pH	4.6 ± 0.2
Conductivity, mS/cm	5.3 ± 0.2
TSS, g/L	6.8 ± 0.7
COD, g/L	40.3 ± 1.0

According to "Regulation on the Procedures and Principles of Operation and Inspection of Poultry Meat and Meat Products Production Facilities" in Turkey, treated slaughterhouse wastewater having this feature can be reused for steam generation, cooling of refrigeration equipment, fire extinguishing, and flotation of fur and debris generating from the slaughter facility.

3.4 Olive Mill Wastewater (OMWW)

OMWW used in this study was obtained from a factory, in which three-phase continuous system is used for olive oil production, in Ayvalik, Turkey. The characteristics of raw wastewater are given in Table 7 (Coskun et al. 2010). Laboratory studies were carried out with different types and configuration of membranes such as only-NF, UF + NF, and UF + RO after centrifugation. Three types of NF membrane such as NP010, NP030, and NF270, and two types of RO membrane such as XLE and BW30 were used for the treatability studies of OMWW. Membrane properties used in the studies are given in Table 8.

For the treatment of OMWW, investigations were carried out using only-NF, UF + NF, and UF + RO membranes, while centrifugation (CENT) process was used as pretreatment in all studies. For NF options, the best result was observed with the CENT + UF + NF configuration and about 85.5% COD removal efficiency was achieved with NF270 membrane type, but the effluent COD concentration (around 6000 mg/L) was not even dischargeable according to the regulations in Turkey.

Two types of RO membranes such as XLE and BW30 were used for reuse study of OMWW with the CENT + UF + RO configuration. Even very high and similar removal efficiencies, which were higher than 97%, were obtained using both RO membranes, COD values (around 1100 mg/L) was not acceptable to reuse of the effluent. It is considered that the high COD value in the effluent water is due to the fermentation products resulting from the storage of the wastewater. For three-phase continuous system, some researchers also reached almost the same effluent quality in the literature (Ochando-Pulido et al. 2014). Results obtained in this study showed that RO membranes treating the OMWW are usable to meet the limits to sewage discharge according to Turkish legislation.

4 Conclusions

The results indicate that if the membrane processes are applied after the pre-treatment in the purification of the poultry slaughterhouse wastewater and textile wastewater, it is possible to obtain adequate water quality to reuse it. However, the olive mill wastewater has very high

Table 8 Characteristics of the membranes used for OMWW treatment

Membrane type	Manufacturer	Material	Property	MWCO[a], kDa
UC 010	Macrodyn Nadir	Cellulose	NA[b]	10
NP 010	Macrodyn Nadir	Polyethersulfone	Hydrophilic	NA
NP 030	Macrodyn Nadir	Polyethersulfone	Hydrophilic	NA
NF 270	DOW Filmtec	Polyamide	NA	0.2–0.3
XLE	DOW Filmtec	Polyamide	NA	NA
BW 30	DOW Filmtec	Polyamide	NA	NA

[a]MWCO: Molecular weight cut-off
[b]NA: Not available

COD concentration and the water quality after treatment doesn't meet the standards for reuse.

The article concludes that the currently applied state-of-the-art technologies are not sufficient to treat industrial wastewaters to the quality standards required wastewater reuse. Instead of following the traditional pollution control approach, the industry should focus on reducing pollutants and water use in the production process.

Acknowledgments I would like to thank DAAD and Exceed SWINDON project for providing financial support to participate in the Water Security and Climate Change Conference 2017.

References

Comodo N, Masotti N, Sacco C, Tedioli G (1993) Wastewater treatment with ozone for industrial and agricultural reuse (in Italian). In: ANDIS biennial conference, Palermo (Italy), pp 21–23

Coskun T, Debik E, Demir NM (2010) Treatment of olive mill wastewaters by nanofiltration and reverse osmosis membranes. Desalination 259(1–3):65–70

Coskun T, Debik E, Kabuk HA, Demir NM, Basturk I, Yildirim B, Temizel D, Kucuk S (2016) Treatment of poultry slaughterhouse wastewater using a membrane process, water reuse, and economic analysis. Desalin Water Treat 57(11):4944–4951

Debik E, Kaykioglu G, Coban A, Koyuncu I (2010) Reuse of anaerobically and aerobically pre-treated textile wastewater by UF and NF membranes. Desalination 256(1–3):174–180

Del Nery V, Damianovic MHZ, Moura RB, Pozzi E, Pires EC, Foresti E (2016) Poultry slaughterhouse wastewater treatment plant for high quality effluent. Water Sci Technol 73(2):309–316

de Nardi IR, Del Nery V, Amorim AKB, dos Santos NG, Chimenes F (2011) Performances of SBR, chemical-DAF and UV disinfection for poultry slaughterhouse wastewater reclamation. Desalination 269(1–3): 184–189

Erdumlu N, Ozipek B, Yilmaz G, Topatan Z (2012) Reuse of effluent water obtained in different textile finishing processes. AUTEX Res J 12(1):23–28

Li XZ, Zhao YG (1999) Advanced treatment of dyeing wastewater for reuse. Water Sci Technol 39:249–255

Malmali M, Askegaard J, Sardari K, Eswaranandam S, Sengupta A, Wickramasinghe SR (2018) Evaluation of ultrafiltration membranes for treating poultry processing wastewater. J Water Process Eng 22:218–226

Na YS, Lee CH, Lee TK, Lee SW, Park YS, Oh YK, Park SH, Song SK (2005) Photocatalytic decomposition of nonbiodegradable substances in wastewater from an acrylic fibre manufacturing process. Korean J Chem Eng 22(2):246–249

Ochando-Pulido JM, Stoller M, Di Palma L, Martinez-Ferez A (2014) Threshold performance of a spiral-wound reverse osmosis membrane in the treatment of olive mill effluents from two-phase and three-phase extraction processes. Chem Eng Process Process Intensif 83:64–70

Rozzi A, Malpei F, Bonomo L, Bianchi R (1999) Textile wastewater reuse in Northern Italy (Como). Water Sci Technol 39:121–128

Sardari K, Askegaard J, Chiao YH, Darvishmanesh S, Kamaz M, Wickramasinghe SR (2018) Electrocoagulation followed by ultrafiltration for treating poultry processing wastewater. J Environ Chem Eng 6(4): 4937–4944

Speece RE (1995) Anaerobic biotechnology for industrial wastewater. Vanderbilt University, Tennessee, USA

Uzal N (2007) Recovery and reuse of indigo dyeing wastewater using membrane technology. The Middle East Technical University, PhD Thesis, Ankara

Wang XJ, Xu QK, Qi LQ (2012) Advanced treatment of textile wastewater for reuse by ozonation-biological and membrane processes. Eco-Dyeing, Finish Green Chem 441:578–583

Responses to Water Scarcity Considering Social Determinants: Groundnut Farmers in the Lawra and Nandom Districts, Ghana

Abdul Rauf Zanya Salifu, Elaine T. Lawson, and Charlotte Wrigley-Asante

Abstract

Households' food security in semi-arid Ghana is under pressure due to water scarcity and climate variability. Climate variability has become the biggest setback to livelihoods and wellbeing in this area, where high dependence on rainfed agriculture coexists with fragile ecosystems. However, the household vulnerability to water scarcity is not only determinate by environmental challenges. Climatic challenges correlate with social, economic, cultural and political inequalities. These factors determine the adaptive responses to water scarcity of different social groups. Focusing on the case study dedicated to the groundnut farmers in semi-arid Ghana, this paper aims at defining the vulnerability determinants to water scarcity and at assessing capacities to adapt to water crises considering the social components (age, gender, education, land ownership). The results show that the reasons for non-adoption and increased vulnerability are directly influenced by social and demographic factors such as age, gender and land ownership. We conclude that to assess the vulnerability and develop successful adaptation strategies, it is critical to identify and articulate not just the climatic drivers, but also the social, economic and political conditions. The socio-economic and political factors contribute to both, to intensifying the vulnerability and to shaping the ability to adapt to changing climatic conditions.

Keywords

Food security · Climate change · Gender · Resilience · Technology · Water insecurity · Women

1 Introduction

Changes in climate have occurred in the past five decades, leading to extreme weather such as droughts, floods, heatwaves and thunderstorms (IPCC 2013). Africa contributes the least to global warming (Fields 2005), yet it is the hardest hit by the impacts of anthropogenic climate change because of the continent's high

A. R. Z. Salifu · E. T. Lawson (✉)
Institute for Environment and Sanitation Studies (IESS), College of Basic and Applied, Sciences, University of Ghana, P.O. Box LG 209, Legon, Accra, Ghana
e-mail: elaine_t@staff.ug.edu.gh

A. R. Z. Salifu
e-mail: arzsalifu@st.ug.edu.gh

C. Wrigley-Asante
Department of Geography and Resource Development, College of Humanities, University of Ghana, P. O. Pox LG 72, Legon, Accra, Ghana
e-mail: CWrigley-Asante@ug.edu.gh

L. Ribbe et al. (eds.), *Towards Water Secure Societies*,
https://doi.org/10.1007/978-3-030-50653-7_9

vulnerability and low adaptive capacity (MESTI 2013; Niang et al. 2014). Africa's high vulnerability relative to other parts of the world is heightened by high levels of poverty, conflicts, poor infrastructure and heavy reliance on climate-sensitive livelihoods among others (Boko et al. 2007; Niang et al. 2014). In West Africa, it is estimated that near-surface temperature increased between 0.5 and 0.8 °C from 1970 and 2010, with future temperatures expected to rise above global average by the end of the century (Niang et al. 2014). Future projections for temperature in Ghana generally show that temperatures are expected to rise by 0.6 °C, 2.0 °C and 3.9 °C by the years 2020, 2050 and 2080, respectively (MESTI 2013).

Food production systems and rural livelihoods in Ghana are projected to be severely impacted by climate variability and change (World Bank 2013). Many areas are characterized by high degrees of exposure to droughts, dry spells, intra- and inter-annual variability, high temperatures (Padgham et al. 2015), high levels of sensitivity due to over-reliance on climate-sensitive economic and livelihood systems (MoFA 2011), as well as limited capacity to adapt or lessen the adverse impacts of climate variability and change (Niang et al. 2014; Antwi-Agyei et al. 2015a). Low yields and crop failure in semi-arid Ghana are linked to droughts, dry spells, truncated length of the growing season and to a decrease in total rainfall (Padgham 2009; Lyimo and Kangalawe 2010; Sarr 2012; Mawunya and Adiku 2013; Mawunya and Adiku 2013; Padgham et al. 2015).

Due to differences in levels of vulnerability and adaptive capacity, the impacts of climate variability and change will vary considerably among regions, countries, sectors as well as social groups, with the world's poorest being the worst affected (Nielsen and Reenberg 2010; Popke et al. 2014; Nyantakyi-Frimpong and Bezner-Ker 2015; Coirolo and Rahman 2014). The impacts of climate variability and change on communities, households or individuals are partly shaped by societies' values, culture, perceptions and power structures (Popke et al. 2014). Thus, vulnerability is not only determined by biophysical factors (floods/droughts) but also by sociocultural, structural and institutional factors that may include gender, age, disability status, ethnicity, level of education and political affiliation among others (Heltberg et al. 2009). These factors determine the adaptive responses of different social groups (Heltberg et al. 2009) by shaping their levels of asset base, access to services and infrastructure, level of dependence on climate-sensitive livelihoods and political agency (Padgham et al. 2015).

This study sets out to isolate the drivers of vulnerability to climate change among groundnut farmers in the Upper West Region and to identify the ways in which the different social groups are adapting to climate variability and change. The region produces about 37% of the total groundnut output of Ghana (Angelucci and Bazzucchi 2013). The differences in vulnerabilities and capacities among female and male farmers have led to gendered vulnerabilities related to climate change and its impacts on communities (Huyer et al. 2015; Rao et al. 2017).

The paper's key objective is to detail the climatic and non-climatic stressors, which intersect to increase vulnerability and shape the kind of responses adopted by groundnut farmers in the Upper West Region of Ghana. It contributes to the ongoing body of empirical studies by providing in-depth knowledge that highlights the role of assets, entitlements, factors and processes in defining differential vulnerability to climate-related shocks and stresses. The results of the study could also serve as a decision-supporting guide for policymakers to develop socially appropriate adaptation policies and responses aimed at reducing vulnerability and improving resilience in groundnut production.

2 Materials and Methods

This paper is based on a study that was conducted in the Upper West Region (1°25″ and 2°45″W; 9°30″and 11° N), located in north-west of Ghana. The region shares borders with Burkina Faso to the north, Northern Region to the south, Upper East Region to the east and La Cote

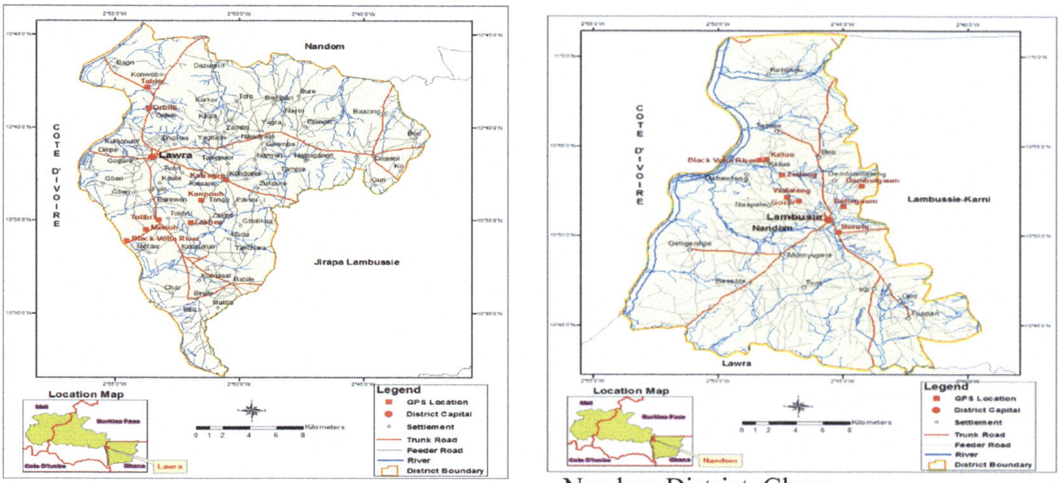

Lawra District, Ghana

Nandom District, Ghana

Fig. 1 Map of the study area. *Source* Ghana Water Research Institute

d'Ivoire to the west. It covers a land size of 18,476 km^2, which is about 12.7% of the total land of Ghana and a population of 702,110 (GSS 2013) (Fig. 1). Farming, mostly subsistent farming, is the main occupation in the study areas (GSS 2013). Crops cultivated include groundnut, maize, millet, sorghum, rice, yam, soya beans and cowpea. Livestock (pigs, cattle, goats, sheep and poultry) rearing is also common in the region (GSS 2013).

A multi-stage sampling procedure was adopted considering the descriptive and interdisciplinary set-up of the study. First, the Upper West Region (UWR) was purposively selected based on its documented vulnerability to climate change (MESTI 2013; Etwire et al. 2013; Rademacher-Schulz et al. 2014; Ndamani and Watanabe 2015; Nyantakyi-Frimpong and Bezner-Ker 2015). Secondly, a simple random sampling was used to select 2 out of 11 districts in the region, namely the Lawra and Nandom Districts. Out of 54 and 62 communities in the Lawra and Nandom Districts, respectively, 2 communities were again randomly selected from each district for the study. They were Kalsagre and Tolibri in Lawra and Goziri and Ketuo in Nandom. The communities selected were part of several communities visited during a reconnaissance survey in September 2015, during which time community members indicated the sort of stresses they faced in their main livelihood activity (farming) and the evidence that they were experiencing the impacts of climate change and variability.

A mixed-method was adopted to better understand the complex social issues in the study areas (Antwi-Agyei et al. 2012). Primary data were obtained using focus group discussions (FGDs), key informant interviews (KIIs) and individual questionnaires. Qualitative data were obtained from FGDs and KIIs. Participants for the FGDs were purposively selected with the aim of understanding perceptions of climate change and variability, the importance of groundnut farming as a livelihood strategy and the factors that increased their vulnerability and as well as their adaptation strategies. The FGDs were conducted in *Dagaare* (the local dialect) and later transcribed into English. A total of eight FGDs were conducted during the study, with two being held in each community. Separate FGDs were held for men and women because women from the study communities are often reluctant to participate in discussions when an elderly man was present (Kuruppu and Liverman 2011). A total of five key informant interviews were selected purposively. The key local stakeholders included officials from the District Department of

Agriculture (DoA) under the Ministry of Food and Agriculture (MoFA) and non-governmental organizations (NGOs) such as Association of Church-Based Development Projects (ACDEP) and Nandom Deanery Integrated Rural Development Programme (NANDRIDEP) and some community leaders. KIIs were conducted in English and sought to gain in-depth information on the roles of these and how their interventions have influenced vulnerability or wellbeing of the different social groups.

Semi-structured individual questionnaires were used to collect quantitative data to complement the data obtained from the FGDs and the KIIs.

A total of 180 respondents took part in the survey, 45 from each study community. Convenience sampling was used to identify respondents (Lawson et al. 2016). Participation in the study was based on availability, accessibility and readiness of targeted respondents to participate in the survey (Marshal 1996; Frey et al. 2000 as cited in Latham 2007). The semi-structured survey covered four major themes, the basic sociodemographic characteristics such as gender, age, marital status and household size, the livelihood information of the respondents particularly in relation to groundnut production (occupation, land size, average annual income), groundnut producers' perception and understanding of the drivers of vulnerability in groundnut cultivation. The questionnaire was prepared after a review of relevant literature and pre-tested for content, context and clarity. The survey was conducted in both English and *Dagaare*.

2.1 Data Analyses

Statistical Package for the Social Sciences (SPSS) Version 21 was used to analyse the demographic and socioeconomic characteristics of respondents. Inferential statistics in the form of chi-square test and binary logistic regression were used to analyse the relationship between some adaptation strategies and social groups and incomes of farmers, respectively. Binary logistic regression was run to predict how the adoption of some adaptation strategies influences farmers' incomes from groundnut sales. Finally, chi-square tests were run to determine if the adoptions of some adaptation strategies were socially differentiated (gender, age and land ownership). The results from the FGDs and KIIs were transcribed, categorized into themes and interpreted in line with research objectives. The implication of adaptation interventions on the wellbeing or vulnerability of different social groups was analysed thematically from the qualitative data from the KII. A qualitative software (Atlast.ti) was used in the qualitative data analysis and to detail the perceived drivers of vulnerability among groundnut farmers in study areas.

3 Results

3.1 Background Characteristics of Surveyed Farmers

Majority of the respondents (57%) were women and the mean age was 47.5 years (Table 1). Sixty-eight per cent of the farmers had no formal education, with more women (47.2%) than men (21.1%) had no formal education. About 15% of respondents (8.9% males and 6.1% females) had primary school education and 14.5%, (10.6% men and 3.9% women) had secondary education.

3.2 Land Access and Ownership

About 62.2% of the respondents did not own land. Almost all the groundnut farmers who owned land were men (Fig. 2). Renting land was uncommon among the farmers and those who did not own land farmers farmed on lands borrowed from a relative.

Sixty-nine per cent of farmers described themselves as "facing problems with land tenure". In their view, this impacted groundnut production by affecting the capacity of farmers to increase production (22.8%), invest in the land to improve soil fertility (26.7%), and crop yield (20%). The mean farm size was about 2.29 acres (0.93 ha), with a minimum size of 0.5 acres

Table 1 Background characteristics of surveyed groundnut farmers

Characteristic	Mean	Min.	Max.	Frequency	Percentage (%)
Gender					
Male				77	42.8
Female				103	57.2
Age	*47.5*	*24*	*80*		
<25				1	0.6
26–35				29	16.1
36–45				50	27.8
46–55				64	35.6
56–65				22	12.2
> 66				14	7.8
Marital status					
Single				3	1.7
Married				144	80
Widowed				33	18.3
Household size	*7.26*	*2*	*21*		
Under 4				18	10
5–8				115	63.9
9–12				41	22.8
13 and Above				6	3.3
Ethnicity					
Dagaaba				176	97.8
Waala				4	2.2

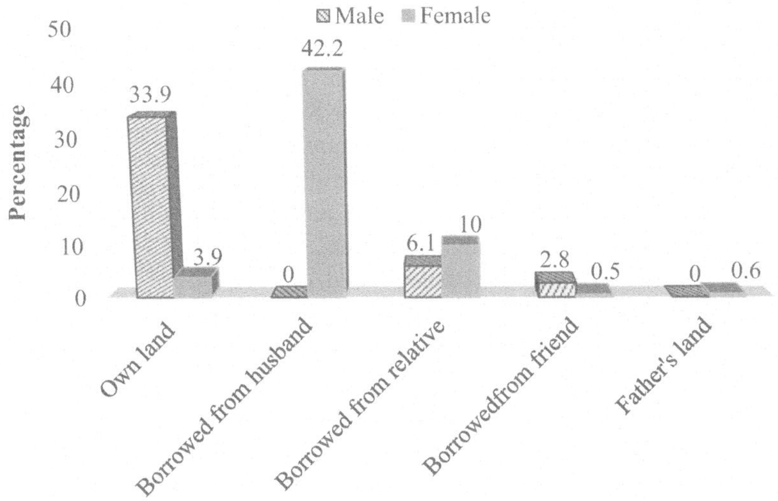

Fig. 2 Land access and ownership among groundnut farmers in study areas

(0.20 ha) and a maximum of 6 acres (2.43 ha). Women had relatively smaller fields, with farm sizes ranging between 0.5 and 1 acres (0.20–0.40 ha), with a distribution of 6.1% and 29.4%, respectively (Fig. 3). The results showed a significant effect of gender on land ownership

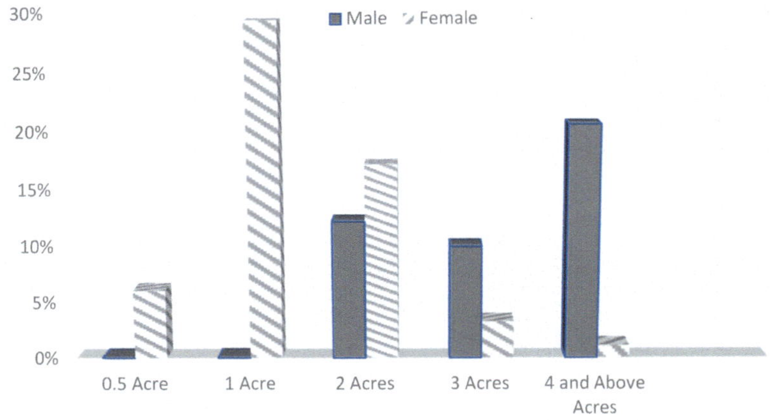

Fig. 3 Distribution of farm size based among male and female groundnut farmers in study areas

($X^2 = 98.321$, $p = 0.000$). Only 6.8% of the women-owned land as compared 79.2% of the males.

3.3 Perceptions of Drivers of Vulnerability

Groundnut production is vulnerable to climatic drivers, as well as the social, economic and political conditions that contribute to intensifying vulnerability (Rao et al. 2017; Lawson et al. 2019). The results in this section were largely obtained through qualitative data from the surveys and focus group discussions. The perceived drivers of vulnerability cited by the respondents can be found in Fig. 4 and are similar to the findings of Nyantakyi-Frimpong and Bezner-Ker (2015) and Padgham et al. (2015). Figure 4 highlights the complexity of unpacking the drivers of vulnerability and the climatic and non-climatic factors that affect groundnut farmers. In each community, the different issues raised (in middle) were linked to the drivers (on right). Issues such as poor seed germination, pests and diseases, withering crops, poor soil, nutrients were mentioned in all the communities. Noteworthy is the fact that the drivers interact with each other to increase the vulnerability of the farmers. For example, increasing temperatures worsened hardships in the everyday tasks for women farmers, driving increased poor health among farmers, withering crops and increasing poverty. Farmers identified the current rainfall pattern as the main threat to groundnut production. A decreasing trend was noted by 61%, and 32% noted a variable/unpredictable pattern. Erratic rainfall led to poor seed formation and yield losses, increasing farmers' vulnerability (Table 2).

Most of the respondents (78.9%) did not perceive the relationship between floods and their own vulnerabilities. Only 6.7% of them, particularly in Ketuo mentioned that they frequently experienced floods because their farms were near the Black Volta as explained by a 59-year-old man:

> … Some years ago, we used to start planting our crops between April/May and the rains ended in October/November but these days the rains usually start in June and end by October. We sometimes lose our groundnut seeds after planting, when there is drought or long dry spells after early rains. If we are not able to buy new seeds we cannot farm that season. The rainfall pattern has changed, and it is a big problem here. (A 59-year-old man from Goziiri-Nandom).

The increasing temperature was identified by 69.4% of the respondents as affecting their yield. Rising temperature was linked by respondents to several problems such as crop failure, poor germination and pest and disease infestation. Over 80% of the respondents identified changing temperatures as negatively affecting groundnut production. It caused crops to wither (34.4%), poor germination and seed loss after planting (18.9%) and stunted growth leading to low yields

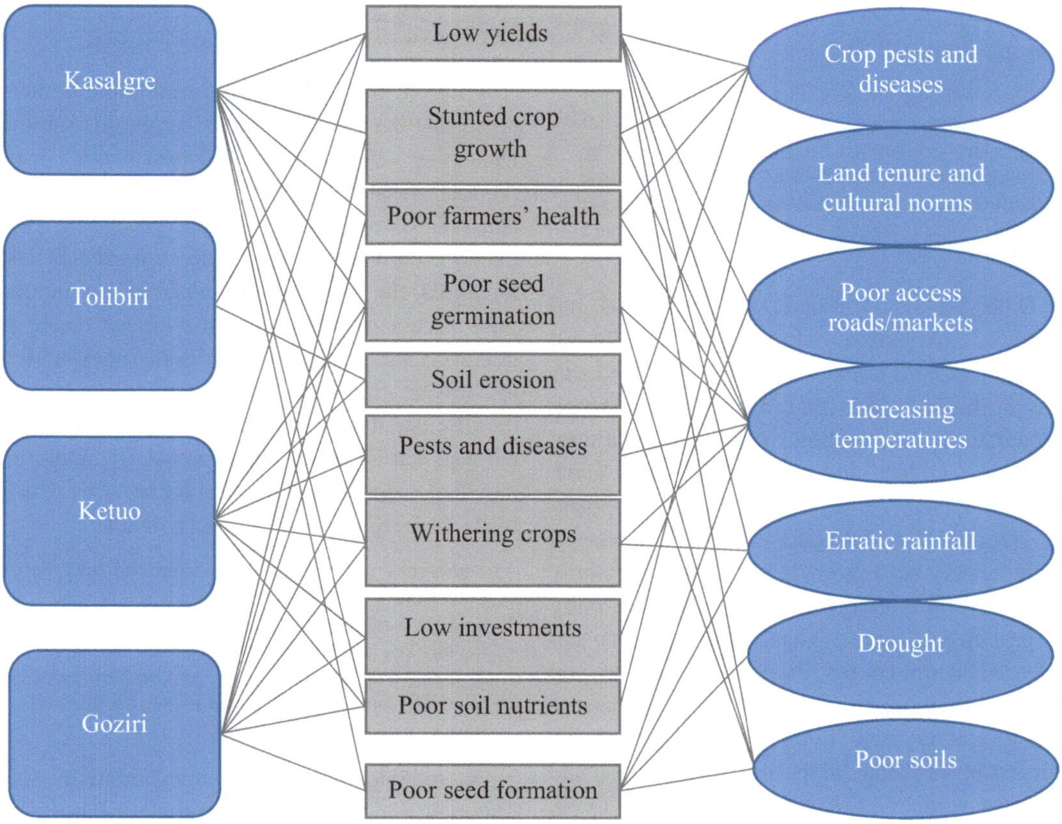

Fig. 4 Perceived drivers of vulnerability among groundnut farmers in study areas

Table 2 Farmers' perceptions of the effects of rainfall variability on groundnut production

Effects	Percentage of respondents
Yield losses due to poor rainfall during crop growing period	31.7
Poor seed formation attributable to low rainfall during pegging	30.0
Seed loss as a result of poor rainfall after planting	11.7
Crop failure due to late onset and early cessation of rain	13.9
Promotes pest and disease infestation (Aflatoxins)	1.7

(14.4%). Most of the farmers (34.4%) identified crop withering as major stress from rising temperatures. A 65-year-old woman in Kalsagre-Lawra noted in the FGD:

> The temperatures are very high, as compared to two or three decades ago. When I go to my farm on a hot day I cannot work because I will fall sick….

About 89% of the respondents identified soil infertility as a problem for groundnut production. Declining soils were linked to bush burning and tree cutting (48.3%), over-cropping and excessive cultivation on the same land (31.7%), excessive use of agrochemicals (9%). A 46-year-old woman (Ketuo-Nandom) noted in the FGD:

> …our soils are not good that is why we don't get good yields. Farming on 1 acre (0.2 ha) decades ago is not the same as 1 acre (0.2 ha) now because now the yields are low due to low soil fertility. The same farm is cultivated continuously so the nutrients get depleted and if you can't afford fertilizer- you can't get more yields.

A 46-year-old woman (Ketuo-Nandom) also said during the FGDs:

> …our soils are not good that is why we don't get good yields. Farming 1 acre (0.2 ha) before is not the same as 1 acre (0.2 ha) now because now the yields are low due to low soil fertility. The same farm is cultivated continuously so the nutrients get depleted and if you can't afford fertilizer- you can't get more yields.

Other threats identified were crop pests and diseases. Aflatoxin contamination affected groundnuts before and after harvesting, and this was attributed to high rainfall variability and high temperatures. According to the Agricultural Extension Officer and the Head of an NGO, Nandom Deanery Integrated Rural Development Programme (NANDIRDEP) the main weakness of the widely used groundnut variety (commonly called '*China*' variety) was its high susceptibility to pest and disease. Other common pests and diseases identified by respondents include pod borers, grasshoppers, beetles, leaf spot disease and groundnut rust. Forty-one per cent of the farmers observed an increasing trend in pests and diseases while 32.8% noted a declining trend. Respondents maintained that pests and diseases made it difficult to store groundnuts for longer periods (25.6%), did not encourage increased cultivation (19.4%) and caused low yields (13.3%). A 40-year-old man (Tolibri-Lawra) said:

> …pests and diseases affect our yields and income. Insects spoil our crops on the farms and groundnut seeds when we store it. Sometime when you even get good harvest you are afraid of losing it because when you store it for some time it gets ruined. So you don't have any option than to sell your crops soon after harvest.

Women farmers especially had poor access to land. According to the respondents, land tenure insecurity limited the ability of groundnut farmers to expand production (22.8%), invest on land to improve soil fertility (26.7%) and this led to low crop yield due to poor soil fertility (20%).

> …some husbands or landowners will take back their land when they realize you have improved the land fertility. So the land on which we farm is not secure. It can be taken away from you at any time by the owners. Some women may also lose the land they cultivate on after the death of their husbands. (A 36-year-old woman from Tolibri- Lawra).

Fifty-one per cent of the groundnut farmers noted that the decreasing local support (from the District Assembly, Department of Agriculture/Ministry of Food and Agriculture and NGOs) negatively affected groundnut production. Participants, especially from the FGDs, identified the removal of subsidies on inputs, poor access to inputs and machinery (such as tractors) as the problems that limited their ability to cultivate groundnuts on a large scale. Some of the discussants argued that some years ago they enjoyed subsidies on farm inputs. In addition, they had access to cheaper tractor services which made farming activities more profitable.

> Sometimes you will have the money to hire the tractor but it is difficult to get it when you need it because there are many farmers competing for one tractor. The government and NGOs are not helping us to solve this problem, so we end up farming on relatively smaller pieces of land. (A 56-year-old farmer from Goziiri-Nandom).

A reliable means of transportation was important for 52.8% of the groundnut farmers as many of them could not afford to buy donkey carts, bicycles or motorbikes and relied commercial tricycles ("*motor-king*"), which were sometimes expensive. The generally bad roads worsened transportation problems.

> …we don't have lorries (buses or taxies) here. So I sometimes walk from here (Ketuo) to Nandom (about 7 kms) to sell and buy my items. Occasionally on market days there is a motor-king that transports people to Nandom Market but not all people can afford because it is costly (Gh¢ 3.00 per head). (A 48-year-old widow, Ketuo-Nandom).

Poor markets were highlighted in the FGDs and the questionnaire survey. About 92% of the respondents sold part of their cultivated groundnuts for income. Farmers, however, complained that poor market pricing of groundnut negatively affected their incomes and wellbeing. The prices of groundnut were usually determined by buyers who bought them at the lowest prices possible. They also mentioned a lack of unity among groundnut producers as a huge barrier to food security in the area.

> …We are not able to bargain for better prices at the markets because groundnut farmers are not united. That is why the buyers always cheat us. Sometimes when you send your produce to the market if the prices are bad we do not sell. But at times we have no option than to sell it at the low price just to get something home. (A 52-year-old man in Tolibri-Lawra).

3.4 Differences of Adaptation Strategies Across Different Social Groups

The adaptation strategies to the impacts of climate change can be divided into two groups: planned and autonomous adaptation strategies used to reduce vulnerability. As shown in Fig. 5, 85% of respondents used early maturing varieties, 73% used compost or manure and 96% reared livestock. All the farmers indicated that they had changed their planting dates to adapt to the current changes in rainfall patterns. Off-farm jobs as adaptation strategies were adopted by 37.1% of the respondents. Agrochemicals (fertilizers/pesticides) (2.8%), drought-tolerant varieties (5.6%) and seasonal migration (7%) to other areas to work were the least adopted. None of the respondents practised irrigation farming, rainwater harvesting and drought insurance (Fig. 5).

The most cited reasons for non-adoption were poor knowledge about a strategy (33.5%), lack of infrastructure and facilities to sustainably implement the strategy (21.9%) and inability to invest in the strategy due to low-income levels (19.3%). The study sought to determine the impact of strategies adopted on the lives of the groundnut farmers. A binary logistic regression was used because income data was collected on a categorical scale involving four ranges. However, based on the findings of this study, two ranges were barely acknowledged and were collapsed to form part of the highly indicated ones. Sweet (1999) and Hosmer et al. (2013) postulate that logistic regression is the most appropriate tool for a dichotomous dependable variable and measurements of varying levels (Table 3).

The characteristics of the model include the Exp. (B) which denotes the odds of the outcome event, the Wald and the significance (P), which shows the power that each independent variable has on the entire model and B represents the

Fig. 5 Adaptation practices of groundnut farmers

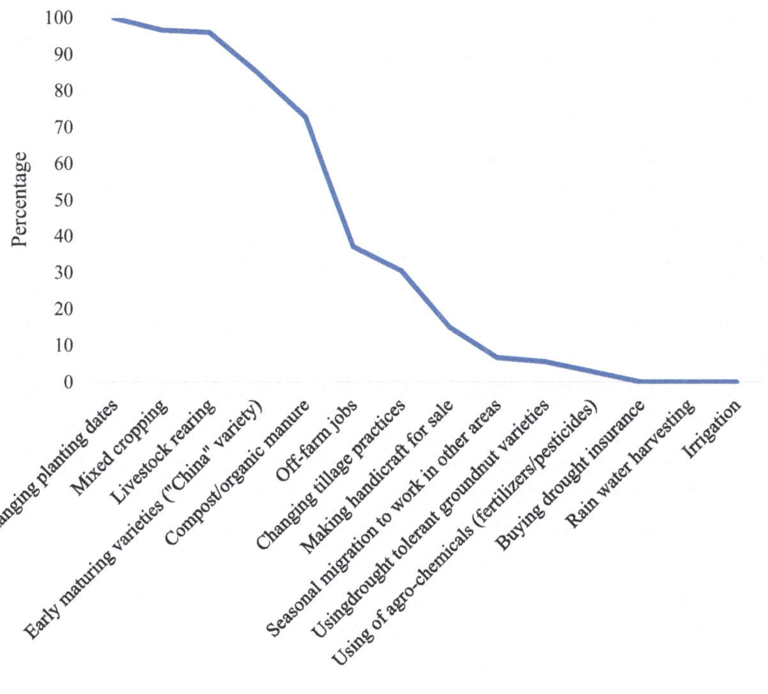

Table 3 Impact of adaptation strategies on annual income from groundnuts

Strategies	B	Odds	Sig.	S.E	Wald
Early maturing varieties	19.90	1.96	0.01*	0.73	6.54
Rearing livestock	19.20	1.72	0.00*	0.15	4.22
Composting	1.82	1.63	0.01*	0.67	7.34
Seasonal migration	−1.16	0.31	0.12	0.76	2.40
Off-farm jobs	1.28	3.61	0.00*	0.45	8.25
Constant	40.43	3.61	0.01	0.16	10.42

Nagelkerke $R^2 = 0.30$; Hosmer and Lemeshow Test: $\chi^2 = 8.429$, df = 4, $P = 0.07$; Omnibus Tests of Model Coefficients: $\chi^2 = 36.60$, df = 5 $P = 0.00$.
Significant at $*p < 0.05$

unstandardized beta. To be considered significant in the logit model, a predictor variable should have odds of more than 1 and a $P < 0.05$ (Sweet 1999). An odds ratio less than 1 means the increasing value of the variable is parallel to decreasing odds of the event's occurrence and the reverse is true.

With an Omnibus tests model coefficient of ($\chi^2 = 36.60$, $P < 0.00$) and Hosmer and Lemeshow test of $\chi^2 = 8.429$, $P > 0.07$), a statistically significant relationship was observed between the adoption of adaptation strategies and famers' annual income from groundnut. The model was fitted at a Hosmer and Lemeshow P-value greater than 0.05 (Hosmer et al. 2013; Pallant 2005). The Nagelkerke R^2 or the Pseudo R-Square for the model is about 0.30. Hence, adopted adaptation strategies impacted 30% of the annual income of groundnut farmers.

However, not all the adaptation strategies influenced farmers' income in the same way. For instance, off-farm jobs had a major influence on the income of farmers (Odds = 3.61, $P = 0.00$). This means that farmers who engaged in off-farm jobs were 3.6 times more likely to experience an increase in income from groundnuts as compared to farmers without off-farm jobs. Also, the use of early maturing varieties (Odds = 1.96, $P = 0.01$) had a significant influence on the income of groundnut farmers. Also, respondents who used early maturing varieties were about two times likely to increase their income. Similarly, those who practised composting (Odds = 1.63, $P = 0.01$) and livestock rearing (Odds = 1.72, $P = 0.00$) were twice more likely to earn higher incomes than those who did not practice. However, seasonal migration (Odds = 0.31, $P = 0.12$) did not show any significant association as those who migrated seasonally to work in other areas were about three times less likely to increase their income.

Three variables (age, gender and land ownership) were selected based on the results of the FGDs and interviews, to understand how the adaptation strategies affected some social groups (Table 4). Chi-square for the goodness of fit was used to present results. The adoption of strategies such as using early maturing varieties ($\chi^2 = 16.235$, $P = 0.00$), mixed cropping ($\chi^2 = 4.640$, $p = 0.03$) and composting ($\chi^2 = 16.389$, $P = 0.00$) of farmlands significantly varied across gender.

More women adopted early maturing varieties (51%) and used mixed cropping (55.7%). Younger farmers (56.3%) engaged in off-farm economic activities. There were no significant gender differences in the adoption of livestock rearing ($\chi^2 = 0.60$, $P = 0.44$) and off-farm jobs ($\chi^2 = 3.02$, $p = 0.08$) as adaptation strategies. However, the adoption of off-farm strategies differed significantly with age ($\chi^2 = 5.293$, $P = 0.02$) with younger farmers (56.3%) engaging in off-farm activities such as *pito* brewing, Shea butter processing, petty trading, making handicraft, sale of livestock and artisanal fishing. The adoption of early maturing varieties ($\chi^2 = 2.824$, $p = 0.09$), livestock keeping ($\chi^2 = 2.683$, $p = 0.10$), mixed cropping ($\chi^2 = 1.940$, $p = 0.16$) and composting ($\chi^2 = 1.179$, $p = 0.28$) did not vary significantly with age.

Table 4 Main adaptation strategies by social groups (based on chi-Square test)

Adaptation strategies (%)										
Social groups	Early maturing varieties		Livestock rearing		Mixed cropping		Composting		Off-farm job	
Gender	Yes	No	Yes	No	yes	no	Yes	No	Yes	No
Men	49	7.4	43.4	28.6	44.3	0.0	51.9	18.4	34.8	48.2
Women	51	92.6	56.6	71.4	55.7	100	48.1	81.6	65.2	51.8
	$X = 16.235$, $P = 0.00$*		$X = 0.60$, $p = 0.44$		$X = 4.640$, $p = 0.03$*		$X = 16.389$, $p = 0.00$*		$X = 3.02$, $p = 0.08$	
Age groups										
Young farmers	41.8	59.3	45.7	14.3	45.4	16.7	42	51	56.3	38.4
Older farmers	58.2	40.7	54.3	85.7	54.6	83.3	58	49	43.9	61.6
	$X = 2.824$, $p = 0.09$		$X = 2.683$, $p = 0.10$		$X = 1.940$, $p = 0.16$		$X = 1.179$, $p = 0.28$		$X = 5.293$, $P = 0.02$*	
Land ownership										
Own land	43.1	7.4	38.2	28.6	39.1	0	44.3	20.4	18.2	50
Borrow land	56.9	92.6	61.8	71.4	60.1	100	55.2	79.6	81.8	50
	$X = 12.464$, $p = 0.00$*		$X = 0.263$, $p = 0.61$		$X = 3.768$, $p = 0.05$*		$X = 8.642$, $p = 0.03$*		$X = 17.809$, $p = 0.00$*	

The use of early maturing varieties ($\chi^2 = 12.464$, $p = 0.00$), mixed cropping ($\chi^2 = 3.768$, $p = 0.05$), composting ($\chi^2 = 8.642$, $p = 0.03$) and off-farm jobs were found to vary significantly across land ownership types. Respondents who borrowed land were more likely to engage in the use of early maturing varieties of groundnut, mixed cropping, composting and off-farm jobs. However, livestock rearing ($\chi^2 = 0.263$, $p = 0.61$) was not found to vary across land ownership types.

4 Discussion

Groundnut farmers in the Upper West Region face a combination of climatic and non-climatic stressors, which intersect to increase their vulnerability and shape the kind of responses they adopt. Drivers highlighted in this study included erratic and variable rainfall patterns, increasing temperature, poor and infertile soils and poor access to land. To adapt to the impacts of a changing climate groundnut farmers had adopted several on-farm and off-farm strategies, to manage these risks and to sustain livelihoods. The uptake or adoption of strategies differed with factors such as gender, land ownership and age. For example, younger farmers were more likely to engage in off-farm activities and farmers who borrowed land were more likely to adopt adaptation strategies such as early maturing varieties of groundnut, mixed cropping, composting and off-farm jobs. Most of the farmers who borrowed land were women (Antwi-Agyei et al. 2015b). In nearly a third of developing countries, men and women do not have the security of land tenure which affects their access to credit and inputs (Huyer 2016). In a model simulation of adaptation policy response among smallholder farmers in northern Ghana, Wossen and Berger (2015) underscored the importance of off-farm income in reducing poverty and food insecurity under both price and rainfall variability. The study indicated that off-farm activities will yield a greater impact if credit facilities are made easily available, alongside off-farm income activities.

The study also identified a direct link between the farmers' annual income from groundnut and their adoption of certain adaptation practices, particularly, the use of early maturing varieties, composting, livestock keeping, off-farm jobs and seasonal migration. The main barriers to adaptation as highlighted by research participants in the KIIs and FGDs included sociocultural practices, gender-related barriers, market failures, transportation, financial and logistical constraints.

The reasons for non-adoption of technologies included a lack of infrastructure and finances to invest in the strategy. With respect to technology, poor farmers in sub-Saharan Africa, especially women, lack access to irrigation infrastructure and technologies (Jost et al. 2015; Perez et al. 2015), while women are less likely to receive information on and adopt improved agricultural technologies in comparison to men (Huyer 2016). Rainfall variability was highlighted as one of the key drivers of vulnerability. However, minimal emphasis was placed on strategies used for adapting to climate-induced changes in water resources. Yet, the greatest scope for action in semi-arid regions is in improving adaptive capacity and responding to changes in water demands, and this calls for the sustainable management and development of water resources (Yidana 2016). According to Kay (2001), the types of irrigation likely to succeed in the northern savannah ecological zone of Ghana are water harvesting, irrigation of river plains and exploitation of shallow groundwater all year round. At the household level, adoption of irrigation determined by factors such as credit, labour, off-farm income, farm equipment, distance to irrigation facility and distance to market. (Wossen et al. 2014; Yilma et al. 2008). In addition, Nyantakyi-Frimpong and Bezner-Kerr (2014) reported low adoption of improved varieties among smallholder farmers in the Upper West region with the reason that hybrid seeds were sensitive, weak and required extra care. They noted that improved seeds needed to be purchased every planting season and required high fertilizer use. The socio-economic implications of these choices were apparent in the reasons respondents gave for not adopting some strategies.

5 Conclusion

Research on vulnerability to climate change has sought to identify the means through which wellbeing can be enhanced through reducing risk and promoting resilience (Adger 2006). In unpacking the complex issue of vulnerability, it is critical to identify and articulate not just the climatic drivers but also the social, economic and political conditions that contribute to intensify vulnerability and shape the ability to adapt to current and future climate change (Rao et al. 2017). The findings of this study showed that groundnut production in the Upper West Region has increased because land hitherto used to cultivate other crops is being converted into groundnut farms. Groundnut cultivation is also a means of coping with the low rainfall and water scarcity, as well as increasing vulnerability. This paper has shown that groundnut farmers in the Upper West Region face a combination of climatic and non-climatic stressors, which intersect to increase their vulnerability and shape the kind of responses they adopt. In addition, the uptake of adaptation strategies differed with gender, land ownership and age.

Land access and tenure insecurity are still big issues for women and non-natives within the study areas (Ahmed et al. 2016). To improve the status quo, more education, advocacy and policy intervention are needed. The Government, local NGOs, religious and traditional leaders must help remove these cultural barriers. These efforts should be also backed by the requisite laws and policies (Lawson 2016). Areas of inequality in relation to climate adaptation include women's limited access to and control of land, high household work burdens that include the responsibility for water and fuelwood collection, high levels of responsibility for agricultural production and lack of access to formal education (Rao et al. 2017). Ensuring that effective strategies are adopted includes a consideration of the socially differentiated reasons for non-adoption. Finally, achieving food and water security will involve building long-term resilience through stronger institutions, investing in cost-effective and adaptive water management and promoting technology transfer among women.

References

Ahmed A, Lawson ET, Mensah A, Gordon C, Padgham J (2016) Adaptation to climate change or non-climatic stressors in semi-arid regions? Evidence of gender differentiation in three agrarian districts of Ghana. Environ Dev 20:45–58. https://doi.org/10.1016/j.envdev.2016.08.002

Adger WN (2006) Vulnerability. Global Environ Change 16:268–281. https://doi.org/10.1016/j.gloenvcha.2006.02.006

Angelucci F, Bazzucchi A (2013) Analysis of incentives and disincentives for groundnuts in Ghana. Technical notes series, MAFAP, FAO, Rome

Antwi-Agyei P (2012) Vulnerability and adaptation of Ghana's food production systems and rural livelihoods to climate variability. Unpublished Doctoral Thesis. The University of Leeds School of Earth and Environment, UK

Antwi-Agyei P, Dougill AJ, Stringer LC (2015a) Land use policy impacts of land tenure arrangements on the adaptive capacity of marginalized groups: the case of Ghana's Ejura Sekyedumase and Bongo districts. Land Use Pol 49:203–212. https://doi.org/10.1016/j.landusepol

Antwi-Agyei P, Dougill AJ, Stringer LC (2015b) Barriers to climate change adaptation in sub- Saharan Africa: evidence from northeast Ghana & systematic literature review. Climate Dev 7(4):297–309. https://doi.org/10.1080/17565529.2014.951013

Arora-Jonsson S (2011) Virtue and vulnerability: discourses on women, gender and climate change. Global Environ Change 21(2):744–751. https://doi.org/10.1016/j.gloenvcha.2011.01.005

Boko M, Niang I, Nyong A, Vogel C, Githeko A, Medany M et al (2007) Africa. In: Parry ML, Canziani OF, Palutikof JP, van der Linden PJ, Hanson CE (eds) Climate change 2007: impacts, adaptation and vulnerability. Contribution of working group II to the fourth assessment report of the intergovernmental panel on climate change. Cambridge University Press, Cambridge, pp 433–467

Coirolo C, Rahman A (2014) Power and differential climate change vulnerability among extremely poor people in Northwest Bangladesh: lessons for mainstreaming. Climate Dev 6(4):336–344. https://doi.org/10.1080/17565529.2014.934774

Etwire PM, Al-Hassan RM, Kuwornu JKM, Osei-Owusu Y (2013) Application of livelihood vulnerability index in assessing vulnerability to climate change and variability in Northern Ghana. J Environ Earth Sci 3 (2):157–170

Fields S (2005) Continental divide: why Africa's climate change burden is greater. Environ Health Pers 113(8): A534–A537

Frey LR, Botan CH, Kreps GL (2000) Investigating communication: an introduction to research methods, 2nd edn. Allyn and Bacon, Boston

Ghana Statistical Service (2013) 2010 Population and housing census: Regional Analytical Report: Upper West Region. Ghana Statistical Service. https://s3-us-west-2.amazonaws.com/new-ndpc-static1/CACHES/PUBLICATIONS/2019/01/21/2010_PHC_Regional_Analytical_Reports_Upper_West_Region.pdf

Heltberg R, Siegel PB, Jorgensen SL (2009) Addressing human vulnerability to climate change: toward a "no-regrets" approach. Global Environ Change 19(1):89–99. https://doi.org/10.1016/j.gloenvcha.2008.11.003

Hosmer DW Jr, Lemeshow S, Sturdivant RX (2013) Applied logistic regression. Wiley-Blackwell, Oxford

Huyer S, Twyman J, Koningstein M, Ashby J, Vermeulen S (2015) Supporting women farmers in a changing climate: five policy lessons. Agriculture and Food Security Programme. Copenhagen, Denmark: CGIAR Climate Change. https://cgspace.cgiar.org/rest/bitstreams/60479/retrieve. Accessed 10 July 2017

Huyer S (2016) Closing the gender gap in agriculture. Gender Tech Dev 20(2):1–12

IPCC(2013) The physical science basis, summary for policymakers. https://www.ipcc.ch/pdf/assessment-report/ar5/wg1/WGIAR5_SPM_brochure_en.pdf. Accessed 21 Sept 2016

Jost C, Kyazze F, Naab J, Neelormi S, Kinyangi J, Zougmore R, Aggarwal P, Bhatta G, Chaudhury M, Tapio-Bistrom M, Nelson S, Kristjanson P (2015) Understanding gender dimensions of agriculture and climate change in smallholder farming communities. Climate Dev 8(2):1–12. https://doi.org/10.1080/17565529.2015.1050978

Kay M (2001) Smallholder irrigation technology: prospects for Sub Sahara Africa. IPRTRID, FAO, Rome

Kuruppu N, Liverman D (2011) Mental preparation for climate adaptation: the role of cognition and culture in enhancing adaptive capacity of water management in Kiribati. Global Environ Change 21(2):657–669. https://doi.org/10.1016/j.gloenvcha.2010.12.002

Lawson ET, Ayivor JS, Ohemeng F, Ntiamoa-baidu Y (2016) Social Determinants of a potential spillover of bat-borne viruses to humans in Ghana. https://doi.org/10.5539/ijb.v8n2p66

Lawson ET (2016) Negotiating stakeholder participation in the Ghana national climate change policy. Int J Climate Change Stra Man 8(3):399–417. https://doi.org/10.1108/IJCCSM-04-2015-ss0041

Lawson ET, Alare RS, Salifu ARZ, Thompson-Hall M (2019) Dealing with climate change in semi-arid Ghana: understanding intersectional perceptions and adaptation strategies of women farmers. GeoJournal, 1–14. https://doi.org/10.1007/s10708-019-09974-4

Lyimo JG, Kangalawe RY (2010) Vulnerability and adaptive strategies to the impact of climate change and variability. The case of rural households in semi-arid Tanzania. Environ Econ 1(2):89–97

Marshal MN (1996) Sampling for qualitative research. Fam Pract 13(6):522–525

Mawunya FD, Adiku SGK (2013) Implications of climate change for agricultural productivity in Ghana: an agrometeorological perspectives. Ghana Pol J 5:7–26

MESTI (2013) Ghana National Climate Change Policy. Ministry of Environment, Science, Technology and Innovation Ghana

MoFA (2011) Agriculture in Ghana: facts and figures (2010). Ministry of Food and Agriculture. Issued by the Statistics, Research and Information Directorate. Accra

Ndamani F, Watanabe T (2015) Farmers' perceptions about adaptation practices to climate change and barriers to adaptation: a micro-level study in Ghana. Water 7(9):4593–4604. https://doi.org/10.3390/w7094593

Niang I, Ruppel OC, Abdrabo MA, Essel A, Lennard C, Padgham J, Urquhart P (2014) Africa. In: Barros VR, Field CB, Dokken DJ, Mastrandrea MD, Mach KJ, Bilir TE, Chatterjee M, Ebi KL, Estrada YO, Genova RC, Girma B, Kissel ES, Levy AN, MacCracken S, Mastrandrea PR, White LL (eds) Climate change 2014: impacts, adaptation, and vulnerability. Part B: regional aspects. Contribution of working group II to the fifth assessment report of the intergovernmental panel on climate change. Cambridge University Press, Cambridge, United Kingdom and New York, NY, USA. https://www.ipcc.ch/pdf/assessment-report/ar5/wg2/WGIIAR5-Chap22_FINAL.pdf. Accessed 12 Nov 2017

Nielsen JØ, Reenberg A (2010) Cultural barriers to climate change adaptation: a case study from Northern Burkina Faso. Global Environ Change 20(1):142–152. https://doi.org/10.1016/j.gloenvcha.2009.10.002

Nyantakyi-Frimpong H, Bezner-Kerr R (2015) The relative importance of climate change in the context of multiple stressors in semi-arid Ghana. Global Environ Change 30:40–56. https://doi.org/10.1016/j.gloenvcha.2015.03.003

Nyantakyi-Frimpong H, Kerr RB (2014) A political ecology of high-input agriculture in northern Ghana. Afr Geog Rev 34(1):13–35. https://doi.org/10.1080/19376812.2014.929971

Owusu K, Waylen P, Qui Y (2008) Changing rainfall inputs in the Volta basin: implications for water sharing in Ghana. sournal 71:201–210https://doi.org/10.1007/s10708-008-9156-6

Padgham J, Abubakari A, Ayivor J, DietrichK, Fosu-Mensah B, Gordon C, Habtezion S, Lawson E, Mensah A, Nupkezah D, Ofori B, Piltz S, Sidibe A, Sissoko M, Totin E, Traore S, Daze A, Echeverria D (2015) Vulnerability and adaptation to climate change in semi-arid areas in West Africa. ASSAR working paper, ASSAR PMU, South Africa. https://www.assar.uct.ac.za/sites/default/files/image_tool/images/138/RDS_reports/West_africa/West%20Africa%20Regional%20Diagnostic%20Study.pdf. Accessed 20 Sep 2016

Padgham J (2009) Agricultural development under a changing climate: opportunities and challenges for adaptation. The World Bank. Joint Departmental Discussion Paper Issue 1. World Bank, Washington, DC

Pallant J (2005) SPSS Survival Manual: a step by step guide to using SPSS for Windows (version 12). Allen & Unwin, New South Wales, Australia

Perez C, Jones EM, Kristjanson P, Cramer L, Thornton PK, Forch W, Barahona C (2015) How resilient are farming households and communities to a changing climate in Africa? A gender-based perspective. Global Environ Change 34:95–107. https://doi.org/10.1016/j.gloenvcha.2015.06.003

Popke J, Curtis S, Gamble DW (2014) A social justice framing of climate change discourse and policy: adaptation, resilience and vulnerability in a Jamaican agricultural landscape. Geoforum. https://doi.org/10.1016/j.geoforum.2014.11.003

Rademacher-Schulz C, Schraven B, Mahama ES (2014) Time matters: shifting seasonal migration in northern Ghana in response to rainfall variability and food insecurity. Climate Dev 6(1):46–52. https://doi.org/10.1080/17565529.2013.830955

Rao N, Lawson ET, Raditloaneng WN, Solomon D, Angula M (2017) Gendered vulnerabilities to climate change: insights from the semi-aridregions of Africa and Asia. Climate Dev, 1–13. https://doi.org/10.1080/17565529.2017.1372266

Sarr B (2012) Present and future climate change in the semi-arid region of West Africa: a crucial input for practical adaptation in agriculture. Atmospheric Science Letters 13(2):108–112. https://doi.org/10.1002/asl.368

Sweet S (1999) Data analysis with SPSS. Allyn and Bacon, Massachusetts

World Bank (2013) Turn down the heat: climate extremes, regional impacts, and the case for resilience. A report for the World Bank by the Potsdam Institute for Climate Impact Research and Climate Analytics. World Bank, Washington, DC

Wossen T, Berger T, Swamikannu N, Ramilan T (2014) Climate variability, consumption risk and poverty in semi-arid Northern Ghana: adaptation options for poor farm households. Environ Dev 12:2–15. https://doi.org/10.1016/j.envdev.2014.07.003

Wossen T, Berger T (2015) Climate variability, food security and poverty: agent-based assessment of policy options for farm households in Northern Ghana. Environ Sci Pol 47:95–107. https://doi.org/10.1016/j.envsci.2014.11.009

Yilma T, Berg E, Berger T (2008) The agricultural technology-market linkage under liberalisation in Ghana: evidence from micro data. J Afr Econ 17 (1):62–84. https://doi.org/10.1093/jae/ejm005

Yidana AA (2016) Social differentiation in climate change adaptation strategies of smallholder farmers in the Upper West Region of Ghana MPhil Thesis, Climate Change and Sustainable Development, University of Ghana

Temporal Climate Variability and Staple Cereals in Southern Burkina Faso

Mwenda Borona, Cheikh Mbow, and Issa Ouedraogo

Abstract

Farming activities are the primary source of livelihood for agro-pastoral communities in the Sahel biogeographic zone in Burkina Faso. The climate variability and change have a direct impact on performance, quantity, and quality of staple cereals yields. Understanding how the crops respond to climate variability is crucial for ensuring adequate crops management and food security. Hence, this paper aims at investigating how different staple cereals respond to certain indicators of climate variability at inter-annual and inter-decadal levels in southern Burkina Faso. In our study, we use regression models and scatter plots. The model results show that climate variability has mixed effects on cereal yields. The study demonstrates the interdependencies among district-scale cereal yields in the study area including maize, sorghum, and millet casually associate with the inter-annual variation of selected climate variability indicators. Maize yields, for example, relate strongly to the rainfall amount variation ($R^2 = 51.8\%$) showing high moisture dependence during critical growth stages. Conclusions emphasize the adoption of efficient water conservation and utilization platforms, especially those that have evidently increased yields and strengthening of forecasts dissemination and related climate information services.

Keywords

Climate variability and change · Staple cereal yields · Seasonality · Rainfed farming · Southern Burkina Faso · Rainfall

1 Introduction

Climate variability refers to short-term spatial–temporal changes in temperature and rainfall conditions (Borona et al. 2016). These changes are largely driven by a complex mix of land–sea–atmospheric changes and interactions, among them variations in sea surface temperature,

M. Borona (✉)
Intergrative Research Institute on Transformations of Human-Environment Systems (IRI THESys), Humboldt-Universität zu Berlin, Quartier Stadtmitte, Friedrichstraße 19110117, Berlin, Germany
e-mail: pius.mwenda.borona@hu-berlin.de; mwebo2@gmail.com

C. Mbow
Future Africa, University of Pretoria, South St, Koedoespoort 456-Jr, Pretoria 0186, South Africa
e-mail: cheikh.mbow@up.ac.za

I. Ouedraogo
The CGIAR Research Program on Climate Change, Agriculture and Food Security (CCAFS), ICRISAT West & Central Africa Regional Office, BP 320, Bamako, Mali
e-mail: i.ouedraogo@cgiar.org

L. Ribbe et al. (eds.), *Towards Water Secure Societies*,
https://doi.org/10.1007/978-3-030-50653-7_10

resulting in the ENSO (El-Niño Southern Oscillation) (Porter and Semenov 2005; FAO 2014b; Rojas et al. 2014). Climate variability has varying direct and indirect effects on crop yield variability (Semenov and Porter 1995; Van der Linden and Xiaousu 2001; Porter and Semenov 2005; Slingo et al. 2005; Osborne and Wheeler 2013). Unfavorable rains have immediate and severe effects on the supply of subsistence commodities including water and food (FAO 2014b). Drought spells in particular, including their frequency and duration, are a driver of water scarcity largely affecting food supply (Parry et al. 1999; Lobell and Field 2007). In sub-Saharan Africa, for example, crop failure and low crop yields have been widely related to rainfall failure (Blanc 2012). Related events such as uncertain rains, also associated with climate change and variability, are a common trait in semi-arid regions particularly the Sudano-Sahelian agro-ecological zone of the Sahel (FAO 1993) and include aspects of unpredictable, off-season and irregular rainfall (Simelton et al. 2011).

In addition, water stress resulting from various forms of droughts and their frequency is a major limiting factor of crop growth, yield quantity, and quality as well as an array of other crop development processes (Semenov and Porter 1995; Rosenzweig et al. 2001; Rao and Okwach 2005; Lin 2011). Instances of crop failure linked to rainfall failure and high temperatures eventually lead to impacts such as malnutrition and even famine or severe food shortage in extreme cases (Haile 2005). Crops and mostly cereals respond differently to variation in climate conditions with effects occurring when certain thresholds are exceeded. Rötter and Van de Geijn (1999), for example, highlight the positive effects of higher moisture on seasonal crops in arid regions. Grain production is subjected to rainy season profiles beyond the rainfall aggregates where various climate factors have different influence on the plant growth (IRRI 1976; Rötter and van de Geijn 1999). Several studies outline other influence associated with variations in climatic factors including temperature effects on the functioning of photosynthetic apparatus, evapotranspiration on available plant moisture, rainy days, and rainfall intensity on available soil moisture (Porter and Semenov 2005; Challinor et al. 2007; Tshiala and Olwoch 2010). The resulting net effects determine available soil moisture balance and subsequent crop performance including yield quality and quantity.

When harsh climate conditions co-occur with economic challenges, it becomes a tremendous challenge to effectively manage and benefit from food production systems. The economy of Burkina Faso largely depends on agriculture (Jalloh et al. 2011; Lodoun et al. 2013) that accounts for 40% of the gross domestic product and 60% of the countries aggregated exports (Somé et al. 2013). In principle, while agriculture forms the core of the economy, it is vulnerable to unpredictable climate events. The climate in most of the country includes a Sahelian profile with high temperatures and unreliable seasonal rainfall (Haggett 2002; Ouedraogo et al. 2011). Farming in most of the country is also based on the use of rudimentary techniques (Somé et al. 2013; Lodoun et al. 2013; FAO 2014a). Sorghum (*Sorghum bicolor*) and millet (*Pennisetum glaucum*) varieties are widely cultivated in the Sahel (Sanginga and Woomer 2009; Mishra et al. 2008) and are major staples in Burkina Faso, whereas maize (*Zea mays*) is an important food crop (Somé et al. 2013). In the study area, these cereals constitute an important source of energy, protein, vitamins, and minerals and their yields are likely to be affected when seasons get shorter or there is a strong variability in the seasonal rainfall.

Few studies are referring to derived and fine time climatic indicators and how they influence staple cereal yields in Burkina Faso. The lack of information is not unique to this area, as daily climate data at fine temporal scale (Semenov and Porter 1995; Lodoun et al. 2013) is rare and hence not often used for analyzing changes in crop yields. This presents an enormous challenge because these derived indicators adequately define seasonality including onset, end, and subsequent length, all key factors in influencing crop growth, development, and yield quality and quantity.

Accordingly, this study aims at investigating how individual staple cereals casually respond to certain indicators of climate variability.

This analysis refers to the existing relationship between crop growth and climate variability induced stresses including their frequency and extremity. Understanding of presented climate variability trends and their impact on crop yields is essential to appropriate crop management decisions and crop choices. Initially, instances of variability as depicted by key representative derivatives in the time series including rainy days, length of the growing period, dry spells, and crop evapotranspiration estimates are revealed. This is followed by regression models and scatter plots outlining the casual association between this variability and annual cereal anomalies. These descriptors are chosen as we view them as adequately describing moisture balance within the growing season that influences crop growth and development. Indeed, under limited moisture conditions, for example in semi-arid environments, it is reasonable to associate crop yields with rainfall deficiency (Bannayan et al. 2010).

2 Data and Methodology

2.1 The Study Area

The study was carried out in Cassou, Ziro Province (11° 16′N to 11° 45′N and −2° 10′W to −1° 48′W), southern Burkina Faso (Fig. 1). The area is characterized by low relief with an average altitude of 300 m a.s.l. Ziro Province lays within the south Sudanian ecological zone (Fontès et al. 1995) and receives between 900 and 1200 mm of rainfall annually. The unimodal rainy season lasts for about 6 months, May–October. The dominant farming system is the traditional cultivation of cereals (sorghum, millet, and maize), tubers (yam and sweet potatoes), and animal husbandry. In the complementary work, Borona et al. (2016), the locations natural vegetation profile, soil type distribution, population dynamics, and ethnic community distribution are detailed.

2.2 Cereal Yield Data

Historical production (Metric tons) and harvested area (Hectares) data (for the years 1984–2013) for sorghum, millet and maize for the Sissili–Ziro Province was obtained from the Agricultural Ministry in Ouagadougou in the year 2013 (INSD 2015). Production and harvested area records were used in the computation of respective cereal yields (tons per hectare) for each year in the time series as presented in Eq. 1. Cereal yields in the study area (Sissili–Ziro Province) (Fig. 2) indicate an increasing trend along with the time series as

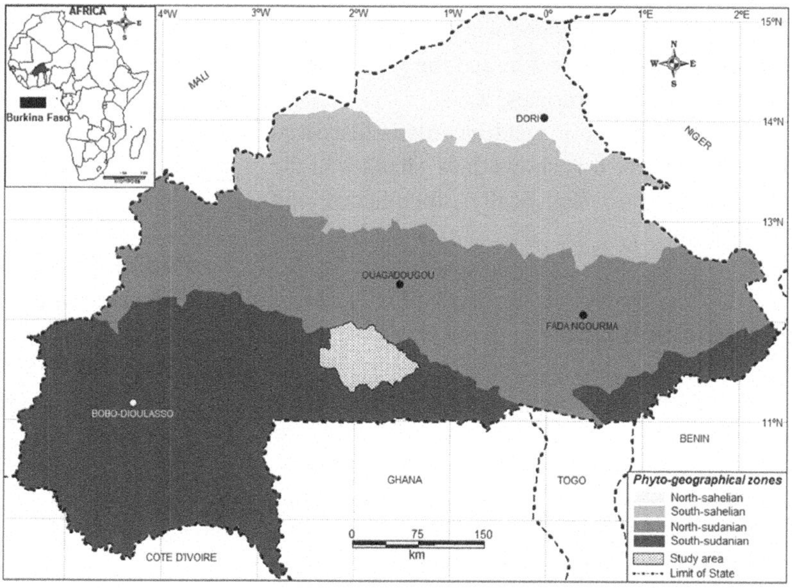

Fig. 1 Map of Burkina Faso showing the study area, Cassou, overlaid on North–South Sudanian or Sahelian agro-ecological zones (Source, Authors)

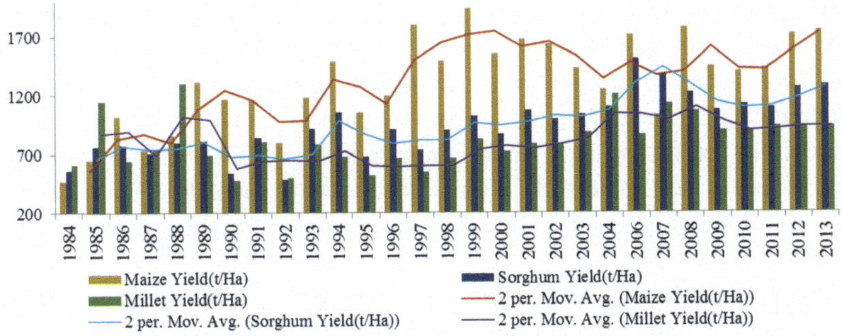

Fig. 2 Staple cereal yields trend, in Sissili–Ziro Province, Burkina Faso, shown as moving averages for every 2 years (2 per.mov. Avg.)

demonstrated by the respective moving average trend line. The three cereal yield's 29-year historical records are independently regressed against selected climatic descriptors in the respective years. Further, the generalized additive model (GAM) technique, Eq. 2 (Wood 2006), was used in interpolating for years with missing yields by fitting the available yield records in the function. The GAM applied in this study was implemented in the R software package *mgcv ()*.

$$\text{Yield}\left(\frac{T}{Ha}\right) = \frac{\text{Annual cereal Production (Tons)}}{\text{Cereal harvested area (Ha)}} \quad (1)$$

$$g(\mu_i) = X*_i\theta + f_1(x_{1i}) + f_2(x_{2i}) + f_3(x_{3i}, x_{4i}) \quad (2)$$

where $\mu_i \equiv \mathbb{E}(Y_i)$ and $Y_i \sim$ an exponential family distribution, Y_i is a response variable, $X*_i$ is a row of the model matrix for any strictly parametric model components,θ is the corresponding parameter vector, and the f_j are smooth functions of the covariates, X_k.

Further analysis of yield records is discussed in Sect. 3 where it is shown how changes in the mean yields are statistically smoothened and normalized.

2.3 Climate Data

Temperature and precipitation historical records covering 35 years were obtained from the National Meteorology Service of Burkina Faso for the nearest station, Po, and analyzed to come up with an array of climatic descriptors. This station falls within the agro-ecological zone of the study area. Initially, the daily climatic data were checked for missing daily values. In this study, gaps were only filled in temperature data including instances of missing years since rainfall records exhibited completeness in the 5-month growing seasons (May–September).

To this end, occurrences of missing maximum and minimum temperature data were estimated using multiple imputations (Markov Chain Monte Carlo) (Geyer 2011) for the years running from 2008 to 2013 as well as gaps within the period 1979–2007. This approach is detailed in complementary work, Borona et al. (2016) and a related study by Hadgu et al. (2013).

Precipitation is the principal factor in agricultural production and hence has been the primary meteorological variable in this analysis. Daily rainfall data are used to compute a wide range of descriptors such as number of rainy days, crop evapotranspiration estimates, dry spells, and the length of the growing period, to effectively describe variability in climate at a multidimensional scale. In the present study, comprehensive data from one synoptic station are referred to avoid spatial extrapolation and to maintain local focus.

3 Definition of Computed Climatic Descriptors

3.1 Annual Rainfall and Number of Rainy Days in the Study Period

Daily rainfall annual records are aggregated by summation of precipitation values where precipitation is greater than 0 mm. In this analysis, a rainy day is defined after Mathugama and Peiris (2011) as instances of daily rainfall above 0 mm. Intuitively the number of rainy days is arrived at by counting of instances where ppm > 0 mm in the respective years.

3.2 Length of the Growing Period (LGP)

Initially, daily precipitation climatic records were unstacked/rearranged and loaded onto INSTAT + for analysis of seasonality descriptors. To define the OGP, we referred to Sivakumar (1992), as applied in recent studies such as Lodoun et al. (2013). The OGP is defined as the time of the year when rainfall over three consecutive days is at least 20 mm after 1 May without instances of dry spells (in this case 7 days) in the next 30 days. The cessation or end of the growing period (CGP) is defined after Maikano (2006) and Traoré (2000) as the date after 1 September when the available soil water or the soil water content at 60 cm depth is 0 mm with daily evapotranspiration of 5 mm. The water holding capacity threshold is explored after Hengl et al. (2014) as detailed in the complementary paper by Borona et al. (2016). The LGP to this end is computed as the difference between the onset of the growing period (OGP) and the cessation of the growing period (CGP).

3.3 Dry Spells, Consecutive Dry Days (CDD), and Sum of Dry Days (TDD)

Dry spells are defined as the maximum number of consecutive days with minimal precipitation (0–0.85 mm) from the onset of rains (OGP) for the season for each year in the study period. This threshold is mentioned in accordance with related studies such as Mathugama and Peiris (2011). Dry spells are computed up to the fifth month from the onset of the season (150 days), i.e., May–September. This length of days also corresponds to the critical growing stages for the cereals considered in this analysis and also the LGP for the cereals as reported by Wang et al. (2008. The sum of dry days in each year (TDD), implying total dry days, was computed by aggregating the maximum dry spells computed for each month (CDD).

3.4 Crop Evapotranspiration (ETc) Estimates

Evapotranspiration refers to the combined evaporation or moisture loss from all surfaces as well as transpiration from plants on exposure to temperature (Chang 1974) and is an important component of the hydrological cycle (Goyal and Harmsen 2013). The evapotranspiration concept, reference evapotranspiration (ET_o) in this study is computed from weather data and denotes evapotranspiration from a reference surface without water scarcity (Allen et al. 1998).

In the present study, a reduced data set constituting radiation estimates, daily rainfall, and air temperature was used to estimate daily ET_o using the Modified Hargreaves method, Eq. 3, adopted from Droogers and Allen (2002) and Farmer et al. (2011);

$$ET_0 = 0.0013 * 0.408(T_m + 17) * (T_{\max} - T_{\min} - 0.0123P)^{0.76} * R_a \quad (3)$$

where T_m is the daily mean air temperature in $_0$C, $T_{\max}$ and T_{mins} represent the daily maximum and minimum air temperature, respectively. R_a is the extraterrestrial radiation in $MJ\text{m}^{-2}\text{day}^{-1}$. The coefficient 0.408 is used in converting $MJ\text{m}^{-2}\text{day}^{-1}$ into mm/day. In Hargreaves equation, mean air temperature is an average of the maximum and minimum temperature, while the R_a is calculated with reference to the location of the site (latitude) and time of the year (month).

The daily ET_o estimates computed with reference to the above equation were then aggregated to represent total ET_o for respective months in the growing season. The generated monthly ET_o estimates were compared with previous estimates in the study area by Sivakumar (1988) and confirmed as near equal values with no significant variation, which validates the applied estimation method. Monthly ET_o values were then aggregated to come up with seasonal ET_o total for the period May–September for each year in the time series. Seasonal ET_o values were further modified to reflect the crop water requirements for the different cereals at initial, development, and mid-season growth stages by referring to modified crop coefficients from Wang et al. (2008). Crop evapotranspiration, Eq. 4, was as such computed as a function of reference evapotranspiration (ET_o) and crop coefficients (Kc).

$$ETc = EToXKc \tag{4}$$

The present study emphasizes on the ETc at the development growth stage of the cereals, as this is the phase most sensitive to water shortages in a wide range of crops (Brouwer et al. 1985).

3.5 Cereal Yields Smoothing and Regression Models

The study recognizes that as Chen et al. (2004) conveys, other factors such as CO_2 concentration, market forces, management practices, e.g., irrigation, pests, and diseases control as well as soil fertility may also influence the trend and performance of crop yields. This study, however, emphasizes on precipitation and temperature derivatives as key limiting factors as Osborne and Wheeler (2013) also accentuate. In addition, as Schneider et al. (2000) add, while farmers institute adaptation means, these adaptations may lag behind instances of climate variability due to the difficulty in distinguishing climate trends and natural variability. As such, the present study does not attempt to capture aspects of crop physiology and farm management but pays attention to a whole range of seasonal descriptors that play a role in influencing crop development. Assessment of climatic effects on crop yields is a vital topic but the application of any particular technique constitutes uncertainties and deficiencies (Hengl et al. 2014).

In this study, a Gaussian smoothing function is employed to absorb and/or minimize the effects of slowly changing non-climatic factors on the annual crop yield trends. We initially evaluated the sensitivity of results by using other approaches such as the median filter and first differences where we observed that the Gaussian filter provided reliable results. After Lobell and Field (2007), climatic derivatives are also subjected to the Gaussian filter but these variables are only applied in the following regression models. In their analysis, they, however, applied the first differences approach to detrend their pool of variables.

The one-dimensional Gaussian function used to remove noise or extreme deviations from the mean is presented in Eq. 5, adopted from Davies (2012), where σ denotes the standard deviation of the yield distribution.

$$G(x) = \frac{1}{\sqrt{2\pi\sigma}} e^{-\frac{x^2}{2\sigma^2}} \tag{5}$$

Detrended annual yield values were then standardized using Eq. 6 to come up with the annual yield anomalies.

$$\boldsymbol{Z} = \frac{(\boldsymbol{x} - \boldsymbol{\mu})}{\boldsymbol{\delta}} \tag{6}$$

where Z is the standardized yield, x is the respective year's yield, μ is the time series mean, and δ is the time-series standard deviation.

We subsequently used regression models to explore the casual association between climatic derivatives and cereal yield variables. Prior to linear regression, we ran curve estimations (results not presented here) to detect the nature and strength of the relationship between cereal yields and climatic derivatives. In these estimations, we sought to find out whether the variables exhibited a linear, cubic, quadratic, or power relationship. From these estimates, we noted a dominant linear relationship between cereal yield variables and climatic derivatives. A multiple linear regression model was as such applied to explore the casual relationship between cereal yields and selected climatic parameters for the period 1984–2013. Climate descriptors, running from the year

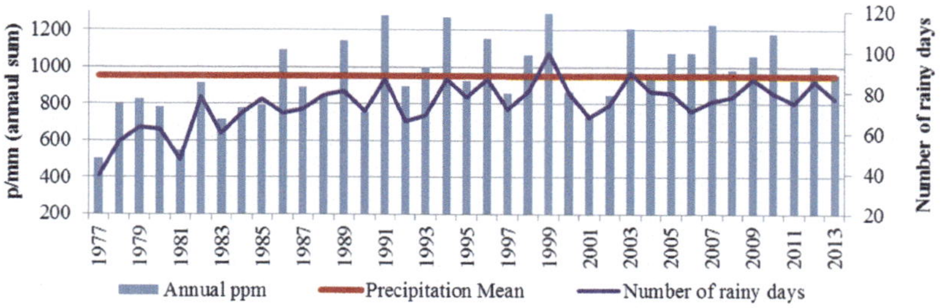

Fig. 3 Variability in annual rainfall and number of rainy days in the study area

1977–1983, were clipped to correspond to the yield records period of 1984–2013.

A multiple linear regression model allows prediction of a continuous dependent variable (Y) based on several continuous or categorical variables ($X_1\_X_p$) (Afifi et al. 2003) and is an extension of linear or bivariate regression (Tabachnick and Fidell 2001), Eq. 7.

$$y = \beta_0 + \beta_1 x_1 + \beta_2 x_2 + \ldots + \beta_m x_{m+} \mathrm{E}, \quad (7)$$

where y is the dependent, target, or response variable, in our case cereal yield anomalies.

$X_{j,}\ j = 1,\ 2\ \ldots\ m,$ represent m different independent or explanatory in our case climatic descriptors.

β_O is the intercept value when all predictors are 0, also denoted as α in other cases.

$\beta_j,\ j = 1,\ 2,\ \ldots,\ m,$ denote the respective m regression coefficients.

E is the random error or disturbance term, usually assumed to be normally distributed with mean zero and variance. Also denoted as μ in other cases.

The model was tested for assumptions such as the presence of influential points or outliers, multicollinearity as well as the independence of errors to ensure the validity of the presented results. These tests among other reasons facilitate the exclusion of non-influential factors and retaining of strong predictors. In this analysis, yields data for selected common and staple cereals in Sissili–Ziro Province of Burkina Faso and individual climatic descriptors are further presented in scatter plots to graphically explain the direction of the relationship along the best-fit line.

4 Results

4.1 Rainfall Trends in Cassou Area

Figure 3 presenting total annual rainfall in the study period shows evidence of variation in the trend of inter-decadal rainfall amounts over the years with the second decade experiencing the highest annual rainfall records. The mean annual precipitation (952 mm) acting as a threshold shows several instances of years with above and below average rainfall effectively exhibiting inter-annual total annual rainfall variability such as 1281 mm, 1268 mm, and 1290 mm in the years 1991, 1994, and 1999, respectively.

The secondary axes representing the total number of rainy days (ppm $\geq$ 0) gives evidence of a positive relationship between the total annual rainfall and the number of rainy days in the respective year and the contribution of rainy days in defining the total annual rainfall. Further, there is characteristic evidence of improved instances of rainy days in the recent years (2003 onward).

4.2 Length of the Growing Period (LGP) and Dry Spells Distribution

Successful onset dates are defined as instances where the onset, including and excluding the dry spell coincide or share the same date in the vertical axis. An instance of an unsuccessful date is characterized by an abrupt dry spell lasting 7 days just after onset of the season. Results

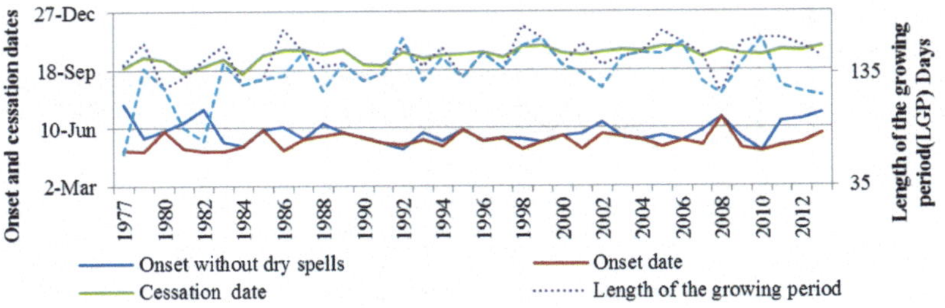

Fig. 4 Annual variation of onset, cessation (Dates), and length of the growing period (Days) in Cassou Area

show instances of the unsuccessful season starts in more than half of the years, 58.3%, of the study period. Figure 4 shows inter-annual variability in the onset, cessations as well as instances of false starts where the onset without spell dates vary from the "onset date."

This variation leads to changes in the LGP when the dry spell condition is included in the computation. In the recent decade, we observe instances of late cessation of the growing period subsequently leading to a longer LGP. In this analysis, the LGP including the dry spell at onset is presented in Fig. 4 and shows improvement (longer season) in most of the years in the last decade (2000s).

In Fig. 5, the existing proportionality of instances of total dry days to the LGP is shown. The results show that years experiencing longer dry days are characterized by shorter season lengths and vice versa. Primarily, we reveal that dry days, which are derived from dry spells, are a determinant of the LGP. The results, however, do not reveal a standard trend in the distribution of dry days in the study period since there is a distinct inter-annual variation in the proportionality of dry days to the LGP.

4.3 Smoothened Cereal Yields to Cater for Non-climatic Influences

Results in Fig. 6 indicate the accommodation of spikes along with the annual cereals time series. In the results, spikes include instances where cereal yields sharply increase or decrease from subsequent or previous years. The following Fig. 6 indicates the inter-annual variation of cereal yields against the respective study period mean. From smoothened yield anomalies, (Fig. 7) the 1988 millet yields represent the highest yield records with records of above average yield records also appearing from the year 2002 through 2013. Sorghum yields show a yield trend characterized by years of below average yields from the period 1984 through 1993. The recent decade notably from the year 2001 breakpoint is characterized by mostly above average sorghum yields with a characteristic

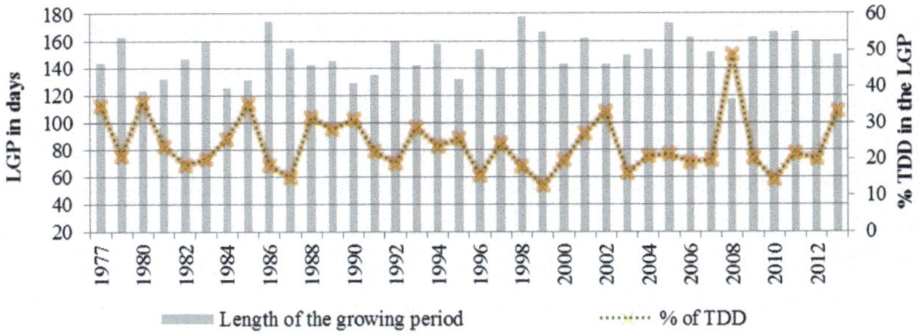

Fig. 5 Annual distribution of total dry days against the LGP in Cassou Area

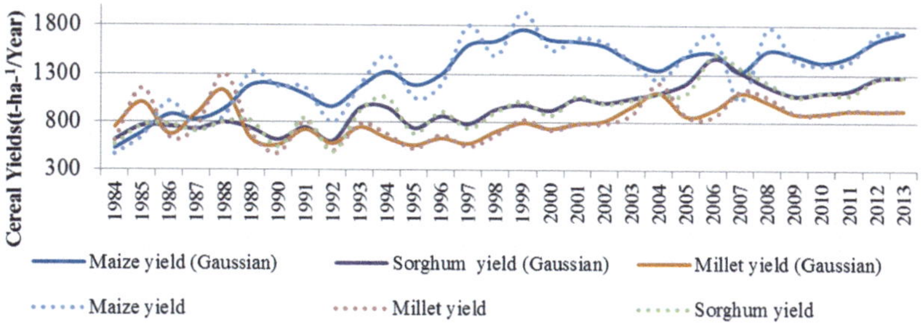

Fig. 6 Detrended cereal yields according to computed cereal yields, Sissili–Ziro Province

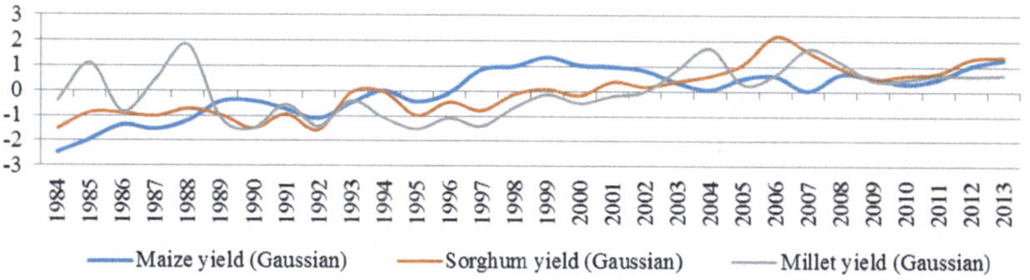

Fig. 7 Detrended cereal yields anomalies showing annual variation against a mean yield value, Sissili–Ziro Province

steadily increasing trend. Maize yields show a clearer yield trend in the time series with the larger part of the records characterized by above average yields in the second decade, with subsequent years after 2004 indicating lower records.

5 Regression Model Parameters and Plots

In the regression plots, we relied on detrended annual cereal yield anomalies run against selected climatic descriptors, which are also detrended. The summary statistics of the multiple regression models, Table 1, and bar charts, Fig. 8, reveal beta coefficients that indicate the relative contribution of climatic descriptors in influencing crop yields in the regression model. This is followed by scatter plots, Fig. 9, showing the direction of the casual relationship between individual predictors and dependent variables. The scatter plots further indicate observed values with the line of best fit showing variable directions of the casual relationships between individual climatic derivatives and cereal yield anomalies.

In the multiple regressions summary (Table 1), we emphasize on the R^2, the coefficient of determination, (Dewberry 2004) which indicates the percentage of the variance of the cereal yield that is predicted on the basis of climatic variables. From the model p-values, we observe a non-significant prediction of the dependent variables implying while climatic factors have an influence there could be other non-climatic drivers contributing to these effects. In the next Sect. 5, we dwell further on the observed relationships between cereal yields and climatic variables while referring to related work.

6 Discussion

6.1 Trends and Variability in Climate Descriptors and Annual Cereal Yields

Results demonstrate variability in seasonal and annual rainfall, where the mean annual rainfall acts as a threshold, there are several instances of years with above and below average annual rainfall in the study period. These instances of

Table 1 Summary of multiple regression models, cereal yields against climatic descriptors, 1984–2013

		Maize yield model	Sorghum yield model	Millet yield model
Model R^2		51.8%	45.9%	34.1%
Model *p*-value		0.001	0.004	0.134
95% C.I		−26.4, −9.7	−19.2, −2.7	−14.4,4.5
Selected standardized coefficients				
ETc	*p*-value	0.012	0.001	0.012
	t	2.725	3.627	2.725
Rainy days	*p*-value	0.013	0.027	0.938
	t	2.669	2.351	−0.079
LGP	*p*-value	0.17	0.042	0.753
	t	1.415	−2.143	−0.318
CDD	*p*-value	0.996	0.342	0.514
	t	−0.043	−0.969	−0.662

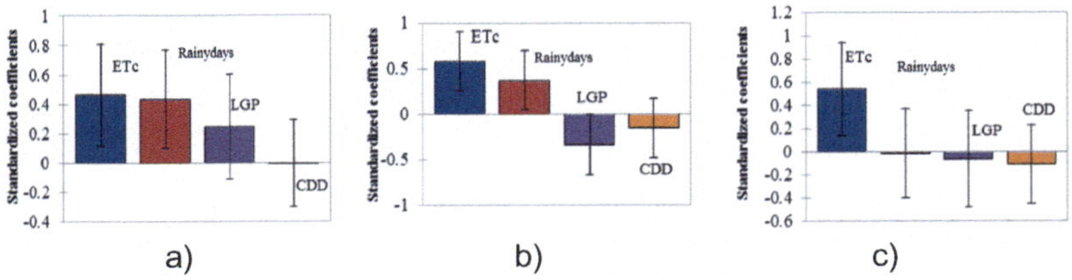

Fig. 8 Color differentiated bar charts showing beta/standardized coefficients for; (**a**) Maize, (**b**) Sorghum, and (**c**) Millet yield anomalies, against selected climatic descriptors

inter-annual and inter-decadal changes relate to a study in central Burkina Faso by West et al. (2008). Annual rainfall shows a strong relationship with the number of rainy days recorded in the year. This to a certain extent indicates the predictability of the total amount of rainfall recorded in the year based on the number of rainy days. The number of rainy days similarly demonstrates the occurrence of inter-annual variation subsequently altering the deviation of the total annual rainfall from the study period mean rainfall.

Dry spells and drought occurrences have a close relationship; dry spells of more than 40 consecutive dry days (CDD) can be effectively termed as a drought instance when this occurs within the growing season (Mathugama and Peiris 2011). From the results, instances of dry spells are a key concern because such dry spells tend to shorten the length of the growing period (Fig. 5) while also widely contributing to false season starts (Fig. 4). These events are indeed documented in perception study by Roncoli et al. (2001) in the central area of Burkina Faso where farming households identify aspects such as late onsets, early cessation, and frequent dry spells at the beginning and within the growing season.

In arid and semi-arid environments, soil moisture availability is to a great degree influenced by the duration and persistence of dry spells (Kisaka et al. 2015) due to the dry ecology of these environments. These instances of dry spells pose great risk among crop farming as they influence crop water deficit during key growth stages subsequently raising the probability of crop failure (Narasimhan and Srinivasan 2005).

The observed yield anomalies in the study area reveal the generally increasing trend over

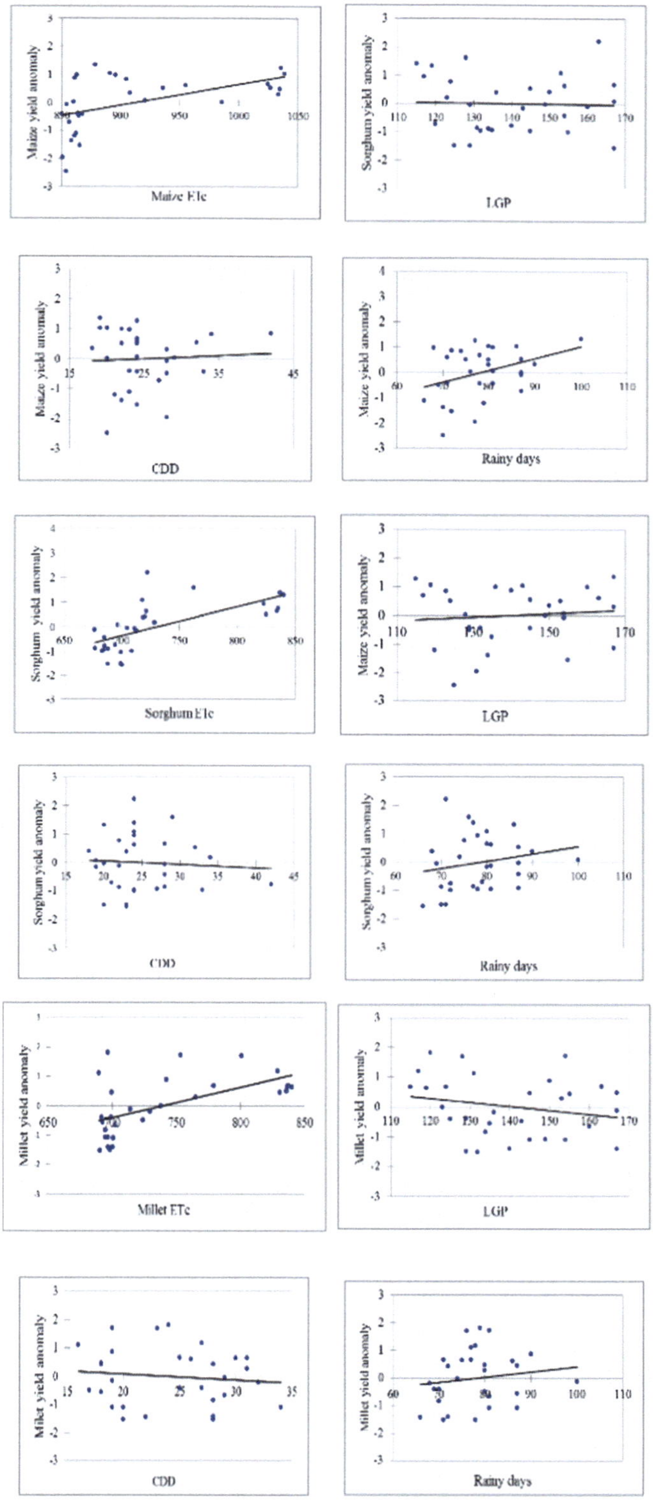

Fig. 9 Scatter plots showing the casual relationship between detrended cereal yields anomalies (Ton/ha) (*Y*-Axis) and development stage ETc, LGP, Dry days, and rainy days, during the growing season (*X*-axis) along the best fit line

the years and characteristic inter-annual shifts. These inter-annual contrasts and the characteristic increasing trends are to a certain extent driven by a blend of climatic and non-climatic drivers. This mix of influences may be complex and difficult to demystify but the adopted statistical techniques attempt to explore the role and contribution of climatic influences in defining yield trends as detailed in the following section.

6.2 Relating Cereal Yields to Climatic Derivatives

The coefficients of variation (R^2) show that the variance in maize and sorghum yields is strongly explained by climatic descriptors, where $R^2 = 51.8\%$ and 45.9%, respectively. These results show that the cereal yields respond near similarly to certain climate variations. When we review the crop–climate variability regression model coefficients (Fig. 8), a dominant positive relationship of cereal yields with the crop evapotranspiration at development stages of crop growth and the number of rainy days is observed. Maize and sorghum yields, for example, are significantly predicted by respective development stage crop evapotranspiration, where $p = 0.012$ and $p = 0.001$, respectively. These results by the bar charts further show the stronger contribution to yield variability by the development stage ETc among all cereals compared to other climatic parameters. The scatter plots, Fig. 9, display a similar and more conspicuous trend where individual cereal yield anomalies exhibit a positive relationship with respective crop evapotranspiration estimates. In semi-arid environments, evapotranspiration influences soil moisture status which is important in maintaining crop growth and subsequent crop yields. This effect on available water has an effect on crop growth and development since evapotranspiration is a function of temperature and rainfall changes driven by the frequency and extent of climate variability (Porter and Semenov 2005). In the present study, we argue that from our results, evapotranspiration falls below seasonal rainfall during the growing season, hence cereals yields respond positively perhaps due to a favorable or optimal soil moisture balance.

The number of rainy days made a positive and significant relationship with maize yield anomalies, where $p = 0.012$ implying the cereal is highly responsive and to a greater extent reliant to rainfall distribution. Maize is indeed very sensitive to hydrous stress and/or high temperatures during the flowering (Ingram et al. 2002; Kambire et al. 2010) and grain filling stages (Parry et al. 1999). Studies in the Sahel show best harvests occur when there are long rainy seasons without dry spells during flowering (Barbier et al. 2009). On the other hand, the explanatory variable indicates a non-significant and weakly positive prediction of sorghum and millet yield anomalies. This weak relationship with the latter cereals is further exhibited by the weaker gradients exhibited by the rainy days versus cereal yield anomalies scatter plots in Fig. 9. Since the number of rainy days is our closest measure of rainfall distribution, results reveal the hardiness of these cereals where rainfall is limited or expresses variability. The lower standardized/beta coefficients Fig. 8c also demonstrate that millet is a hardy crop compared to sorghum, as Behnassi et al. (2013) also discuss. Millet is more efficient in utilization of soil moisture due to a better root configuration; the cereal can hence thrive in much drier areas. The crop is, however, susceptible to waterlogging when compared to sorghum (Ingram et al. 2002).

Cereal yields show varying responses to the CDD and the LGP in the season with, for example, the CDD exhibiting a weakly positive relationship with variation in maize yields as displayed in Figs. 8a and the scatter plot in Fig. 9. This direction is also more conspicuous in the millet and sorghum models and scatter plots where we observe a negative relationship between CDD and the respective cereal yield anomalies. This demonstrates that, while these cereals are highly resistant to extreme climatic conditions, they are susceptible to drier conditions perpetuated by prevailing rainfall variability that could perhaps also affect their growth and development and resulting yields. In addition, while these cereals are resistant and highly adaptable to moisture and temperature stress they exhibit certain limits. Changes in the mean and variability of temperature do affect crop processes, for example, enhancing photosynthesis and respiration up to certain thresholds (Porter and Semenov 2005).

On the same vein, the LGP shows a stronger negative association with cereal yield anomalies of sorghum, where $\beta = -0.343$, t (1) = −2.143, $p = 0.042$ and millet $\beta = -0.065$, t (1) = −0.318, $p = 0.753$. These observations indicate the severity and contribution of unstable climatic conditions that lead to widespread false season onset and early end of the season. The eventual season shortening and uncertainty in the LGP affects even drought-tolerant crops. Other studies, for example, Rowhani et al. (2011) have also reported this observation. Variation in the LGP is a key aspect in on-farm decision making. As Schneider et al. (2000) state, a longer LGP enables farmers to cultivate varieties that require longer maturity periods and further facilitates longer grain filling and higher cereal yields. This argument emphasizes the critical role that seasonality characteristics play in defining and influencing crop yield quantity and quality.

Results show variation in response of cereal yields to climatic derivatives implying crops respond variably to climatic factors with some climatic parameters such as rainy days and crop evapotranspiration exhibiting prominence. Further, the importance of precipitation as demonstrated by a large pool of precipitation derived variables shows the importance of rainfall in explaining crop growth, development and subsequent yields. Similar studies have indeed mentioned experienced effects of rainfall failure and/or variation on yields. For example Barbier et al. (2009), do mention these instances of drop in cereal yields in Burkina Faso in certain years due to climate variation. Specifically, during the 2004 low rainfall event, farmers harvested minimal sorghum and maize with minimal sales to markets. In 2006, the situation was similar, although in this case, farmers suffered from the effects of floods. These effects are further compounded by poor soils West et al. (2008) and desertification (Kandji et al. 2006).

The applied detrending approach does not exactly handle yield changes associated with interventions such as technological inputs, but we view it as an appropriate method to exclude non-climatic influences. Results in the present study, therefore, demonstrate that half to two-thirds or even more of the variance in the cereal yields remained unexplained by the climatic descriptors. This output reflects the influence of non-climatic factors such as the prevailing economic environment (Lobell and Field 2007) and their primary importance. Nevertheless, the proportion of variance in cereal yields at the inter-annual scale explained by the climatic variables provides substantial evidence on the role of local climatic variations in shifting yields. This direction of effects could get severe with more extreme climatic events, subsequently aggravating low yields and eventually leading to food and nutrition deficiency.

7 Summary and Conclusion

This paper reveals instances of climate variability based on inter-annual rainfall variations across the study period (1984–2013). This variability is expressed by the variation in rainfall derived descriptors from year to year against a long-term average. The analysis reveals that cereal yields are characterized by an increasing trend over the years with characteristic oscillations along the period mean. From the results, millet yields show less variation to climate derivatives including instances of dry spells compared to Sorghum and Maize yields. Another interesting observation is the strong relationship between maize and rainfall amount, where maize yield anomalies exhibit a strong relationship with rainfall measured as the number of rainy days. Further instances of dry spells and total dry days also indicate the widespread influences of these events on drought-hardy crops. These results are evidence of climate variability and trends in the study area which has mixed effects on cereal yields. This variability also constitutes a variation in the onset of the growing season, which we equate to the start of sowing that is likely to influence the farming calendar due to the difficulty in prediction.

It is imperative to underline the importance of non-climatic factors since the regression models indicate that climatic factors explain a certain percentage of inter-annual crop yield anomalies variation. This leaves a proportion of yield variance probably defined by an unknown blend of slowly changing factors such as adaptation interventions. We mentioned the uncertainties surrounding the contribution of these factors in varying of cereal yields and emphasize that presented statistical models may not adequately capture these effects. Moreover, attempts in related work by Lobell and Field (2007) encountered challenges in attempts to estimate inter-annual CO_2 changes. This was attributed to minimal inter-annual variations in CO_2 variations. After results from Chap. 4 (Sect. 4.5), we further indicate that in instances there was an observed relationship between two variables such relationship can be explored further by the use of experimental techniques.

While recognizing the complexity of intervening dry spells among other intra-seasonal rainfall indicators and their effect on season length, it is necessary to understand and develop specialized responses to predict and monitor these key calamities affecting crops at moisture-sensitive growth stages. At the same time, it is of importance to enhance access of weather-based information and/or forecasts and beneficial climate research through pathways targeting vulnerable groups, individual farmers and policymakers, on the importance and merits of relying on such information. These could include custom made weather information that incorporates probabilities of dry spell occurrence and suggested response strategies to institute precautionary planning among farmers. In instances where such information is availed to farmers, follow-up strategies are imperative to monitor usefulness and perceptions. An elaborate climate information system could include these elements. Recognizing the inter-annual and decadal rainfall variations and the existent uncertainty exhibited by climate variability, response aimed at managing crop failure including mechanisms aimed at maintaining soil water content is imperative. These could include adequate and climate information constituting timely warning systems, water harvesting facilities backed by policies and irrigation as well as access to drought-tolerant varieties and income diversification.

Acknowledgments This work was funded by the Finish Ministry of Foreign Affairs through the World Agroforestry Centre under the BIODEV Project (Building Biocarbon and Rural Development in West Africa). We thank Dr. Richard Coe (ILRI/ICRAF) for input in computation and adjustment of variability descriptors and Dr. David Stern (University of Reading) for clarification in the interpretation of growing season descriptors. We also would like to acknowledge Dr. Jorge De Jesus of ISRIC for his assistance in the extraction of soil type information. We thank the National Meteorology Service of Burkina Faso for providing long-term daily climate data.

References

Afifi A, May S, Clark VA` (2003) Computer-aided multivariate analysis. CRC Press

Allen RG, Pereira LS, Raes D, Smith M (1998) Crop evapotranspiration-Guidelines for computing crop water requirements-FAO Irrigation and drainage paper 56 FAO, Rome 300

Bannayan M, Sanjani S, Alizadeh A, Lotfabadi SS, Mohamadian A (2010) Association between climate indices, aridity index, and rainfed crop yield in northeast of Iran. Field Crops Res 118:105–114

Barbier B, Yacouba H, Karambiri H, Zoromé M, Somé B (2009) Human vulnerability to climate variability in the sahel: farmers' adaptation strategies in Northern Burkina Faso. Environ Manage 43:790–803. https://doi.org/10.1007/s00267-008-9237-9

Behnassi M, Pollmann O, Kissinger G (2013) Sustainable food security in the era of local and global environmental change. Dordrecht

Blanc É (2012) The impact of climate change on crop yields in Sub-Saharan Africa

Borona M, Mbow C, Ouedraogo I (2016) Unstacking high temporal resolution meteorological data for multidimensional analysis of climate variability in southern Burkina Faso. Geogr Tidsskrift-Danish J Geography 116(2):176–189. https://doi.org/10.1080/00167223.2016.1212668

Brouwer C, Goffeau A, Heibloem M (1985) Irrigation water management: training manual No. 1-introduction to irrigation. IILRI and FAO, Rome

Challinor A, Wheeler T, Garforth C, Craufurd P, Kassam A (2007) Assessing the vulnerability of food crop systems in Africa to climate change. Clim Change 83:381–399

Chang JH (1974) Climate and agriculture: an ecological survey. Aldine

Chen C-C, McCarl BA, Schimmelpfennig DE (2004) Yield variability as influenced by climate: a statistical investigation. Clim Change 66:239–261
Davies ER (2012) Computer and machine vision: theory, algorithms, practicalities. Elsevier, Waltham
Dewberry C (2004) Statistical methods for organizational research: theory and practice. Psychology Press
Droogers P, Allen RG (2002) Estimating reference evapotranspiration under inaccurate data conditions. Irrig Drain Syst 16:33-45
FAO (1993) Soil Tillage in Africa: needs and challenges, vol 69. Food and Agriculture Organization of the United Nations: Land Water Development Division, Rome
FAO (2014a) Burkina Faso Country fact sheet on food and agriculture policy trends. FAO. https://www.fao.org/docrep/field/009/i3760e/i3760e.pdf. Accessed 16 June 2015
FAO (2014b) Understanding the drought impact of El Niño on the global agricultural areas; An assessment using FAO's Agricultural Stress Index (ASI). FAO, Rome
Farmer W, Strzepek K, Schlosser CA, Droogers P, Gao X (2011) A method for calculating reference evapotranspiration on daily time scales. MIT Joint Program on the Science and Policy of Global Change
Fontès J et al. (1995) Carte de la végétation et de l'occupation du sol du Burkina Faso: notice explicative. Ministère de la coopération française, Projet Campus, Toulouse, France
Geyer CJ (2011) Handbook of markov chain monte carlo. In: Brooks S, Gelman A, Jones G, Meng XL (eds). CRC Press
Goyal MR, Harmsen EW (2013) Evapotranspiration: principles and applications for water management. Apple Academic Press
Hadgu G, Tesfaye K, Mamo G, Kassa B (2013) Trend and variability of rainfall in Tigray, northern Ethiopia: analysis of meteorological data and farmers' perception. Acad J Agric Res 1:088–100
Haggett P (2002) Encyclopedia of world geography, vol 17. Marshall Cavendish, New York
Haile M (2005) Weather patterns, food security and humanitarian response in sub-Saharan Africa Philos Trans R Soc B Biol Sci 360:2169–2182
Hengl T et al (2014) SoilGrids 1 km—global soil information based on automated mapping. PLoS One 9:e105992
Ingram KT, Roncoli MC, Kirshen PH (2002) Opportunities and constraints for farmers of west Africa to use seasonal precipitation forecasts with Burkina Faso as a case study. Agric Syst 74:331–349. https://doi.org/10.1016/S0308-521X(02)00044-6
INSD (2015) Annuaire statistique 2014. Institut National des Statistiques et de la Démographie (INSD), Ouagadougou, Burkina Faso. http://www.insd.bf/n/contenu/pub_periodiques/annuaires_stat/Annuaires_stat_nationaux_BF/Annuaire_stat_2014.pdf
IRRI (1976) Proceedings of the symposium on climate & rice. Irri, Los Banos
Jalloh A, Rhodes ER, Kollo I, Roy-Macauley H, Sereme P (2011) Nature and management of the soils in West and Central Africa: a review to inform farming systems research and development in the region. CORAF/WECARD, Dakar
Kambire H, Abdel-Rahman G, Bacyé B, Dembele Y (2010) Modeling of maize yields in the south-Sudanian zone of Burkina Faso-West Africa American-Eurasian. J Agric Environ Sci 7:195–201s
Kandji ST, Verchot L, Mackensen J (2006) Climate change and variability in the Sahel region: impacts and adaptation strategies in the agricultural sector. UNEP and ICRAF, Nairobi
Kisaka OM, Mucheru-Muna M, Ngetich F, Mugwe J, Mugendi D, Mairura F (2015) Seasonal rainfall variability and drought characterization: case of eastern arid region, Kenya. In: Leal Filho W, Esilaba AO, Rao KPC, Sridhar G (eds) Adapting African agriculture to climate change. Climate change management. Springer International Publishing, pp 53–71. https://doi.org/10.1007/978-3-319-13000-2_5
Lin BB (2011) Resilience in agriculture through crop diversification: adaptive management for environmental change. Bioscience 61:183–193
Lobell DB, Field CB (2007) Global scale climate–crop yield relationships and the impacts of recent warming. Environ Res Lett 2:014002
Lodoun T, Giannini A, Traoré PS, Somé L, Sanon M, Vaksmann M, Rasolodimby JM (2013) Changes in seasonal descriptors of precipitation in Burkina Faso associated with late 20th century drought and recovery in West Africa. Environ Dev 5:96–108. https://doi.org/10.1016/j.envdev.2012.11.010
Maikano I (2006) Generate prototype WCA recommendation maps for selected sorghum (8) and millet (8) cultivars based on updated end-of-season dates (PRODEPAM, activity). ICRISAT, Bamako
Mathugama SC, Peiris TSG (2011) Critical evaluation of dry spell research. Int J Basic Appl Sci 11:153–160
Mishra A, Hansen JW, Dingkuhn M, Baron C, Traoré SB, Ndiaye O, Ward MN (2008) Sorghum yield prediction from seasonal rainfall forecasts in Burkina Faso. Agric For Meteorol 148:1798–1814
Narasimhan B, Srinivasan R (2005) Development and evaluation of soil moisture deficit index (SMDI) and evapotranspiration deficit index (ETDI) for agricultural drought monitoring. Agric For Meteorol 133:69–88
Osborne T, Wheeler T (2013) Evidence for a climate signal in trends of global crop yield variability over the past 50 years. Environ Res Lett 8:024001
Ouedraogo I, Savadogo P, Tigabu M, Cole R, Oden PC, Ouadba J-M (2011) Trajectory analysis of forest cover change in the tropical dry forest of Burkina Faso. West Afr Landsc Res 36:303–320. https://doi.org/10.1080/01426397.2011.564861
Parry M, Rosenzweig C, Iglesias A, Fischer G, Livermore M (1999) Climate change and world food security: a new assessment. Glob Environ Change 9

(Supplement 1):S51–S67. https://doi.org/10.1016/S0959-3780(99)00018-7
Porter JR, Semenov MA (2005) Crop responses to climatic variation. Philos Trans R Soc Lond B: Biol Sci 360:2021–2035
Rao K, Okwach G (2005) Enhancing productivity of water under variable climate. In: International water management institute conference papers.
Rojas O, Li Y, Cumani R (2014) Understanding the drought impact of El Niño on the global agricultural areas: an assessment using FAO's Agricultural Stress Index (ASI): Food and Agriculture Organization of the United Nations (FAO)
Roncoli C, Ingram K, Kirshen P (2001) The costs and risks of coping with drought: livelihood impacts and farmers' responses in Burkina Faso. Clim Res 19:119–132
Rosenzweig C, Iglesias A, Yang X, Epstein PR, Chivian E (2001) Climate change and extreme weather events; implications for food production, plant diseases, and pests. Glob Change Hum Health 2:90–104
Rötter R, van de Geijn SC (1999) Climate change effects on plant growth, crop yield and livestock. Clim Change 43:651–681. https://doi.org/10.1023/a:1005541132734
Rowhani P, Lobell DB, Linderman M, Ramankutty N (2011) Climate variability and crop production in Tanzania. Agric For Meteorol 151:449–460
Sanginga N, Woomer PL (2009) Integrated soil fertility management in africa: principles, practices, and developmental process. Tropical Soil Biology and Fertility Institute of the International Centre for Tropical Agriculture
Schneider SH, Easterling WE, Mearns LO (2000) Adaptation: sensitivity to natural variability, agent assumptions and dynamic climate changes. Clim Change 45:203–221
Semenov MA, Porter JR (1995) Climatic variability and the modelling of crop yields. Agric For Meteorol 73:265–283. https://doi.org/10.1016/0168-1923(94)05078-K
Simelton E et al. (2011) African farmers' perceptions of erratic rainfall Sustainability Research Institute Paper
Sivakumar M (1988) Predicting rainy season potential from the onset of rains in Southern Sahelian and Sudanian climatic zones of West Africa. Agric For Meteorol 42:295–305
Sivakumar MVK (1992) Empirical analysis of dry spells for agricultural applications in West Africa. J Clim 5:532–539. https://doi.org/10.1175/1520-0442(1992)005%3c0532:eaodsf%3e2.0.co;2
Slingo JM, Challinor AJ, Hoskins BJ, Wheeler TR (2005) Introduction: food crops in a changing climate. Philos Trans R Soc Lond B Biol Sci 360:1983–1989
Somé L, Jalloh A, Zougmoré R, Nelson GC, Thomas TS (2013) Burkina Faso. In: Jalloh A, Nelson GC, Thomas TS, Zougmoré RB, Roy-Macauley H (eds) International food policy. Research Institute, Washington
Tabachnick BG, Fidell LS (2001) Using multivariate statistics
Traoré S (2000) Drought adaptation of Malian local sorghum ecotypes. Secheresse 11:227–237
Tshiala M, Olwoch JM (2010) Impact of climate variability on tomato production in Limpopo Province, South Africa
Van der Linden P, Xiaousu D (2001) Contribution of working group I to the third assessment report of the intergovernmental panel on climate change. Cambridge University, Cambridge
Wang Y-M, Traore S, Kerh T (2008) Computing and modeling for crop yields in Burkina Faso based on climatic data information. WSEAS Trans Inform Sci Appl 5:832–842
West CT, Roncoli C, Ouattara F (2008) Local perceptions and regional climate trends on the central plateau of Burkina Faso. Land Degrad Dev 19:289–304
Wood S (2006) generalized additive models: an introduction with R. Taylor & Francis

Multivariate Standardized Drought Indices to Identify Drought Events: Application in the Maipo River Basin

Melanie Oertel, Francisco Javier Meza, and Jorge Gironás

Abstract

Timely and reliable drought detection is the basis for sustainable drought management, which aims to keep negative impacts to a minimum. Droughts are natural and recurring events and as such cannot be avoided. Ensuring future preparedness relies on knowledge of past events. Drought indices are commonly used to detect drought events, among them standardized indices are often utilized due to their comparability across time and space. We propose a multivariate approach based on four standardized indices; the Standardized Precipitation Index (SPI), the Standardized Precipitation Evapotranspiration Index (SPEI), the Standardized Streamflow Index (SSI), and the Standardized Soil Moisture Index (SSMI), allowing us to evaluate drought patterns by each variable of interest as well as to investigate river basin specific drought propagation patterns. We apply the multivariate approach in the Maipo river basin in Chile. Detection of starting and end points identified by SPI and SPEI with three different thresholds permits us to see how and when triggering processes occur in soil moisture and streamflow. Drought propagation patterns are identified, as SSI values respond to SPI (SPEI) values with a minimum lag of three months, proving that it is not sufficient to rely only on one index for operational drought management. Further, we develop a multivariate autoregressive model (MAR1 model) for operational forecast based on SPEI, SSI, and SSMI. Model results show the applicability to obtain forecasts for each index and for jointly marginal probability density functions.

Keywords

Drought propagation · Standardized indices · Multivariate autoregressive model · Semi-arid river basin

M. Oertel (✉) · F. J. Meza · J. Gironás
Centro Interdisciplinario de Cambio Global. Pontificia, Universidad Católica de Chile, Santiago, Chile
e-mail: moertel@uc.cl

M. Oertel · F. J. Meza
Departamento de Ecosistemas Y Medio Ambiente. Pontificia, Universidad Católica de Chile, Santiago, Chile

F. J. Meza
Aquasec. IAI Center of Excellence for Water Security, Tucson, USA

J. Gironás
Departamento de Ingeniería Hidráulica y Ambiental, Pontificia Universidad Católica de Chile, Santiago, Chile

Centro de Desarrollo Urbano Sustentable CEDEUS CONICYT/FONDAP/15110020, Santiago, Chile

Centro de Investigación Para La Gestión Integrada del Riesgo de Desastre CIGIDEN CONICYT/FONDAP/15110017, Santiago, Chile

L. Ribbe et al. (eds.), *Towards Water Secure Societies*,
https://doi.org/10.1007/978-3-030-50653-7_11

1 Introduction

Droughts are recurring natural events which can have devastating impacts on all levels of society. As droughts cannot be avoided, the aim of successful drought management is to minimize negative impacts and facilitate the recovery of affected systems. What is considered a drought event differs between climatic zones and depends on the vulnerability of a region (Wilhite and Buchanan-Smith 2005; Kallis 2008). Although droughts can occur in any climatic zone, some regions are more prone to droughts and their impacts than others. That is the case for semi-arid regions with growing populations, due to a combination of driving factors such as seasonality of rainfall and high evapotranspiration rates (Beguería et al. 2014), as well as intensive use of water resources. Further, to declare a drought event depends on the context, for example, the severity of a passed drought event is often measured by the sum of monetary losses due to occurred negative impacts, whereas drought detection is based on thresholds, thus monitoring of drought-related variables, such as precipitation and streamflow. Minimizing negative drought impacts is feasible if drought events are detected as early as possible. Early drought detection is an ongoing challenge in drought research, and several improvements in detection methods have been presented in recent years. This advancement is explained by the greater availability and accessibility of data (e.g. satellite images, open online source), larger computational capacity and the more active exchange of knowledge, as drought management is discussed on several levels of societies nation—and worldwide. Just a few examples are: The Integrated Drought Management Programme in Central and Eastern Europe (Kindler and Okruszko 2014), the National Drought Observatories in Chile, Peru, and Honduras in Latin-America of the MWAR-LAC Project (Latin American Flood and Drought Monitor), the State Drought Plans in the US (State Drought Plans), and the Australian national drought policy (Botterill and Cockfield 2013). Finding appropriate detection methods has a long history, mainly based on the analysis of precipitation records. Since the introduction of the Palmer Drought Severity Index (Palmer 1965) drought indices have evolved to more than 100 today (see for example Niemeyer 2008 and Zargar et al. 2011). Drought indices are tools used to synthesize information and present it as a single value, which can be easily used for communication purposes. The diversity of indices varies from simple threshold methods to copula-based calculation procedures, as the joint deficit index presented by Kao and Govindaraju (2010), or the multivariate drought index by Hao and Agha-Kouchak (2013). A complete review of different types of indices can be found in the work of the following authors: Hao and Singh (2015), Hayes (2007), Heim (2002), Keyantash and Dracup (2002), Mishra and Singh (2010), Vicente-Serrano et al. (2012b), and Zargar et al. (2011).

The variety of drought indices reflects the diversity of drought definitions and classifications, hence there is no unique drought index satisfying all requirements, similar to the case of a universal drought definition (e.g. Heim 2002; Niemeyer 2008; Zargar et al. 2011). Conceptual definitions are mainly used to separate drought events from other terms such as aridity or water scarcity (Oertel et al. 2015). Operational definitions are preferred for drought management purposes, as they allow us to differentiate between meteorological, agricultural, hydrological, and groundwater droughts (Wilhite 1993; Heim 2002; van Loon and Laaha 2015). Such definitions are used to describe the type of drought, but inappropriate for operational drought management, as starting and end points of drought types are interwoven and may lead to confusion among different stakeholders (Wilhite and Buchanan-Smith 2005; Smakhtin and Schipper 2008). Given the number of different indices and the complexity of the phenomenon, multivariate approaches are needed to better describe the dynamics of involved variables, such as the triggering process through the hydrological cycle, and drought propagation. Hence, a groundwater drought may be the result of a prolonged meteorological drought, which occurs due to a precipitation deficit (Beyene et al. 2014). Numerous drought indices are including several variables and can offer a more comprehensive picture, for example the Surface Water Supply Index (SWSI) (Shafer and Dezman 1982), the Aggregate Drought Index (ADI) (Keyantash and Dracup 2004), or the Self-Calibrated PDSI (Wells et al. 2004). In this context, jointly monitoring and documenting is recommended as a first step to improve regional drought management (Wilhite et al. 2007; Maliva and Missimer 2012).

Approaches with multiple variables, which consider characterizations and predictions, are not easy but necessary (Kao and Govindaraju 2010). But stand-alone indices including several variables, require input data of the same quality, are difficult to compare against other indices, and often lack replicability as their calculation procedure is more complex (Keyantash and Dracup 2002; Kallis 2008; Zargar et al. 2011). These are common critics found in literature and explain the preference for indices such as the Standardized Precipitation Index (SPI), or the use of percentiles of the variable of interest.

The World Meteorological Organization (WMO 2012) confirmed the application of the SPI after McKee et al. (1993) as an international common denominator. Standardized indices have gained importance in recent years. Independently of the input variable, they follow mainly the same calculation procedure and are easily replicable. Their main advantage is the comparability of results among time and regions. This facilitates communication as it offers a common base for discussion, and hence index results can be discussed broadly to share insights of drought events. Tools to calculate standardized indices such as the software from the World Meteorological Organization (WMO 2012), "DrinC" (Tigkas et al. 2015) or the R-package "SCI" (Gudmundsson and Stagge 2016; Stagge et al. 2015, 2016) are freely available. Nonetheless, caution must be taken when preparing the input variables and selecting the probability distributions to obtain reliable results.

Because drought detection should be easy and replicable, our objective is to evaluate the use of standardized indices to understand drought propagation. Thus, we address three research questions: (1) How do the calculation procedure and distribution fitting of standardized indices influence results? (2) How can we use results gained by standardized indices to identify drought propagation processes? and (3) How can we use standardized indices for operational forecast?

Following the idea of learning from the past to identify drought patterns, we present in this study a multivariate approach based on four standardized indices to evaluate drought characteristics in the Maipo river basin, a semi-arid basin located in central Chile. We take a methodological perspective on drought event detection with the application of the SPI, the Standardized Precipitation Evapotranspiration Index (SPEI), the Standardized Streamflow Index (SSI), and the Standardized Soil Moisture Index (SSMI). Evaluation of drought propagation patterns with different indices is presented elsewhere (e.g., Van Loon et al. 2015; Barker et al. 2016). The study of Barker et al. (2016) also uses standardized indices, applying the SPI and SSI to identify drought relationships. To investigate several variables, individual and jointly, we extend in the number of applied indices, including temperature (SPEI) and soil moisture, to obtain a more comprehensive view on drought propagation.

2 Methods

2.1 Standardized Indices

The calculation of the standardized indices—independent of the variable under analysis—follows the same structure:

1. Obtain the time series of monthly values of the desired variable,
2. Fitting a probability distribution function to monthly data,
3. Converting the data to a cumulative distribution function,
4. Use an inverse function to obtain standardized gaussian values (i.e. values that follow a normal distribution with mean 0 and standard deviation of 1).

A crucial step in the calculation of standardized indices is the selection of an adequate probability distribution function. There is no unique distribution which fits best for all variables at the same time. To identify the most suitable distributions we used EasyFit 5.6 software developed by MathWave© Technologies. The software compares more than 50 distributions and offers goodness of fit tests (Kolmogorov-Smirnov and Anderson-Darling), to evaluate the fitting. We applied two fittings; best fit for each month and recommended distributions, to evaluate the influence on results of the indices. To ensure comparability among the indices we applied the same calculation procedure for each index, following recommendations provided in the literature (McKee et al. 1993; Vicente-Serrano et al. 2012b; Beguería et al. 2014; Stagge et al. 2015, 2016; Vicente-Serrano and Beguería 2016).

2.2 Study Area

The Maipo river basin lies in central Chile and its boundaries are nearly coinciding with the administrative boundaries of the Metropolitan Region of Santiago. The Maipo river originates in the Andes and flows to the Pacific Ocean, with a catchment area of approximately 15,000 km^2. Santiago is the most important city in Chile in terms of population and economic activities, with an Urban areas of cover around 640 km^2. The city has a population of 7 million, equivalent to 40% of the Chilean population (Bonelli et al. 2014; Meza et al. 2014; Bustos and Meza 2015).

The river basin has a Mediterranean climate with a mean annual temperature of 14 °C and an annual mean precipitation of 350 mm in the valley. Precipitation falls concentrated during winter months (June–September) with snow accumulation above 1,500 m (Meza et al. 2012). Given the Mediterranean climate, agricultural activities depend on irrigation, which is taken mainly from surface waters. Agriculture is the main water user in the river basin, accounting ~80% of freshwater demand. Main crops are fruits, vegetables, wine grape, and cereals. Natural vegetation is grass and shrublands in the lower part of the river basin as well as some native forests (Bustos and Meza 2015). Streamflow is snowmelt-driven, hence high streamflow values occur during summer months (September–March). This strong seasonality makes the basin vulnerable to the impacts of droughts.

Based on its characterization the river basin can be considered drought-prone by different factors. Santiago´s population is steadily increasing, accounting for higher drinking water demands, and as urban areas are growing they are competing with agricultural areas. Hence the land use in the river basin is changing and so with its properties (Puertas et al. 2014).

The recent megadrought in the Maipo river basin illustrate the interwoven impacts and driving forces of droughts. Staring in 2009 the drought has remained for several years (Garreaud et al. 2017). Climate change projections reassure increasing vulnerability for the region to drought events (Bonelli et al. 2014). Accordingly, the study of drought events in the Maipo river basin is representative for other semi-arid river basins, facing similar challenges of keeping negative impacts to a minimum while population is growing, agricultural activities are intensified, and more extreme weather events are expected. Figure 1 shows the location of the Maipo river basin and the stations used in this study.

2.3 Data and Data Sources

We used data freely available from governmental institutions, which are subject to quality control procedures. Data include daily precipitation (Pp), maximum and minimum temperature (T_{max}, T_{min}) as well as monthly streamflow (Q). Data were obtained from the Dirección General de Aguas (DGA), which belongs to the Ministry of Public Works. Because of the influence of human

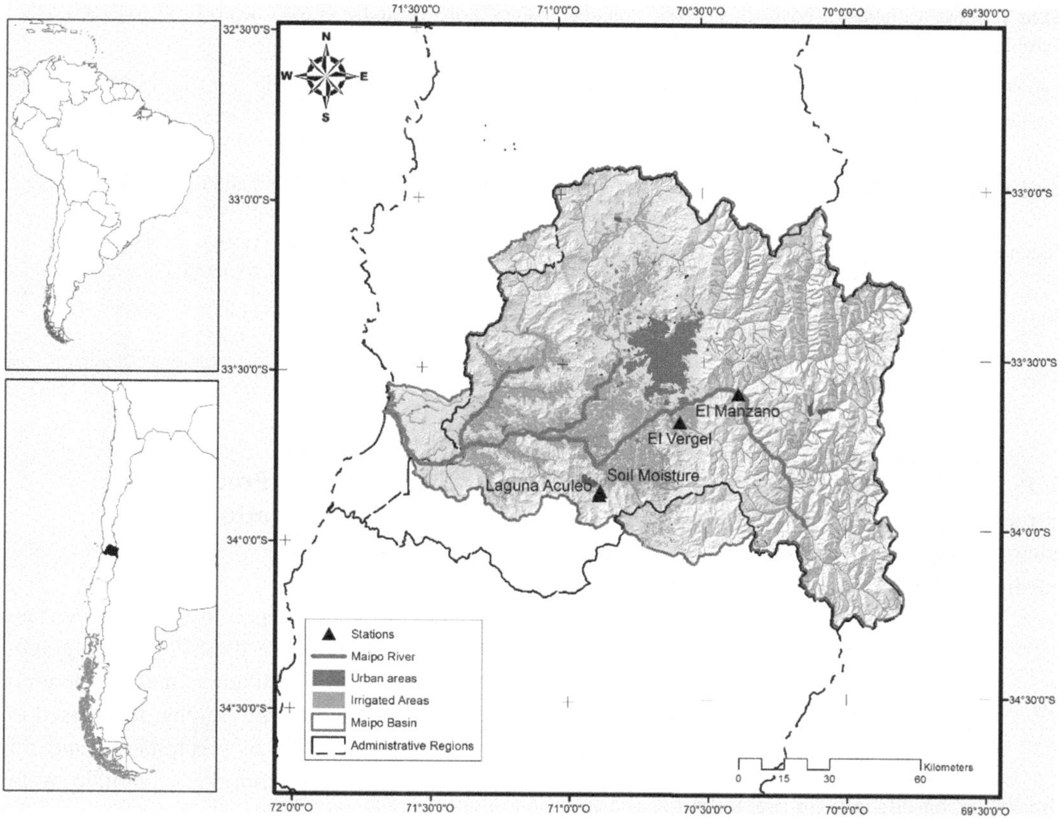

Fig. 1 Location of the Maipo River Basin and measuring stations used in this study

activities, we have selected a streamflow station with a natural regime, in the upper part of the basin. Satellite data was used as an indirect source to obtain soil moisture, which was derived from the ESA CCI SM v03.2 data set, provided by the Climate Change Initiative (CCI) program of the European Space Agency. This dataset is developed by merging various active and passive microwave products into one record (Liu et al. 2012; Dorigo et al. 2017; Gruber et al. 2017). The version was released in February 2017 and provides daily soil moisture retrievals from 1978 until 2015 with a spatial resolution of 25 km^2 for the upper centimeters of soil. There are three different packages of data available: the active, the passive and the combined, which merges fusing scatterometer and radiometer products (Liu et al. 2012; Dorigo et al. 2017). For our study, we chose the combined package, as it provides the highest number of observations. The data set includes from 1992 on more sensors and hence more data points are available. For this study we used data from 1995–2015, as for this period less missing monthly values are identified, and ensure thereby more reliable results of the calculated index. Previous versions of the soil moisture data set were used to assess agricultural droughts, applying the same calculation procedure as the SPI, using soil moisture data instead of precipitation (Sridhar et al. 2008; Hunt et al. 2009; Carrão et al. 2016). We apply in this study a similar approach, using soil moisture as input variable for a standardized index, further called Standardized Soil Moisture Index (SSMI).

More details on the source, time period used, and missing values for each variable are summarized in Table 1. Further Table 2 displays the monthly mean, maximum, and minimum values

Table 1 Input variables used in this study, *missing values were filled by closest neighbor ($R^2 = 0.98$) or ** interpolation

Variable	Source	Time period	Station name	Coordinates	Missing values
Precipitation	DGA	Jan 88–Aug 16	El Vergel	Latitude −33.674 Longitude −70.587	0%
Temperature (T_{max}, T_{min})	DGA	May 88–Jul.16	Laguna Aculeo	Latitude −33.886 Longitude −70.878	15%*
Streamflow	DGA	Jan.60–Sep.17	El Manzano	Latitude −33.594 Longitude −70.379	5%**
Soil moisture	ESA CCI	Jan.95–Dec.15	/	Latitude −33.875 Longitude −0.875	8%**

for Pp, T_{max}, T_{min}, Q, and soil moisture. The data mirror the high seasonality of streamflow, as values are inverse to precipitation and soil moisture patterns.

2.4 Standardized Precipitation Index (SPI)

The SPI is based only on precipitation and indicates how dry or wet a certain period (e.g. 1, 3, 6 months) is. The distribution fitting procedure is particularly challenging in semi-arid regions, as zero precipitation for several months is a normal climate pattern (Wu et al. 2007). For precipitation we have chosen the closest station with reasonable observations to the soil moisture point, as soil moisture response quickly to precipitation. This precipitation station provides monthly measurements with no missing values. The covered period is 28 years, even though 30 years is the standard recommendation for this method. We continue with this database in the case study as 20 years are the required minimum (WMO 2012). Based on the data used, we list in Table 3 the best-fit probability distributions for each accumulation period and each month to obtain SPI values. Further monthly precipitation data were fit to the recommended gamma distribution (McKee et al. 1993; Guttmann 1999; Stagge et al. 2015).

2.5 Standardized Precipitation Evapotranspiration Index (SPEI)

The SPEI was developed by Vicente-Serrano et al. (2010) to improve the SPI, by integrating temperature in its calculation, and is therefore more appropriate for arid regions. It is based on the monthly difference of precipitation and reference evapotranspiration, i.e. a simple water balance to estimate surplus or deficit:

$$\mathrm{D} = \mathrm{P} - \mathrm{ET}_0 \tag{1}$$

Originally, ET_0 was obtained by the Thornthwaite approach, but Beguería et al. (2014) recommend the Penman-Monteith equation in a revised version of the SPEI. Alternatively, they recommend the Hargreaves equation (Allen et al. 2004). For this study, we estimated ET_0 (mm/month) using the modified Hargreaves equation after Droogers and Allen (2002):

$$ET_0 = 0.0013 \cdot 0.408\,RA \cdot \left(T_{avg} + 17.0\right) \cdot (TD - 0.0123Pp)^{0.76} \tag{2}$$

where RA (MJ m^{-2} d^{-1}) corresponds to extraterrestrial radiation, T_{avg} (°C) is the average temperature obtained from T_{max} (°C) and T_{min} (°C), TD (°C) is the temperature range (difference of T_{max} and T_{min}), and Pp (mm) is precipitation.

Table 2 Mean, minimum, and maximum values for all variables based on monthly values for available data

Variable	Jan	Feb	Mrz	Apr	Mai	Jun	Jul	Aug	Sep	Okt	Nov	Dez
Precipitation (mm)												
Mean	0.26	1.05	2.91	18.90	57.22	113.99	69.15	62.53	28.06	9.84	4.85	2.81
Max	3.80	13.50	19.00	124.90	212.70	444.70	290.40	156.60	158.40	57.40	50.40	30.50
Min	0.00	0.00	0.00	0.00	0.00	1.50	2.30	0.00	0.00	0.00	0.00	0.00
T_{min} (°C)												
Mean	10.50	9.84	8.29	5.78	3.97	3.17	2.35	3.60	5.12	6.52	7.96	9.63
Max	12.80	11.30	9.76	8.60	6.80	6.90	7.29	7.75	7.05	8.50	10.00	11.20
Min	9.23	8.00	6.10	2.10	0.30	–0.80	–0.10	0.30	2.58	5.10	5.70	7.80
T_{max} (°C)												
Mean	30.72	30.28	28.53	23.77	19.22	16.23	15.56	17.08	19.77	23.39	26.73	29.09
Max	31.80	31.70	31.80	26.90	22.00	19.90	18.38	19.20	23.20	27.10	28.50	30.70
Min	29.40	28.30	26.30	19.70	14.38	13.90	13.10	15.40	17.10	19.80	23.90	26.11
Streamflow (m^3/s)												
Mean	204.40	149.61	104.29	74.39	62.16	60.74	61.31	61.63	69.93	99.73	155.09	209.97
Max	592.84	455.14	261.10	160.67	117.20	106.18	142.69	130.34	144.13	155.46	256.70	539.45
Min	66.66	62.33	51.41	33.36	26.53	24.27	24.59	23.17	27.49	33.82	50.58	43.04
Soil moisture (m^3/m^3)												
Mean	0.17	0.18	0.18	0.20	0.25	0.31	0.30	0.29	0.25	0.21	0.19	0.18
Max	0.22	0.24	0.24	0.26	0.35	0.37	0.34	0.35	0.33	0.31	0.26	0.23
Min	0.15	0.15	0.15	0.15	0.20	0.23	0.24	0.20	0.18	0.17	0.16	015

Missing temperature values of (15%), were filled by correlation with the closest neighbor, given the high correlation of $R^2 = 0.98$. The differences between precipitation and ET_0 were then fit to best fit probability density functions, which are listed in Table 3. Additionally, SPEI values were calculated by applying the recommended Generalized Extreme Value (GEV) (Stagge et al. 2015, 2016) distribution for all months and accumulation periods.

2.6 Standardized Soil Moisture Index (SSMI)

Soil moisture is an important drought indicator. Since field measurements are sporadic, satellite data was used as a proxy. We used data from the ESA CCI SM v03.2 combined dataset and used the period of 1995–2015 for our study as it was nearly complete. The dataset can be freely downloaded at http://www.esa-soilmoisture-cci.org/. The 8% of missing data were filled by linear interpolation. Using soil moisture data to obtain standardized index values is relatively new, hence evaluations of recommended distributions are still missing, even though similar approaches can be found in Hao and AghaKouchak (2013) and Carrão et al. (2016). Our results show that if one distribution is applied to all months, the logistic distribution presented good results for all accumulation periods. Besides, the best fit distributions for each month found are provided in Table 3.

2.7 Standardized Streamflow Index (SSI)

The Standardized Streamflow Index is used for hydrological drought evaluation. Shukla and Wood (2008), and Nalbantis (2008), Nalbantis and Tsakiris (2009) presented independently streamflow indices, based on the SPI procedure—the same applies to this study. The chosen station

Table 3 Best fit distributions for all indices and accumulation periods

Index		Jan	Feb	Mrz	Apr	May	Jun	Jul	Aug	Sep	Oct	Nov	Dec
SPI	1	Gamma	Pareto2	Gamma	Wakeby	Wakeby	Wakeby	Dagum	GenPareto	GenExtreme	GenExtreme	Logistic	Normal
	3	Wakeby	Wakeby	Wakeby	GenLogistic	Exp	Wakeby	JohnsonSB	Wakeby	Gamma	Wakeby	GenExtreme	Wakeby
	6	Wakeby	GenExtreme	Wakeby	Wakeby	Weibull	LogLogistic	JohnsonSB	Wakeby	GenExtreme	Gamma	Wakeby	JohnsonSB
	12	GenGamma	Wakeby	Wakeby	Burr	GenExtreme	GumbelMax	JohnsonSB	Wakeby	JohnsonSB	Gamma	JohnsonSB	Gamma
SPEI	1	Wakeby	GumbelMax	Wakeby	Lognormal	Gamma	Wakeby	Lognormal	Pert	Dagum	GenLogistic	Wakeby	Wakeby
	3	GenLogistic	Wakeby	Laplace	Wakeby	JohnsonSB	LogLogistic	FatigueLife	Erlang	GenExtreme	GenExtreme	GenExtreme	LogLogistic
	6	Normal	GenLogistic	Wakeby	Cauchy	Rayleigh	LogLogistic	GenExtreme	JohnsonSB	JohnsonSB	JohnsonSB	GenGamma	JohnsonSB
	12	Wakeby	Wakeby	Wakeby	GenExtreme	Wakeby	GenExtreme	LogLogistic	PhasedBiExp	JohnsonSB	JohnsonSB	JohnsonSB	Wakeby
SSMI	1	Wakeby	Burr	Gamma	FatigueLife	GumbelMax	Wakeby	JohnsonSB	Hypersecant	Wakeby	Lognormal	Burr	GenLogistic
	3	Frechet	LogLogistic	LogLogistic	Burr	Wakeby	Kumaraswamy	Cauchy	Wakeby	JohnsonSB	Error	JohnsonSB	LogLogistic
	6	Wakeby	GenLogistic	Burr	GenExtreme	Cauchy	Wakeby	Kumaraswamy	GenExtreme	Wakeby	Dagum	Wakeby	Error
	12	GenLogistic	Cauchy	Wakeby	Weibull	Frechet	Wakeby	Cauchy	Cauchy	Cauchy	Wakeby	Wakeby	Wakeby
SSI	1	Lognormal	Lognormal	LogPearson3	GenExtreme	GenExtreme	LogLogistic	Dagum	GenExtreme	Burr	Wakeby	Wakeby	Wakeby
	3	LogPearson3	InvGaussian	InvGaussian	Lognormal	LogGamma	Burr	Wakeby	Pearson5	LogPearson3	JohnsonSB	JohnsonSB	JohnsonSB
	6	JohnsonSB	JohnsonSB	JohnsonSB	JohnsonSB	FatigueLife	Frechet	Wakeby	GenLogistic	LogLogistic	InvGaussian	JohnsonSB	JohnsonSB
	12	GenGamma	Rayleigh	Weibull	Weibull	Gamma	GenGamma	Gamma	GenLogistic	GenGamma	Gamma	GenLogistic	Weibull

is in the upper part of the basin to avoid human-influenced streamflow values, representing a natural flow regime (1960–2015). Missing monthly values (5%) were filled by linear interpolation. For this study we followed the recommendations given by Vicente-Serrano et al. (2012a) and Vicente-Serrano et al. (2012b), hence, we applied best fit for each month (see Table 3) and calculated the SSI as well with GEV and Log-Logistic distributions, as both recommended distributions led to similar results.

2.8 Drought Propagation

Identifying the start and end of drought events is needed for effective and appropriate drought management. However, this is a challenging task as meteorological and hydrological processes are interwoven due to human interventions in a river basin (van Lanen et al. 2008; Lorenzo-Lacruz et al. 2013). To analyze drought propagation in the Maipo river basin, understood as the temporal interdependence through the hydrological cycle, we apply a two-step approach based on the previously presented standardized indices.

Like other studies (Barker et al. 2016) we used the Pearson correlation coefficient as indicator among the four indices with time lags up to 12 months. We correlated values of the SPI with SPEI, SPEI with SSMI, SPEI with SSI, and SSMI with SSI for all accumulation periods (1-, 3-, 6-, and 12-months) based on values obtained by best fit for all indices. Further we compared the duration and magnitude for the different accumulation periods as well as the starting and end of detected events. The set thresholds for our analysis are −0.5, −0.84, and −1 for the SPI and SPEI. We consider a drought event if SPI or SPEI values were below the set threshold for at least three consecutive months. A threshold of −0.5 is low and generally considered to be near normal (McKee et al. 1993; Beyene et al. 2014), however we have chosen this value to see if and how detected events with such a low threshold are detected by soil moisture and streamflow. The value of −0.84 represents the set threshold used for drought detection by the Chilean Water Agency (Dirección General de Aguas (DGA)), while the threshold of −1 is frequently used to determine severely dry periods (Guttman 1998). The choice of thresholds needs to be site specific and must be tested with the data available and previously occurred impacts (Steinemann and Cavalcanti 2006; Tsakiris et al. 2007; Smakhtin and Schipper 2008). The reader is referred to studies of (Wilhite 1993; Tsakiris and Vangelis 2005; Steinemann et al. 2015) for more details on threshold determination.

2.9 Multivariate Autoregressive Model for Standardized Indices

One important feature of the selected indices is that they can be treated as time series of interdependent variables (cross-correlated and auto correlated variables). Synthesizing their information into one single statistical model gives us an opportunity to produce simple forecasts that can be used for drought modelling and detection. Autoregressive models (AR models), are useful to represent time series values based on previous values of same time series. Such models allow the simulation of time series based on random numbers (Wilks 2006). For our study we have chosen to develop an AR-Model including three variables, hence a multivariate AR-Model. This multivariate autoregressive model (MAR-Model) is based on results obtained by the SPEI, SSMI and SSI. We selected the SPEI over the SPI as input variable, as both are highly correlated at all accumulation periods ($R^2 > 0.8$) and because the SPEI is preferable in semi-arid regions to detect droughts at an early stage (meteorological drought) (Vicente-Serrano et al. 2014). Thus, our model includes meteorological (SPEI), agricultural (SSMI), and hydrological (SSI) characteristics to simulate new time series.

MAR models are used to study the linear relationship between variables, where each value is predicted from the same time series values (autocorrelated) as well as of the other time series values (cross-correlated) (Hytti et al. 2006). The multivariate AR-Model of order 1 in matrix notation is defined by Eq. (3) after (Wilks 2006).

$$x_{t+1} - \mu = [\Phi] \cdot (x_t - \mu) + [B]\epsilon_{t+1} \tag{3}$$

where x describes the vector of our input variables (SPEI, SSMI, SSI), and x_{t+1} refers to the simulated value based on previous values of the time series and an added error. Vector μ contains the mean values of the vector elements x. For our model development we can neglect μ, because our input variables are already normally distributed with a mean of 0 and a standard deviation of 1. This simplifies the equation for our case to:

$$x_{t+1} = [\Phi] \cdot (x_t) + [B]\epsilon_{t+1}. \tag{4}$$

Following the terminology used by Wilks (2006), the term $[B]\epsilon_{t+1}$ includes random variables (standard Gaussian) with unit variance ϵ and the matrix $[B]$ (operating on vector ϵ_{t+1}) allows ϵ to have different variances, but to be mutually correlated for each time step. Both matrices $[\Phi]$ and $[B]$ allow to simulate data based on the input time series. To obtain the coefficients of these matrices, concurrent and lag-1 correlations of elements of x are calculated. The matrix $[\Phi]$ represents the autoregressive parameters. We obtain the elements of matrix $[\Phi]$ by multiplying two correlation matrices, $[R_1]$ and $[R_0]^{-1}$. $[R_0]$ is a symmetric matrix of simultaneous correlations of the input time series. In our case the elements are:

$$[R_0] = \begin{bmatrix} 1 & cor(SPEI, SSMI) & cor(SPEI, SSI) \\ cor(SPEI, SSMI) & 1 & cor(SSMI, SSI) \\ cor(SPEI, SSI) & cor(SSMI, SSI) & 1 \end{bmatrix} \tag{5}$$

With cor(X,Y) representing the Person correlation between variable X and Y. The matrix $[R_1]$ is not symmetric, the off-diagonal elements of the matrix are the lag-1 correlations among the variables, and the diagonal contains lag-1 autocorrelations. For our case $[R_1]$ is obtained with:

$$[R_1] = \begin{bmatrix} cor(SPEI, SPEI_{t+1}) & cor(SPEI, SSMI_{t+1}) & cor(SPEI, SSI_{t+1}) \\ cor(SSMI, SPEI_{t+1}) & cor(SSMI, SSMI_{t+1}) & cor(SSMI, SSI_{t+1}) \\ cor(SSI, SPEI_{t+1}) & cor(SSI, SSMI_{t+1}) & cor(SSI, SSI_{t+1}) \end{bmatrix} \tag{6}$$

To obtain the elements of matrix $[B]$ it is necessary to find the matrix:

$$[B][B]^T = [R_0] - [\Phi] \cdot [R_1]^T \tag{7}$$

One solution for this can be found applying Cholesky decomposition which yields:

$$[B][B]^T = \begin{bmatrix} B_{11} & 0 & 0 \\ B_{21} & B_{22} & 0 \\ B_{31} & B_{32} & B_{33} \end{bmatrix} \begin{bmatrix} B_{11} & B_{21} & B_{31} \\ 0 & B_{22} & B_{32} \\ 0 & 0 & B_{33} \end{bmatrix}, \tag{8}$$

In this way the $[B]$ is obtained and the vector x_{t+1} can be calculated, representing $SPEI_{t+1}$, $SSMI_{t+1}$, and SSI_{t+1}.

3 Results

3.1 Distribution Fitting

We apply two approaches for each accumulation period: (1) Selecting the best fit probability distribution for each month, and (2) applying one distribution to monthly data following recommendations from the literature. Table 3 lists all best fit probability distributions for each index and accumulation period, reflecting the variety of suitable distributions, as we use 33 different distributions. The selection of distributions is carried out using the Kolmogorov-Smirnov and Anderson-Darling tests. The most difficult fitting appears for monthly precipitation values, as zero precipitation is a normal pattern in the river basin. For values of e.g. March, July, and August JohnsonSB and Wakeby distributions show best fit for three accumulation periods. Other months

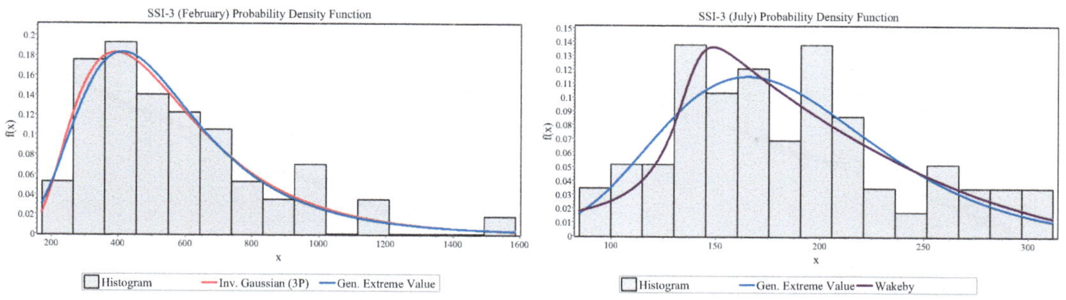

Fig. 2 Examples of monthly fitting for 3-months SSI for February (left) and July (right), graphical comparison of best fit (Inv. Gaussian, Wakeby) and GEV, histogram represents accumulated streamflow values

are represented by three or four different distributions. This is also the case for SPEI, SSMI, and SSI, as there is no clear preference of one distribution. Nonetheless several distributions fit to data show marginally differences in the final index values. Figure 2 shows the example for the SSI 3-months for February and July, comparing graphically the best fit and Generalized Extreme Values (GEV) probability density functions.

Recommended distributions from the literature are: gamma for SPI (Guttmann 1999), GEV for SPEI (Stagge et al. 2015), and SSI (Vicente-Serrano et al. 2012a). For the case of the standardized soil moisture index there is no especially recommended distribution. However, for our data set the Log-Logistic performed well for all months and was used for comparisons with results obtained by best fit distributions.

We compared results for each index and accumulation period obtained with best fit and recommended distributions, and both performed nearly identical with $R^2 > 0.9$. Figure 3 provides examples of the X-Y plots for index values of SPI-1, SPEI-3, SSMI-6, and SSI-12. The higher the accumulation period, the smaller the difference in results between best fit and recommended distributions. Therefore, we if indices are calculated for 1-month period, the best fitting distribution is commended, otherwise, the distributions proposed in the literature work fine. Usually lower accumulation periods of one to three months are used to detect meteorological (precipitation deficit) and agricultural drought events (soil moisture deficit), as they can persist for short periods (weeks). Accumulation periods of six months and more are generally used to describe interlinkages between variables and to detect hydrological or groundwater drought events. Figure 4 displays the SPI 12-months values for the best fit and the gamma distribution; additionally accumulated precipitation is included to exemplify the transformation into an index value.

To demonstrate the influence of a chosen distribution on final index results we selected the period from 1997–2000 for all indices, as this time frame corresponds to the drought event of 1998/99—with one year before and after—to evaluate the influence of different probability distributions on its detection. Figure 5 presents the different graphs associated with drought indices and accumulation periods.

3.2 Drought Propagation

Indices were calculated for the time span of available data for each variable. As soil moisture data is used from 1995 (due to less missing values), we used the 1995–2016 period to analyze drought propagation patterns. Graphical results for visual comparison of the indices are provided in Fig. 6. Given the high correlation of SPI and SPEI without lag ($R > 0.8$, $p < 0.001$), we used SPEI values for correlations among results of SSMI and SSI. Results of Pearson correlations are presented graphically in Fig. 7. The graphs show that with longer accumulation periods correlation values increase. This finding can be used to identify the suitable accumulation period for drought analysis.

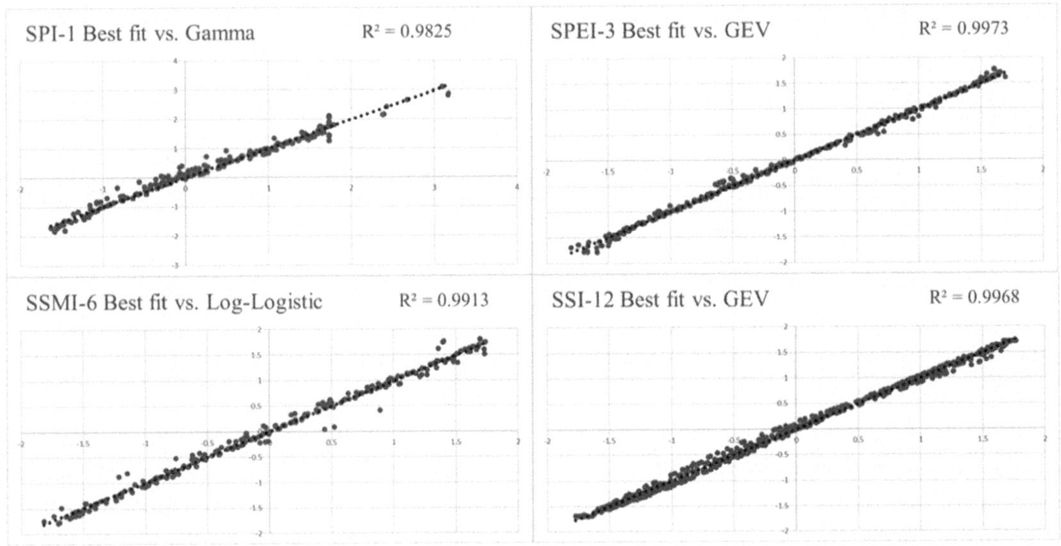

Fig. 3 X-Y plots of index results for best fit and recommended distribution: SPI 1-month, SPEI 3-months, SSI 6-months, and SSMI 12-months

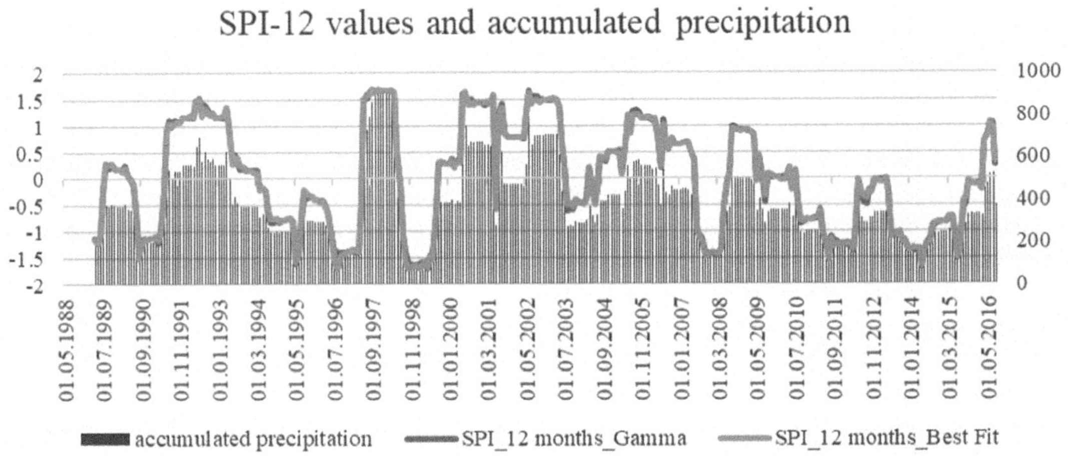

Fig. 4 Results for the SPI 12-months accumulation period, lines represent different distribution fittings and bulks are used to show precipitation

For correlations between SPEI and SSMI highest values were found with 0-lag, hence SPEI and SSMI behave simultaneous as expected. Correlation values between SSMI and SSI are low, highest correlation values are $\sim$0.3 and occur with lag-3 until lag-5. Correlation values remain with in this range for at least five months. Results of correlations between SPEI and SSI increase with accumulation periods and peak values are found with a minimum of lag-3 (1-, and 3-months accumulation period). This result validates the relationship between precipitation and streamflow in the Maipo river basin and prove that relying on one index is not sufficient for operational drought management. Table 4 summarizes highest Pearson correlations found between the indices.

Another approach to detect drought propagation was carried out by setting three different SPI and SPEI thresholds (-0.5, -0.84, and -1) to identify drought events when the index values are below these thresholds at least for three

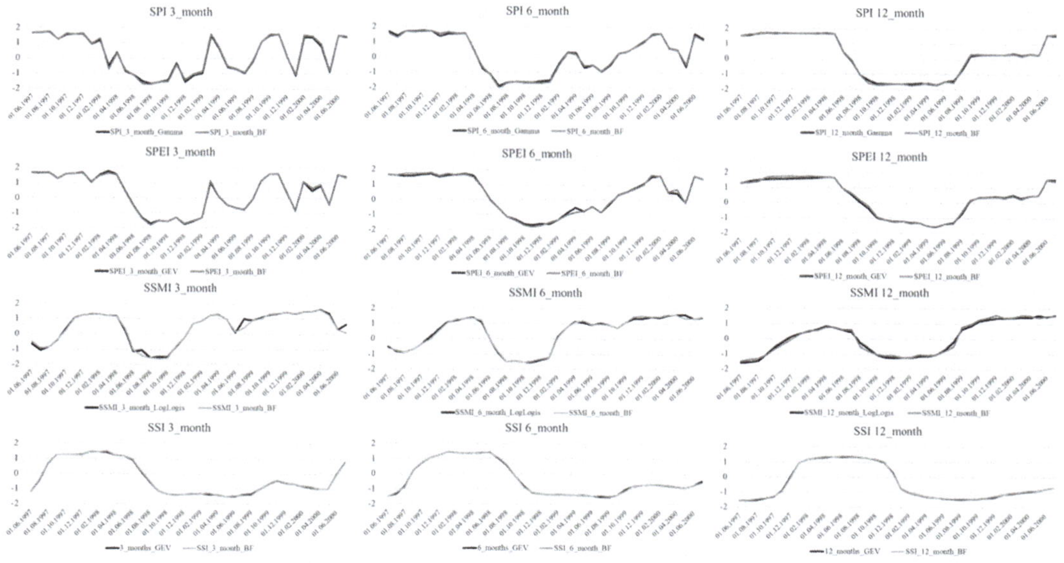

Fig. 5 SPI, SPEI, SSMI, and SSI for 3-, 6-, and 12 months accumulation periods between 1997–2000 with best fit and recommended distributions

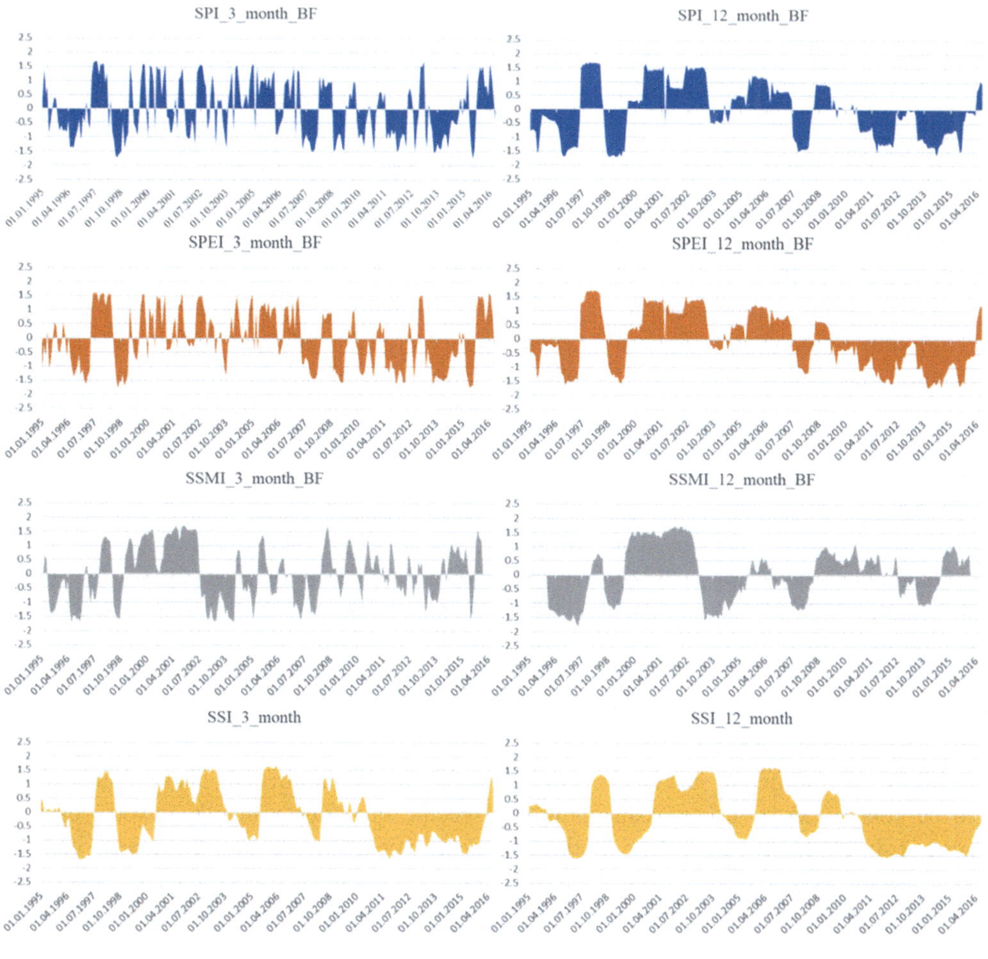

Fig. 6 SPI, SPEI, SSMI, and SSI for 3- and 12-months from Jan. 1995–July 2016

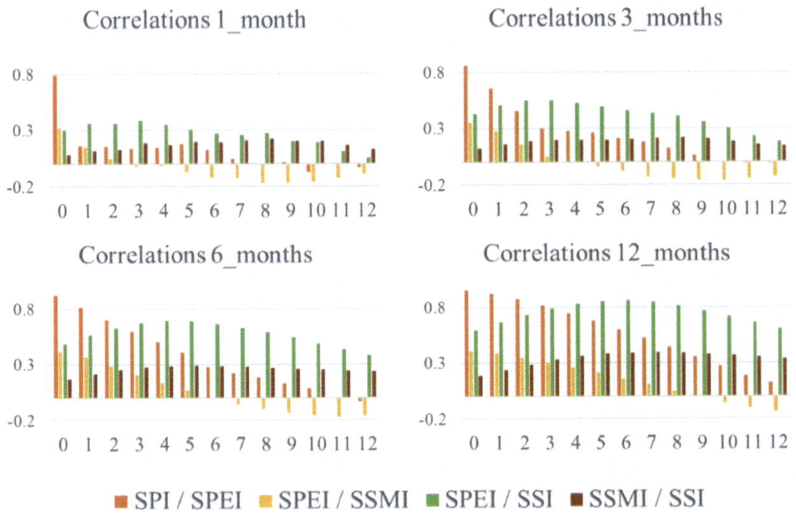

Fig. 7 Correlation results among indices for 1-, 3-, 6-, and 12-months accumulation period with time lags up to 12 months

Table 4 Highest pearson correlations between indices with information on months lagged

Accumulation period	1 month	3 months	6 months	12 months
SPEI/ SSMI	0.33 no lag	0.36 no lag	0.42 no lag	0.41 no lag
SPEI/ SSI	0.38 lag 3 months	0.54 lag 3 months	0.69 lag 4 months	0.86 lag 6 months
SSMI/ SSI	0.22 lag 8 months	0.20 lag 8 months	0.29 lag 5 months	0.40 lag 7 months

months. Detected events were then checked with SSMI and SSI values. Figure 8 represents exemplary detected events by SPEI-3 with SSI-3 values. As expected, the lower the threshold the more detected events, furthermore it is shown that relying only on one index for drought event determination can mislead to capture the whole event.

Additionally, comparing detected events by different thresholds provides the argument for using "warning thresholds", understood as thresholds lower than the ones used for institutional drought detection. For instance, the SSI drought event of 2004/05 was not detected with applied thresholds by SPI or SPEI, as their values were below −1 for (only) two months at the end of 2003. As Fig. 8 demonstrates the relationship between precipitation and streamflow in the Maipo Basin, also different recovery periods need to be considered for a successful drought management. In all cases, streamflow continued for several months below average even when SPI or SPEI values were positive again.

3.3 Multivariate Autoregressive Model (MAR1)

Representing the time series of drought indices with a multivariate autoregressive model of first order creates a unique opportunity to develop simple operational forecasts of the desired variables. A MAR(1) model captures the time dependence among the variables in a way that the future vector values are conditioned on the previous observed. Figure 9 illustrates the evolution of the expected value (mean value) of each index when the latest reported values for 3-months accumulation period are: SPEI = −1.46, SSMI = −0.302, and SSI = −1.64. Since each variable is a Gaussian value with a conditional mean given by the product of the elements of the matrix $[\Phi]$ (values presented in Table 5) and the recent observations, it is also possible to derive specific quantiles using the inverse cumulative distribution function and the proper variance of the indices obtained from the elements of matrix $[B]$ in the model. As expected, the mean values

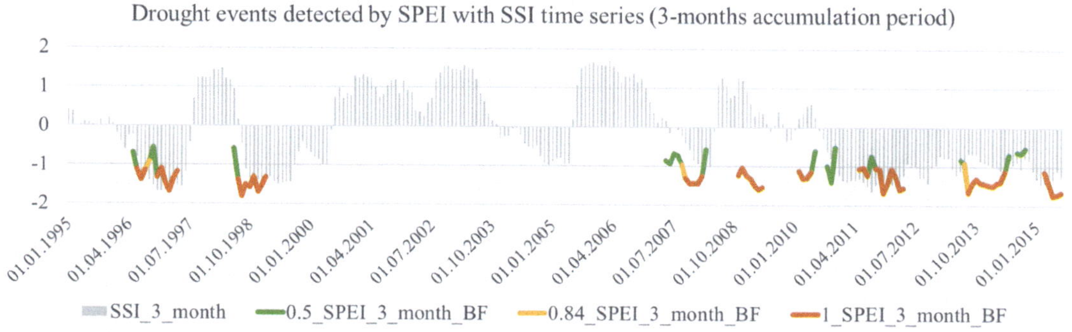

Fig. 8 Drought event detection by SPEI with at least three consecutive months below thresholds of −0.5, −0.84, and −1 and in grey bulks SSI values for 3-months accumulation period

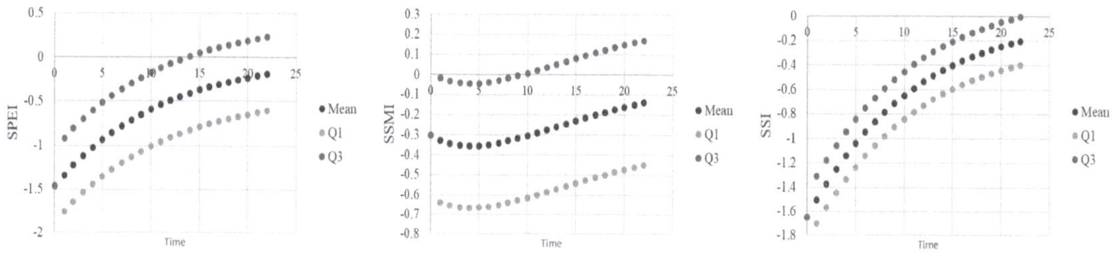

Fig. 9 Evolution of the Mean value and Quartiles (Q1 and Q3) for the drought indices conditioned on previous observations (see text for more details)

Table 5 Parameters for matrices Phi and B for the multivariate autoregressive model

Parameters	3-months	Parameters	3-months
Φ1,1	0.653875033	B1,1	0.619132107
Φ1,2	0.017471646	B2,1	0.167515529
Φ1,3	0.227396259	B2,2	0.460794874
Φ2,1	−0.032297837	B3,1	−0.012267512
Φ2,2	0.872736933	B3,2	−0.007813371
Φ2,3	0.067070836	B3,3	0.288809564
Φ3,1	−0.090281009		
Φ3,2	−0.003437815		
Φ3,3	0.992973291		

slowly approach the values of zero as they become time independent of the latest observation.

Another interesting use of the MAR(1) model for drought indices it the assessment of joint probabilities as a function of different observed values. For instance, one may be interested in evaluating the risk of a severe meteorological, agricultural, and hydrological drought given a set of previously observed initial conditions. To do so, we can obtain the marginal probability density function of a multivariate index and solve analytically the probability of jointly observed specific values of the indices. An alternative, yet

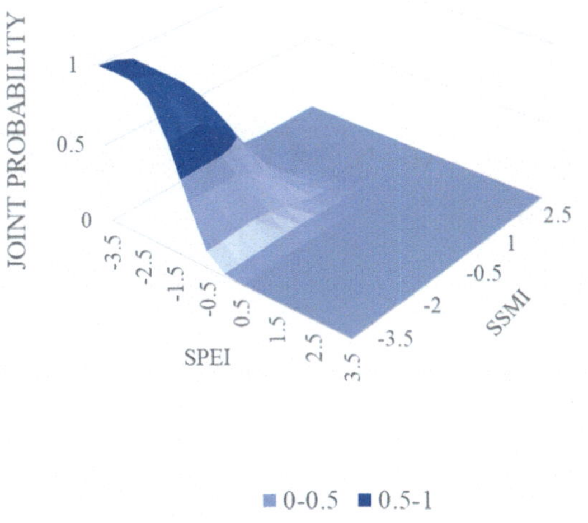

Fig. 10 Estimation of the joint probability of observing future values of SPEI and SSMI equal or smaller than −1 as a function of the latest observed values

very effective approach, is to simulate a sufficiently large number of possible outcomes and empirically obtain values of a specific outcome (Montecarlo model). Figure 10 shows an example of this specific use. For simplicity and to facilitate graphical illustration, we have set the SSI value to −1 and assessed the probability of observing values equal or smaller to −1 in the SPEI and SSMI as a joint distribution by generating 10000 simulations for different initial values of SPEI and SSMI. Given the persistence of the observations, greater probabilities are found when SPEI and SSMI values are smaller than −2.

4 Discussion

4.1 Distribution Fitting

The importance of distribution fitting depends on the accumulation period used. If, for example, a 1-month calculation is carried out, we recommend fitting each month separately. Especially for semi-arid regions where precipitation rarely occurs. Using one distribution with monthly specific parameters can result in unrealistic values, as monthly input values are not supporting a given distribution. Results of our study are in accordance with recommended distributions in literature to obtain reliable values. Besides, the influence of chosen distributions diminishes with higher accumulation periods. Hence, the higher the accumulation period, the lesser the influence of distributions on index results. Following the discussion about recommended distribution for the SPEI, we compared obtained index values for each month to evaluate the performance of GEV (Stagge et al. 2015, 2016) and Log-Logistic (Beguería et al. 2014; Vicente-Serrano and Beguería 2016). For our objective, to identify drought events with standardized indices, both probability distributions lead to nearly the same SPEI values, the minor difference has no influence on the detection, and maximum differences found in magnitude are ~0.1. According to our results we can say that both distributions are providing almost identical results, hence both perform the same for drought event detection.

4.2 Drought Propagation

Setting an appropriate threshold to identify drought events depends on the distribution used and properties of the basin, hence on their propagation through the hydrological cycle. Choosing a very low threshold can lead to an over-alerting management, otherwise it is necessary to follow the development of below-average values as they might indicate an upcoming event. We therefore propose to work for operational drought management with 'warning thresholds', understood as values close to the commonly applied threshold. For instance, if the threshold is set to be −1 for detection, it may be useful to set warning thresholds between −0.9 and −1, for not missing the starting point. Another measure could be to use such warning thresholds for checking the behavior of other variables, e.g. SPI values below −0.9 for two months and to check with the development for streamflow values. This reinforces the importance of site specific thresholds; depending on the input data and suitable transformation procedures.

Results of the propagation analysis show that it can be misleading to rely only on the SPI or

SPEI to identify starting or end points of a drought event. To capture drought events, understood as inclusive events of the hydrological cycle, it is not sufficient to rely on one index for drought detection. This is especially important for operational drought management, and to amplify and evaluate applied practices.

To identify propagation patterns with standardized indices is the aim of this study, by using as less modified input data as possible to ensure replicability, especially where long records of input data are rare. We have chosen a streamflow station in the upper part of the river basin, as those values are less human influenced. Precipitation and soil moisture data are taken from the lower part of the river basin, where main activities take place, but still, as we have chosen a non-irrigated area, can be regarded as less-human influenced. The selected soil moisture point, and gauge station were chosen following criteria of maximum proximity, as soil moisture responds quickly to precipitation. Besides, there are no long-term records for soil moisture in the upper parts of the river basin available. Like this we can identify drought characteristics by using hydro-meteorological data.

4.3 Multivariate Autoregressive Model

Using models for operational forecast is common practice, yet suitable methods for drought forecasts are still missing. We present a multivariate autoregressive model based on three standardized indices. Such a model permits simple operational forecasts for each index, and it can be used to obtain joint probabilities of indices, which are very useful for drought risk management. However, we are aware that it can be challenging to obtain sufficient simultaneous input data; if this is not given, cross—and autocorrelations could not be consistent leading to infeasible solutions (i.e. regular matrices that are nonsingular) of the model.

The application of multivariate models for drought characterization is not necessarily bounded MAR models of first order. One could be interested in exploring higher order autoregressive models to improve forecasting capabilities. For our study case, this was not implemented due to the high persistence of the individual index values.

5 Conclusions

Referring to our opening questions we conclude that (1) we expected more differences in results if different distributions are used. We can state that the higher the accumulation period the lesser the influence of the chosen distribution. This statement must be taken with caution as other input data may lead to different results. For all applied distributions same drought events were detected, and no significant differences in the start/ end points or different magnitudes were found. (2) Our approach is useful to detect drought propagation patterns, confirming the use of standardized indices as powerful tools due to their comparability for different input variables and the simplicity of the calculations. (3) To use standardized indices for operational forecast is suitable as index values are Gaussian values and hence allow straightforward modeling. In this study we use a multivariate autoregressive model, which suits simple forecasts.

Acknowledgements The research of Melanie Oertel is funded by Becas Conicyt, Programa de Formación de Capital Humano Avanzado, Ministerio de Educación de Chile. Further this work was carried out with the aid of a grant from the Inter-American Institute for Global Change Research (IAI) CRN3056 which is supported by the US National Science Foundation (Grant GEO-1128040).

References

Allen RG, Pereira LS, Raes D, Smith M (2004) Crop evapotranspiration: guidelines for computing crop water requirements. FAO irrigation and drainage paper vol 56. FAO, Rome

Barker LJ, Hannaford J, Chiverton A, Svensson C (2016) From meteorological to hydrological drought using standardised indicators. Hydrol Earth Syst Sci 20:2483–2505. https://doi.org/10.5194/hess-20-2483-2016

Beguería S, Vicente-Serrano SM, Reig F, Latorre B (2014) Standardized precipitation evapotranspiration index (SPEI) revisited: parameter fitting, evapotranspiration models, tools, datasets and drought monitoring. Int J Climatol 34:3001–3023. https://doi.org/10.1002/joc.3887

Beyene BS, Van Loon AF, van Lanen HAJ, Torfs PJJF (2014) Investigation of variable threshold level approaches for hydrological drought identification. Hydrol Earth Syst Sci Discuss 11:12765–12797. https://doi.org/10.5194/hessd-11-12765-2014

Bonelli S, Vicuña S, Meza FJ, Gironás J, Barton J (2014) Incorporating climate change adaptation strategies in urban water supply planning: the case of central Chile. J Water Climate Change 5:357. https://doi.org/10.2166/wcc.2014.037

Botterill LC, Cockfield G (2013) Drought, risk management, and policy: Decision making under uncertainty. Drought and water crises vol 2. CRC Press, Boca Raton

Bustos E, Meza FJ (2015) A method to estimate maximum and minimum air temperature using MODIS surface temperature and vegetation data: application to the Maipo Basin, Chile. Theor Appl Climatol 120:211–226. https://doi.org/10.1007/s00704-014-1167-2

Carrão H, Russo S, Sepulcre-Canto G, Barbosa P (2016) An empirical standardized soil moisture index for agricultural drought assessment from remotely sensed data. Int J Appl Earth Observation Geoinf 48:74–84. https://doi.org/10.1016/j.jag.2015.06.011

Dorigo W, Wagner W, Albergel C, Albrecht F, Balsamo G, Brocca L, Chung D, Ertl M, Forkel M, Gruber A, Haas E, Hamer PD, Hirschi M, Ikonen J, de Jeu R, Kidd R, Lahoz W, Liu YY, Miralles D, Mistelbauer T, Nicolai-Shaw N, Parinussa R, Pratola C, Reimer C, van der Schalie R, Seneviratne SI, Smolander T, Lecomte P (2017) ESA CCI soil moisture for improved earth system understanding: state-of-the art and future directions. Remote Sens Environ. https://doi.org/10.1016/j.rse.2017.07.001

Droogers P, Allen RG (2002) Estimating reference evapotranspiration under inaccurate data conditions. Irrigation Drain Syst 33–45

Garreaud R, Alvarez-Garreton C, Barichivich J, Boisier JP, Christie D, Galleguillos M, LeQuesne C, McPhee J, Zambrano-Bigiarini M (2017) The 2010-2015 mega drought in Central Chile: Impacts on regional hydroclimate and vegetation. Hydrol Earth Syst Sci Discuss 1–37. https://doi.org/10.5194/hess-2017-191

Gruber A, Dorigo WA, Crow W, Wagner W (2017) Triple collocation-based merging of satellite soil moisture retrievals. IEEE Trans Geosci Remote Sens 55:6780–6792. https://doi.org/10.1109/TGRS.2017.2734070

Gudmundsson L, Stagge JH (2016) SCI: Standardized Climate Indices such as SPI, SRI or SPEI R-package version 1.0–2

Guttman NB (1998) Comparing the palmer drought index and the standardized precipitation index. J Am Water Resour Assoc 34:113–121

Guttmann N (1999) Accepting the standardized precipitation index: a calculation algorithm. J Am Water Resour Assoc 35:311–322

Hao Z, AghaKouchak A (2013) Multivariate standardized drought index: a parametric multi-index model. Adv Water Resour 57:12–18. https://doi.org/10.1016/j.advwatres.2013.03.009

Hao Z, Singh VP (2015) Drought characterization from a multivariate perspective: a review. J Hydrol 527:668–678. https://doi.org/10.1016/j.jhydrol.2015.05.031

Hayes MJ (2007) Drought Indices: Feature Article From Intermountain West Climate Summary, July 2007. This article originally appeared in a longer form on the National Drought Mitigation Center (NDMC) webpage, http://drought.unl.edu/index.htm. The NDMC works to minimize impacts and vulnerabilities of drought by providing risk management strategies and recommendations. The Intermountain West Climate Summary 3–6

Heim RR (2002) A review of twentieth-century drought indices used in the United States. Bull Am Meteorol Soc 1149–1165

Hunt ED, Hubbard KG, Wilhite DA, Arkebauer TJ, Dutcher AL (2009) The development and evaluation of a soil moisture index. Int J Climatol 29:747–759. https://doi.org/10.1002/joc.1749

Hytti H, Takalo R, Ihalainen H (2006) Tutorial on multivariate autoregressive modelling. J Clin Monit Comput 20:101–108. https://doi.org/10.1007/s10877-006-9013-4

Kallis G (2008) Droughts. In: Reviews Annual (ed) Annual review of environment and resources, vol 33. Palo Alto, CA, pp 85–118

Kao S-C, Govindaraju RS (2010) A copula-based joint deficit index for droughts. J Hydrol 380:121–134. https://doi.org/10.1016/j.jhydrol.2009.10.029

Keyantash J, Dracup JA (2002) The quantification of drought: an evaluation of drought indices. Bull Am Meteorol Soc 1167–1180

Keyantash JA, Dracup JA (2004) An aggregate drought index: assessing drought severity based on fluctuations in the hydrologic cycle and surface water storage. Water Resour Res 40:n/a. https://doi.org/10.1029/2003wr002610

Kindler J, Okruszko T (2014) Integrated drought management programme in central and Eastern Europe (IDMP CEE). Ann Wars Univ Life Sci Land Reclam 46. doi:10.2478/sggw-2014-0014

Latin American Flood and Drought Monitor. http://stream.princeton.edu/LAFDM/WEBPAGE/interface.php?locale=en. Accessed 30 Nov 2017

Liu YY, Dorigo WA, Parinussa RM, de Jeu RAM, Wagner W, McCabe MF, Evans JP, van Dijk AIJM (2012) Trend-preserving blending of passive and active microwave soil moisture retrievals. Remote Sens Environ 123:280–297. https://doi.org/10.1016/j.rse.2012.03.014

Lorenzo-Lacruz J, Vicente-Serrano SM, Gonzalez-Hidalgo JC, Lopez-Moreno JI, Cortesi N (2013) Hydrological drought response to meteorological drought in the Iberian Peninsula. Clim Res 58:117–131. https://doi.org/10.3354/cr01177

Maliva R, Missimer T (2012) Aridity and Drought. In: Maliva R, Missimer TM (eds) Arid lands water evaluation and management. Springer, Berlin, New York, pp 21–39

McKee T, Doesken N, Kleist J (1993) The relationship of drought frequency and duration to time scales. In: American Meteorological Society (Ed.) 1993–Proceedings of the Eight Conference, pp 179–184

Meza F, Wilks D, Gurovich L, and Bambach N (2012) Impacts of climate change on irrigated agriculture in the Maipo Basin, Chile: reliability of water rights and changes in the demand for irrigation. J Water Resour Plann Manage 421–430. https://doi.org/10.1061/(asce)wr.1943-5452

Meza FJ, Vicuña S, Jelinek M, Bustos E, Bonelli S (2014) Assessing water demands and coverage sensitivity to climate change in the urban and rural sectors in central Chile. J Water Clim Change 5:192. https://doi.org/10.2166/wcc.2014.019

Mishra AK, Singh VP (2010) A review of drought concepts. J Hydrol 391:202–216. https://doi.org/10.1016/j.jhydrol.2010.07.012

Nalbantis I (2008) Evaluation of a hydrological drought index. European water 23/24:67–77

Nalbantis I, Tsakiris G (2009) Assessment of hydrological drought revisited. Water Resour Manage 23:881–897. https://doi.org/10.1007/s11269-008-9305-1

Niemeyer S (2008) New drought indices. Options Méditerr Ser A 267–274

Oertel M, Meza FJ, Gironás J (2015) Improving operational drought definitions—taking them to basin scale. In: Joaquin Andreu Alvarez, David Haro-Monteagudo, Henny van Lanen, Javier Paredes-Arquiola, Abel Solera Solera, Andreu J, Solera A, Paredes-Arquiola J, Haro-Monteagudo D, van Lanen H (eds) Drought: research and science-policy interfacing, Online-Ausg. CRC Press, Hoboken, pp 151–159

Palmer WC Meteorological Drought (1965) Research Paper No. 45, p 58

Puertas OL, Henríquez C, Meza FJ (2014) Assessing spatial dynamics of urban growth using an integrated land use model. Application in Santiago Metropolitan Area, 2010–2045. Land Use Policy 38:415–425. https://doi.org/10.1016/j.landusepol.2013.11.024

Shafer BA, Dezman IE (1982) Development of a surface water supply index to assess the severity of drought conditions in snowpack runoff areas

Shukla S, Wood AW (2008) Use of a standardized runoff index for characterizing hydrologic drought. Geophys Res Lett 35. https://doi.org/10.1029/2007gl032487

Smakhtin VU, Schipper EL (2008) Droughts: The impact of semantics and perceptions. Water Policy 10:131. https://doi.org/10.2166/wp.2008.036

Sridhar V, Hubbard KG, You J, Hunt ED (2008) Development of the soil moisture index to quantify agricultural drought and its "user friendliness" in severity-area-duration assessment. J Hydrometeorol 9:660–676. https://doi.org/10.1175/2007JHM892.1

Stagge JH, Tallaksen LM, Gudmundsson L, van Loon AF, Stahl K (2015) Candidate distributions for climatological drought indices (SPI and SPEI). Int J Climatol 4027–4040. https://doi.org/10.1002/joc.4267

Stagge JH, Tallaksen LM, Gudmundsson L, van Loon AF, Stahl K (2016) Response to comment on 'candidate distributions for climatological drought indices (SPI and SPEI)'. Int J Climatol 36:2132–2138. https://doi.org/10.1002/joc.4564

State Drought Plans. http://drought.unl.edu/planning/droughtplans/StateDroughtPlans.aspx. Accessed 30 Nov 2017

Steinemann A, Iacobellis SF, Cayan DR (2015) Developing and evaluating drought indicators for decision-making. J Hydrometeorol 150417113205000. https://doi.org/10.1175/jhm-d-14-0234.1

Steinemann AC, Cavalcanti LFN (2006) Developing multiple indicators and triggers for drought plans. J Water Resour Plann Manag 132(3):164. https://doi.org/10.1061/(asce)0733-9496(2006)

Tigkas D, Vangelis H, Tsakiris G (2015) DrinC: a software for drought analysis based on drought indices. Earth Sci Inform 8:697–709. https://doi.org/10.1007/s12145-014-0178-y

Tsakiris G, Vangelis H (2005) Establishing a drought index incorporating evapotranspiration. Eur Water 9 (10):3–11

Tsakiris G, Loukas A, Pangalou D, Vangelis H, Tigkas D, Rossi G, Cancelliere A (2007) Chapter 7. Drought characterization. Options Méditerr Ser A 85–102

van Lanen HAJ, Kundzewicz ZW, Tallaksen LM, Hisdal H, Fendeková M, Prudhomme C (2008) Indices for different types of droughts and floods at different scales: Technical Report No. 11. WATCH Water and Global Change

Van Loon A, Van Lanen H, Tallaksen LM, Hanel M, Fendekova M, Machlica A, Sapriza G, Koutroulis A, Huijgevoort M, Jódar J, Hisdal H, Tsanis I (2011) Propagation of drought through the hydrological cycle

van Loon AF, Laaha G (2015) Hydrological drought severity explained by climate and catchment characteristics. J Hydrol 526:3–14. https://doi.org/10.1016/j.jhydrol.2014.10.059

Vicente-Serrano SM, Beguería S, López-Moreno JI (2010) A multiscalar drought index sensitive to global warming: the standardized precipitation evapotranspiration index. J Climate 23:1696–1718. https://doi.org/10.1175/2009JCLI2909.1

Vicente-Serrano SM, López-Moreno JI, Beguería S, Lorenzo-Lacruz J, Azorin-Molina C, Morán-Tejeda E (2012a) Accurate computation of a streamflow drought index. J Hydrol Eng 318–331. https://doi.org/10.1061/(asce)he.1943-5584.0000433

Vicente-Serrano SM, Beguería S (2016) Comment on 'candidate distributions for climatological drought indices (SPI and SPEI)' by James H. Stagge et al. Int J Climatol 36:2120–2131. https://doi.org/10.1002/joc.4474

Vicente-Serrano SM, Beguería S, Lorenzo-Lacruz J, Camarero JJ, López-Moreno JI, Azorin-Molina C, Revuelto J, Morán-Tejeda E, Sánchez-Lorenzo A (2012b) Performance of drought indices for ecological, agricultural and hydrological applications. Earth Interact 16:1–27. https://doi.org/10.1175/2012EI000434.1

Vicente-Serrano SM, Van der Schrier Gerard, Beguería S, Azorin-Molina C, Lopez-Moreno J-I (2014) Contribution of precipitation and reference evapotranspiration to drought indices under different climates. J Hydrol. https://doi.org/10.1016/j.jhydrol.2014.11.025

Wells N, Goddard S, Hayes MJ (2004) A self-calibrating palmer drought severity index. J Climate 17:2335–2351

Wilhite DA (1993) Drought assessment, management, and planning: theory and case studies. Kluwer Academic Publishers, Boston, Natural resource management and policy

Wilhite DA, Buchanan-Smith M (2005) Drought as hazard: understanding the natural and social context. In: Wilhite DA (ed) Drought and water crises: science, technology, and management issues. Taylor & Francis Group, Boca Raton, pp 3–30

Wilhite DA, Svoboda MD, Hayes MJ (2007) Understanding the complex impacts of drought: a key to enhancing drought mitigation and preparedness. Water Resour Manage 21:763–774. https://doi.org/10.1007/s11269-006-9076-5

Wilks DS (2006) Statistical methods in the atmospheric sciences: an introduction, 2nd ed. International geophysics series, vol 59. Academic Press, Amsterdam

WMO (2012) Standardized precipitation index user guide (Svoboda, M.; Hayes, M.; Wood, D.). WMO, No. 1090. World Meteorological Organization, Geneva

Wu H, Svoboda MD, Hayes MJ, Wilhite DA, Wen F (2007) Appropriate application of the standardized precipitation index in arid locations and dry seasons. Int J Climatol 27:65–79. https://doi.org/10.1002/joc.1371

Zargar A, Sadiq R, Naser B, Khan FI (2011) A review of drought indices. Environ Rev 19:333–349. https://doi.org/10.1139/a11-013

Monitoring Small Reservoir Volume by Satellite-Based Data for Drought Management in the Central Region of Vietnam

Trung Tuan Luong, Lam Xuan Nguyen, and Trinh Duc Tran

Abstract

Vietnam is one of the countries with a large number of reservoirs. There are around 7000 reservoirs operating currently of which 85% are small reservoirs with volume less than 3 million cubic meter. Those reservoirs have been serving as a major water supply for 75% of the total 4 million ha of agricultural land. During recent years, Central and Highland region of Vietnam has experienced an increasing frequency of drought disasters triggered by heat waves and prolonged dry periods. The El Niño influenced drought in 2014–2016 resulted in 40,000 ha paddy field stopped cultivation and in about 65,000 households lacking water supply. Hence, there is a huge demand for effective drought monitoring and warning systems. However, current systems are based on drought indices such as SPI, PDSI. Those indices have been proven suitable for rainfed area, but not effective for irrigated areas under the service of reservoir. The current advances in public-free-access satellite data offer an effective solution to inform about drought risk in irrigated areas. Based on the information provided by the satellite images about water balances in reservoirs the drought risk can be estimated for the commanding areas. However, it faces difficulties due to low resolution, revisit frequency, cloud, etc. SAR images from Sentinel-1 that has started operation since 2014, is expected to reduce the obstacles since it offers continuous all-weather, day-and-night, 10 m resolution and 6 days revisit-frequency imagery at C-band. The study used Sentinel-1 images to estimate water surface area and reservoir volume in the near-real-time mode. The algorithm was also improved by applying Fuzzy and Frost filter in filtering the speckles. On average, Nash–Sutcliffe coefficient that obtained from the validation process ranged from 0.8 to 0.9 when piloting the algorithm in Binh Dinh province with 168 small reservoirs. The study results show a high potential for application of remote sensing data for reservoir monitoring serving drought management in the region as well as the whole country.

Keywords

Drought monitoring · Remote sensing · Sentinel-1 SAR images · Fuzzy-based filtering method

T. T. Luong · L. X. Nguyen (✉) · T. D. Tran
Vietnam Academy for Water Resources, Hanoi, Vietnam
e-mail: nguyenxuanlamiwe@gmail.com

L. Ribbe et al. (eds.), *Towards Water Secure Societies*,
https://doi.org/10.1007/978-3-030-50653-7_12

1 Introduction

Climate change, environmental degradation, population growth and urbanization have been exerting great pressure on water supplies in many parts of Asian. The pressure is further increased with the prospect of higher extreme weather and hazards frequency (WMO 2016). In 2014–2016 drought spell, it was reported that 240,000 hectares of crops was lost; 338,850 households lacking water, total estimated damages of about 250 million USD (CSCNDPC 2016) in South Central Coast and Central Highland region of Vietnam (SCCCH, Fig. 1).

Small-medium reservoirs play a key role in ensuring water for living life, agriculture, and others in SCCCH region of Vietnam since 75% area of agricultural land is dependent on those reservoirs (MARD 2016) for water supply. There are around 2000 reservoir (of all sizes) located in the SCCCH region and 85% of that are reservoirs with capacity less than 3 million cubic meters. Experience with drought impact in recent years in Vietnam shown that drought usually occurs at the commanding areas of small reservoirs that were constructed for a long time, managed by local governments and people, and usually suffer poor monitoring and operation due to limited capacity and budget (WB8 2014).

Reservoir operation for agricultural water supply in the context of lacking both monitoring system and the technical capacity is a challenging task for a commune based reservoir in a normal climatological condition let alone a context of climate change and drought impact.

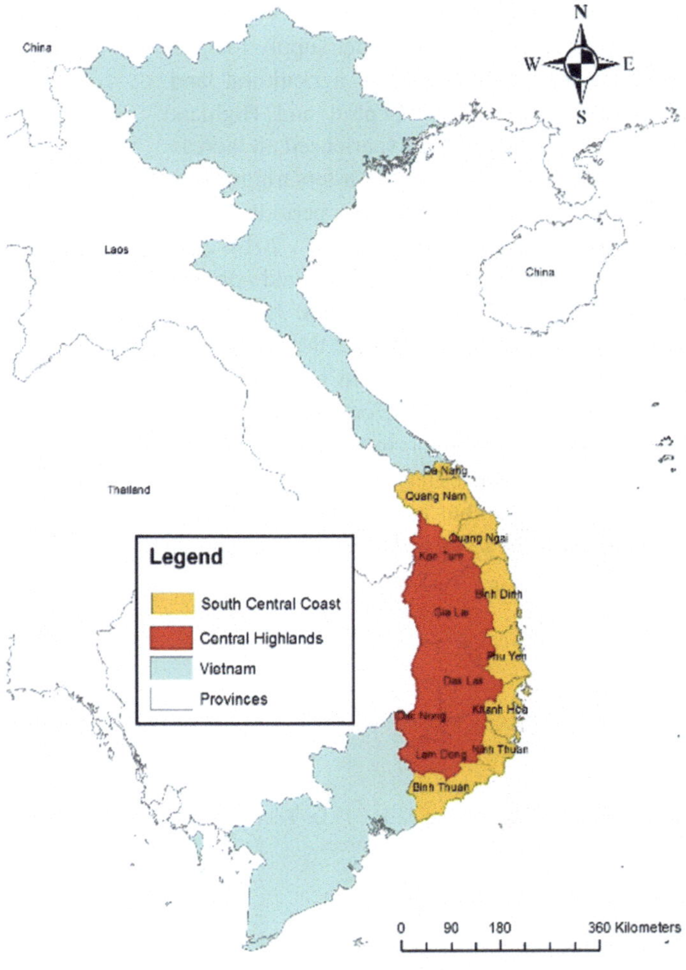

Fig. 1 South Central Coast and Central Highland region and provinces

Additionally, conventional drought monitoring and warning method using meteorological indices (SPI, PDSI, etc.) are usually effective for rain-fed area but not sufficient for the region with extensive irrigation system connected with reservoir water supply. Therefore, for the drought monitoring and warning for irrigated area under service of reservoir just can be achieved with the concurrent monitoring of water in the reservoir together with the commanding area. Recently, remote sensing observations have been used to monitor climatological variables and to assess and quantify impacts of climatological variables on water resources systems. In the field of reservoir monitoring, a wide range of researches has been conducted using optical, radar and altimetry satellite (Amitrano et al. 2014; Abileah and Vignudelli 2011; Magome 2003). Public-free-access satellite data has thus been proven as an effective tool to monitor on large scale, however, it faces difficulties due to low resolution, revisit frequency, cloud interference. SAR images from Sentinel-1A that has started operation since 2014, are expected to help overcome the obstacles when it provides continuous all-weather, day-and-night, 10 m resolution and revisit-frequency of 6 days imagery at C-band. This study is based on the public satellite-based data and expected to provide a cost-effective tool in monitoring small-medium reservoirs for better drought management in the SCCCH region.

2 Methodology

The most important factors for reservoir monitoring are the water level and volume beside the inflow and out flow. To monitor these factors, the methodology of obtaining reservoir volume this paper used is to calculate the storage through water surface area and reservoir volume-area curve. The water surface area can be calculated by extracting the water pixels in satellite radar image (Klemas and Pieterse 2015). The radar image was chosen because radar is not affected by cloud (Musa et al. 2015), which is an important advantage compared to optical satellite data, and also radar image has wider spatial coverage compared to the altimetry data. Synthetic-aperture radar (SAR) reflectance properties, however, are affected by the topography, a pre-processing process was developed to reduce the noise. Besides, the reservoir volume-area curve is a key factor, in case of missing the records, the Digital Elevation Map (DEM) can be used to develop this relation. However, SRTM DEM 30 m was generated in 2000 Feb launch, it is meant that there is no bathymetric data beneath the water-level for reservoirs in operation at that time because the radar of SRTM shuttle was not be able to penetrate the water surface, therefore, an extrapolation from the known data is necessary (Fig. 2).

2.1 Pre-Processing

2.1.1 Numerical and Geometric Correction

SAR calibration converts the pixel values in the SAR image which is qualitatively representative of the biased backscatter signal to pixel values directly related to the radar backscatter. Backscattering coefficient is created from digital number values of SAR image based on the calibration look up tables (LUTs) provided by level-1 products. The geometric correction or terrain correction then is implemented to remove the effects of side looking geometry in SAR images by converting data from ground range geometry without terrain consideration to backscatter coefficient by using digital elevation map (Miranda and Meadows 2015). These pre-processing steps could be conducted by using the Sentinel-1 Toolbox in Sentinel Application Platform (SNAP) software provided by ESA.

2.1.2 Noise Reduction

Speckle noise is the consequence of constructive and destructive interference of various backscattered echoes resulting from random intersection of a radar signal with the small objects on a scale equal to the sound wavelength (Raju et al. 2013). This phenomenon brings about the so called granular pattern in the radar image that make the presence of the image objects significantly degraded (Masoomi et al.

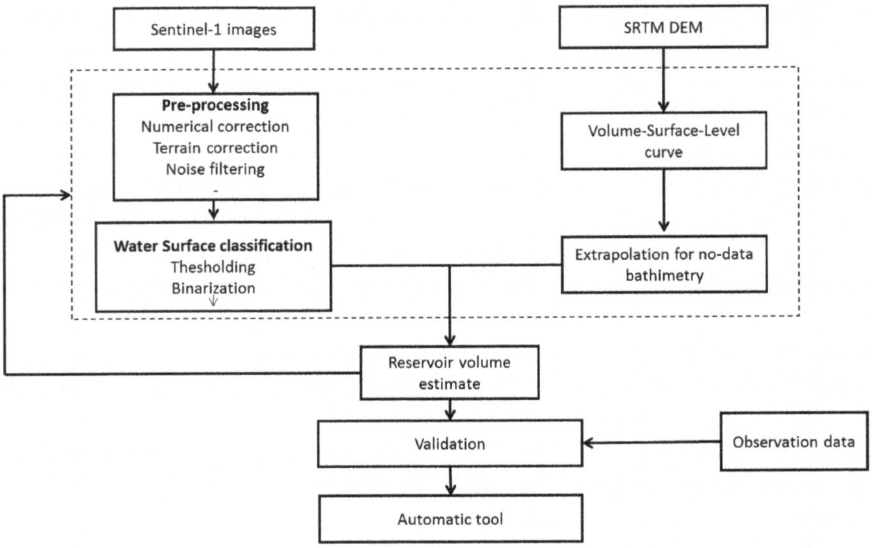

Fig. 2 The approach of remote sensing image-based reservoir storage monitoring

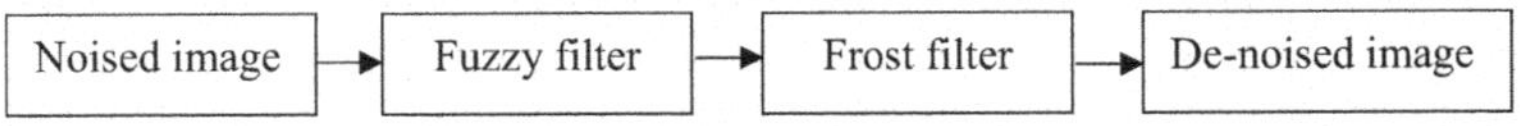

Fig. 3 Method of noise removal for SAR image

2012). The speckle filtering mainly aims at reducing noise occurring in the homogenous region but preserving texture and enhancing the edge by movement of a kernel window over each pixel on the image along the entire image and application of a mathematical algorithm to replace the central pixel value of the window with a new value (Santoso et al. 2016a, b).

For noise reduction in images, linear filtering techniques were often been used because of simple mathematical operation. However, linear filters have limitation in reduction impulse noise and have a bias to darken the edges of an image objects (Keung 2003). In contrast, the fuzzy filters averaging a pixel value by other neighbor pixel but simultaneously conserves important image structures by determination of the pixel having small derivative in specific direction to differentiate between local variation originated from noise and ones from image formation (Van De Ville et al. 2003). In researches of Santoso et al. (2016a, b), a speckle reduction technique based on integration of Frost filter and fuzzy filter was proven to be a favorable solution for suppressing noise while preserving edges of the SAR images. Thus this approach was employed for the noise removal step (Fig. 3).

Frost filter

The Frost filter is a linear filter based on the minimum square error algorithm. The kernel of the filter adapts locally to the coefficient of variation illustrated by the ratio of the local standard deviation to the local average of the noised-affected image. Therefore, it can keep pixels of sharp features unchanged while suppressing the noise (Kupidura 2016). Santoso et al. (2016a, b) express the degraded image and n-size Frost filter respectively by following:

$$\mathrm{I(t)} = \mathrm{R(t)v(t)} \tag{1}$$

$$\mathrm{DN} = \sum_{n\times n} k\alpha e^{-\alpha|t|} \tag{2}$$

where:

R(t) is original image, v(t) is speckle noise, k is normalization constant;

$$\alpha = \frac{4}{n'^2} \times \frac{2}{\bar{I}^2} \tag{3}$$

Therein $\bar{I}$ is local mean, σ denotes the local variance and σ' is image coefficient.

Fuzzy filter

The output image x(i, j) created after applying a 2-D fuzzy filter for input image y(i, j) is illustrated as below (Keung 2003):

$$\mathrm{y(i,j)} = \frac{\sum_{(r,s)A} F[x(i+r, j+s)]x(i+r, j+s)}{\sum_{(r,s)A} F[x(i+r, j+r)]} \tag{4}$$

where F[x(i, j)] is the window function in terms of fuzzy membership functions and A denotes area of the window. For a N x N dimension square window, r and s, respectively, are defined in range of [-R, R] and [-S, S], where N = 2R + 1 = 2S + 1.

There are seven fuzzy filters can be obtained with the different formulation of window functions (Kwan and Cai 2002). Santoso (2016) suggested that combination between Frost filter and Triangular Moving Average (TMAV) filter indicated best performance. The symmetric TMAV fuzzy filter with the moving average value within a window chosen as the center value is defined as (Keung 2003):

$$\mathrm{F_{TMAV}[x(i\ +r,\ j\ +\ s)]} \begin{cases} 1 - \frac{|x(i+r, j+s) - x_{\text{mav}}(i,j)|}{x_{mv}(i,j)} \\ for|x(i+r, j+s) - x_{\text{mav}}(i,j)| \le x_{mv}(i,j) \\ 1 \\ for x_{mv}(i,j) = 0 \end{cases} \tag{5}$$

where:

$$x_{mv}(i,j) = \max[x_{\max}(i,j) - x_{\text{mav}}(i,j), x_{\text{mav}}(i,j) - x_{\min}(i,j) \tag{6}$$

x_{max}(i, j), x_{min}(i, j) and x_{mav}(i, j) are the maximum value, the minimum value and the moving average value of x(i + r, j + s), respectively.

2.2 Water Surface Detection

Segmentation threshold algorithms were used to delineate water surface from satellite images. In this study, we choose to apply Otsu method to classify the satellite image into water and background. The Otsu method is widely acknowledged as successful automatic identification threshold (Fang et al. 2009; Chaki et al. 2014). It is a classification method based the variance maximum value between the two clusters (Xu et al. 2011). In a grayscale image represented by pixels in range of L levels [1, L] with probability of each pixel values defined as P_i, supposing that the image is classified by threshold value t into two classes C_o = [1, t] and C_1 = [t + 1, L] having cumulative probabilities of $P_o(t) = \sum_{i=1}^{t} p_i$ and $P_1(t) = 1 - P_o(t)$, respectively. The mean gray values and variances of C_o and C_1 are presented by μ_o, μ_1, σ_o^2, σ_1^2:

$$\mu o(t) = \sum_{(i=1)}^{t} \frac{(ip_i)}{(P_o(t))} \tag{7}$$

$$\mu 1(t) = \sum_{i=t+1}^{L} \frac{ip_i}{P_1(t)} = 1 - \mu o \tag{8}$$

$$\sigma_0^2(t) = \sum_{i=1}^{t} (i - \mu_o)^2 \frac{p_i}{P_o(t)} \tag{9}$$

$$\sigma_1^2(t) = \sum_{i=t+1}^{L} (i - \mu_1)^2 \frac{p_i}{P_1(t)} \tag{10}$$

The threshold value t^* is estimated by maximizing the variance between two classes σ_b^2 or minimizing the variance inside a class σ_w^2:

$$\sigma_b^2(\mathrm{t}) = \mathrm{P_o(t)}(\mu_o(\mathrm{T}) - \mu)^2 + \mathrm{P_1(t)}(\mu_1(\mathrm{T}) - \mu)^2 \tag{11}$$

$$\sigma_b^2(\mathrm{t}) = Po(t)_o^2(t) + P1(t)\sigma_1^2(\mathrm{t}) \tag{12}$$

where $\mu = P_o(t)\mu_o(T) + P_1(t)\mu_1(T)$ is mean level of the original image.

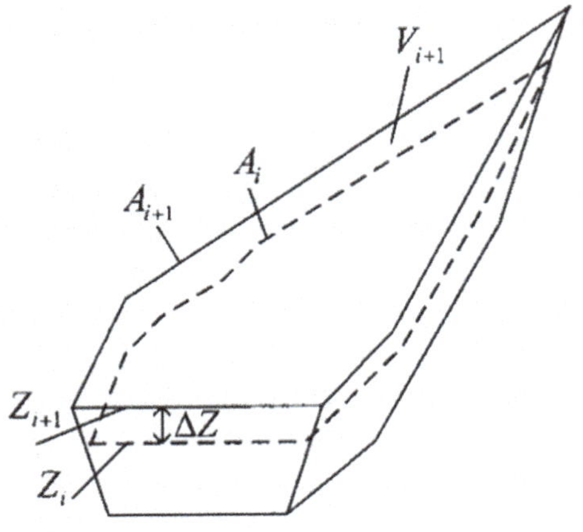

Fig. 4 Sketch map of reservoir storage estimation method

After separation step, the water surfaces include reservoir surfaces and possibly other surrounding small water bodies. In order to estimate more accurately the area of reservoir surfaces, we decide to extract largest part of water pixels, therefore water bodies outside of reservoirs can be removed.

2.3 Reservoir Curve Calculation

The relationship between water surface elevation and surface area was determined by splitting the elevation data into 1 m elevation layers and calculating the surface area of each layer. From that, the storage could be calculated through change in elevation and the water surface. The traditional reservoir storage estimation method sketch map is indicated in Fig. 4 (Peng 2006).

The storage for certain changes in water level ΔZ or the difference between two successive elevations is calculated as follows:

$$V_{i+1} = (A_i + A_{i+1} + \sqrt{A_i A_{i+1}})\Delta Z/3 \quad (13)$$

where A_i is the water surface area at the water level of Z_i (i = 0, 1, 2, 3, …,). The total storage V_t at the water level of $Z_i + 1$ is defined by formula below:

$$V_t = \sum_{k=0}^{i} V_{k+1} \quad (14)$$

Given the water surface and area—storage relationship curve of the reservoir, the storage could be estimated by interpolation. Simulation results are checked with real observation data from monitoring records of the reservoirs (Fig. 5).

2.4 Development of Automatic Tool

The entire procedure of calculation from Sentinel-1 SAR data, updating, processing for

Fig. 5 Rainfall, reservoir and crop pattern in Binh Dinh province

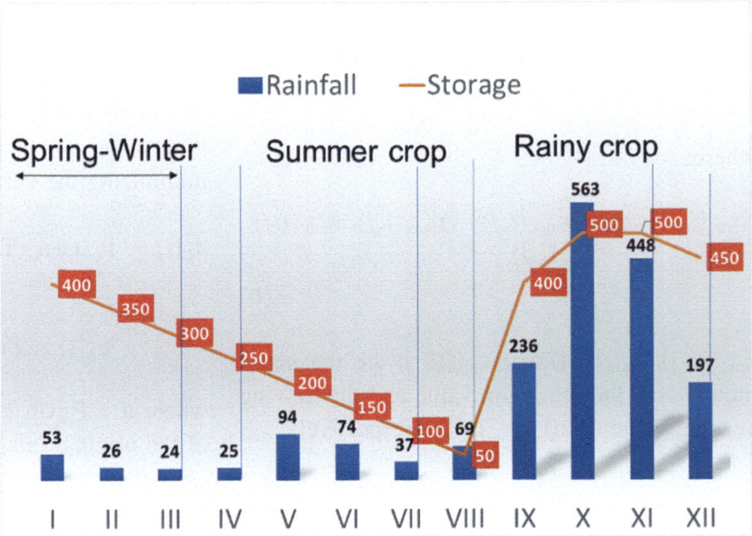

Fig. 6 Automatic tool developed on Matlab

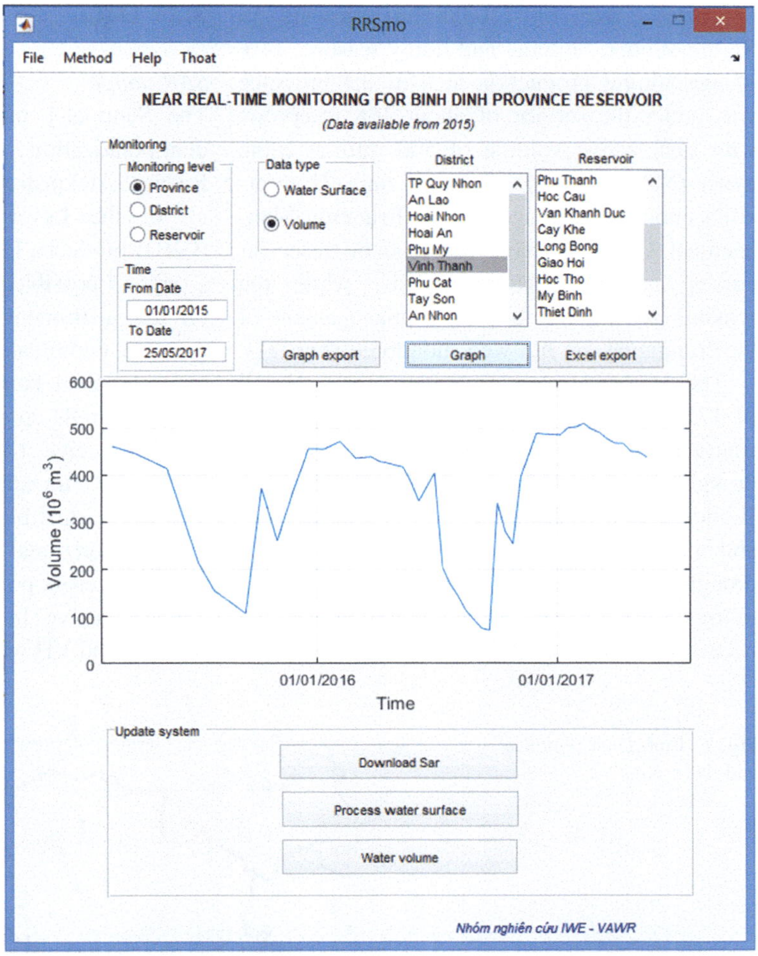

reservoir water surface extraction to storage calculation is programmed using Python and Matlab language to develop an automatic tool. Additional software including ArcGIS and SNAP are linked with Application Programming Interface (API) for the program languages (Fig. 6).

2.5 Assessment of Tool Performance

The research was carried out with a validation of simulated storage results with real observation data by Nash—Sutcliff Coefficient:

$$\text{Nash} = 1 - \frac{\sum_{t=1}^{T}\left(X_s^t - X_o^t\right)^2}{\sum_{t=1}^{T}\left(X_o^t - \overline{X_o}\right)^2} \quad (15)$$

where $\overline{X_o}$ is the average of observation values, X_s^t and X_o^t are respectively simulated and observed values at time t. The algorithm was tested on 21 reservoirs currently under the management of Binh Dinh Irrigation Management Company (IMC).

3 Study Area and Data

3.1 Study Area

Binh Dinh province locates in South Central Coastal, ranging from 13°39'N to 14°42'N latitude, 108°54'E to 108°55'E longitude. The region covers an area of 6851 km^2 and population of 1.9 million. The province has annual rice production area of 110,000 ha, growing in Spring-Winter, Summer and Rainy seasons. 74% of agricultural production area of the province was under the service of about 168 reservoirs with total active volume of 500 million cubic meter (MARD 2016). However, only 21 reservoirs under the management of Irrigation Management Company have full data on reservoir curves and observation records, while the remaining reservoirs under the management of the communes are not well monitored (Fig. 7).

The province has an estimated annual rainfall of 1751 mm but 70÷80% of that rainfall concentrates in September-December period. The drought usually occurs at the end of Spring-Winter season when the reservoirs are dried out and the early rain events have not started thus drought monitoring is really needed at that period to trigger the warnings and communities can take necessary actions (Fig. 5).

3.2 Data

SAR image

The Sentinel-1 mission provides data from a dual-polarization C-band Synthetic Aperture Radar (SAR) instrument. This collection contains all of the Level-1 Ground Range Detected (GRD) products. Each scene contains either 1 or 2 out of 4 possible polarization bands, depending on the instrument's polarization settings. The possible combinations are single band VV or HH, and dual band VV + VH and HH + HV: (1) VV: single co-polarization, vertical transmit/vertical receive; (ii) HH: single co-polarization, horizontal transmit/horizontal receive; (iii) VV + VH: dual-band cross-polarization, vertical transmit/horizontal receive; and (iv) HH + HV: dual-band cross-polarization, horizontal transmit/vertical receive. In this study, the dual polarization VV and VH with 10 m spatial resolution and

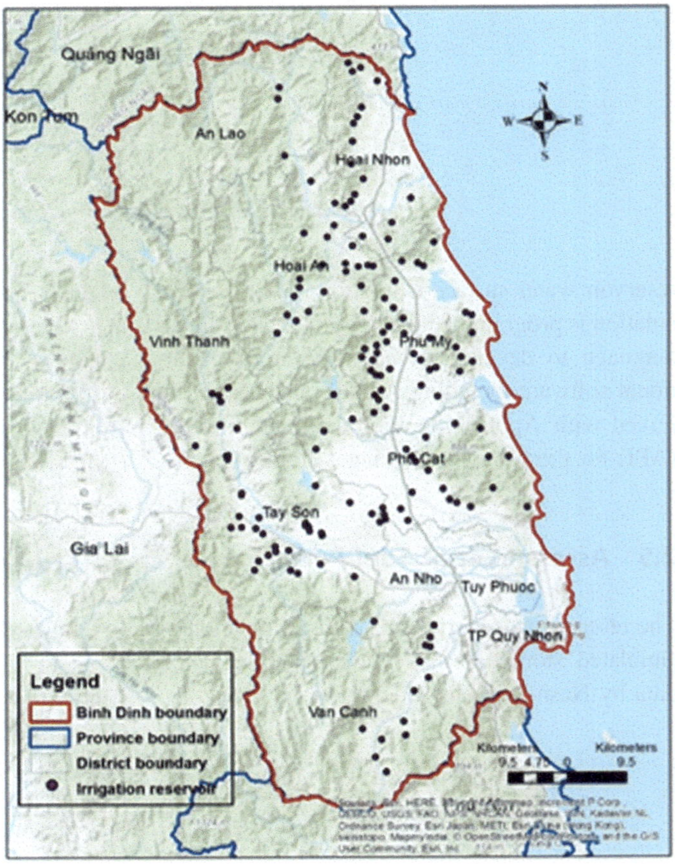

Fig. 7 Binh Dinh province and reservoirs

6-day temporal resolution is applied for water extraction.

DEM data

The relationship between reservoir storage, water elevation and surface area can be derived from 30 m resolution SRTM DEM. However, only reservoirs built after January 2000 can be calculated due to the obtaining time of the DEM, as for the reservoirs built before that, only reservoir storage parts above the water level at the time can be seen.

4 Results and Discussion

Separation of water surface and land

In this study, the crucial step of reservoir estimation is calculating area of water surface of reservoir, in turn, it is detecting water parts on the images. Water and land background separation from image of Sentinel-1 band VV could be accomplished by density slicing. Otsu thresholding analysis was applied to each image from each date to generate threshold value would be used as a point dividing the density values of a satellite image into two groups. The density slicing resulted in a binary image with two classes including land and water. In order to determine the effectiveness of the fuzzy—Frost combined filter, the research was implemented with and without filtering process. As shown in Figs. 8 and 9, there were much noise remained in the non-filtered images, while application of 3 × 3 kernel size filter made the classification of water and land more obvious and the edges of surface of reservoirs more sharp Then extraction of reservoir surfaces from those resulting images would be easier.

In general, SAR images with fine spatial resolution of 10 m are proved as effective data to classify area of water surface of reservoirs in Binh Dinh province. Particularly, Otsu thresholding based—water surface segmentation gives good results in cases of large and medium-scale reservoirs. The step of extract largest water pixel areas also contributed to avoid water parts not belong to the reservoir, hence reservoir surface is more correctly defined. However, due to the existence of a number of small islands in several reservoirs influencing the backscattering coefficient of the water, these parts of water are insufficiently extracted. The effect of undepressed noise and shadow of the mountain or solid particles in water possibly cause some inevitable error in the classification results (Fig. 10).

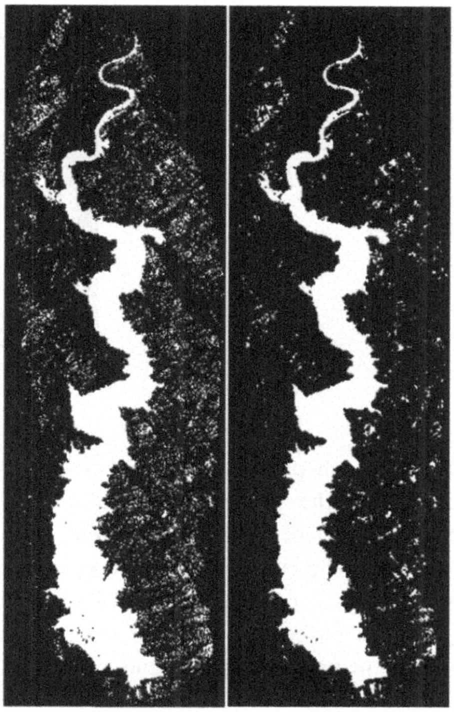

Fig. 8 Binary images of Dinh Binh reservoir on 22/02/2015 before (left) and after (right) filtering application

Nonetheless, since the surface waters and wet areas have almost similar backscattering coefficient, the close humid areas are detected as water bodies by Sentinel-1 remote sensors so they are possibly misidentified as part of reservoir surface. This misidentification would be reason for incorrect estimation of reservoir's surface area and consequently, the storage would be computed inaccurately (Fig. 11). Besides, the difference between geographical characteristics of reservoir areas, which has not been taken into account in the research, should be studied to attain higher performance of water segmentation.

The observed data collected from IMC from January 2015 to December 2016. Nash coefficient was calculated, range from 0.7 to 0.9.

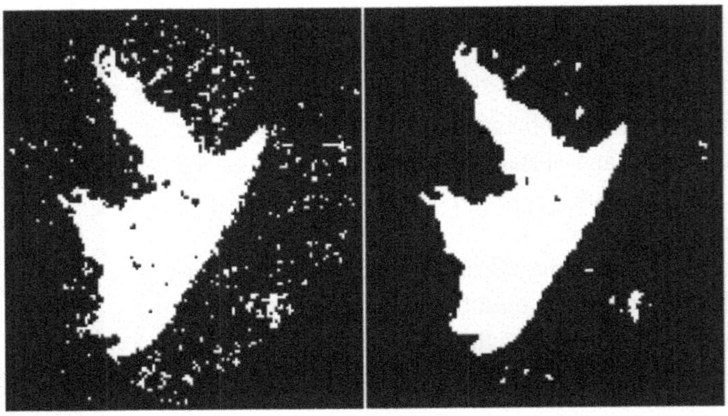

Fig. 9 Binary images of Thuan An reservoir on 22/02/2015 before (left) and after (right) filtering application

Additionally, shape of the storage-change line matches with the general observed trend that decreases during period from beginning months of the year to the end of summer crop season in August and then increases in the rainy months (Fig. 12). Nevertheless, the error exist is quite significant in small reservoirs due to the coarse resolution DEM SRTM of 30 m which caused inaccuracy in the reservoir curve development.

When it comes to seasonal assessment, resulting outputs exhibit higher performance when reservoirs' storage reaches lowest or highest values. During the time of minimum and maximum storage (in the end of summer crop and rainy season respectively), the water volume in reservoirs are kept stable, while on the other hand, throughout spring-winter crop and summer crop seasons, especially from April to June, reservoirs' storage are considerably temporally varied because irrigation is totally dependent on outflow from reservoirs. Because of similarity in patterns of storage and water surface area, when real reservoir's volume decreases its portion of water surface should be also reduced. However, as explained above, the previous reservoir's surface areas which remain wet is possibly identified as water surface and considered as part of reservoir surface, then overestimation took place.

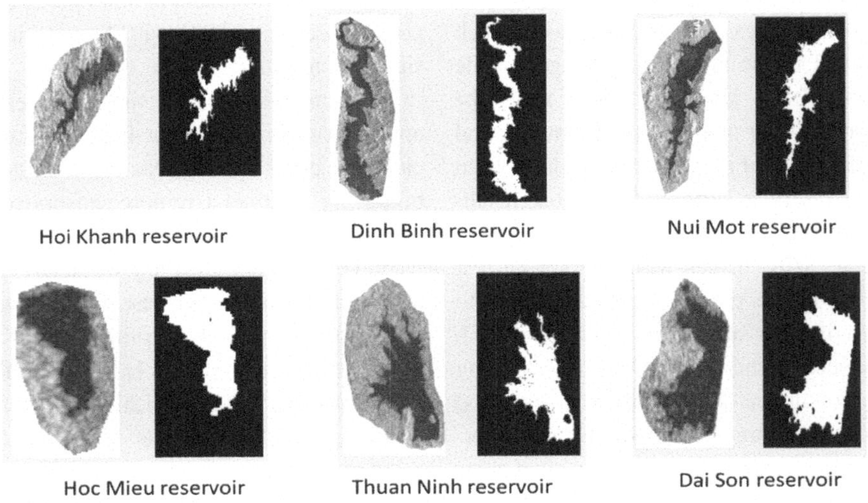

Fig. 10 Water surface classification from background

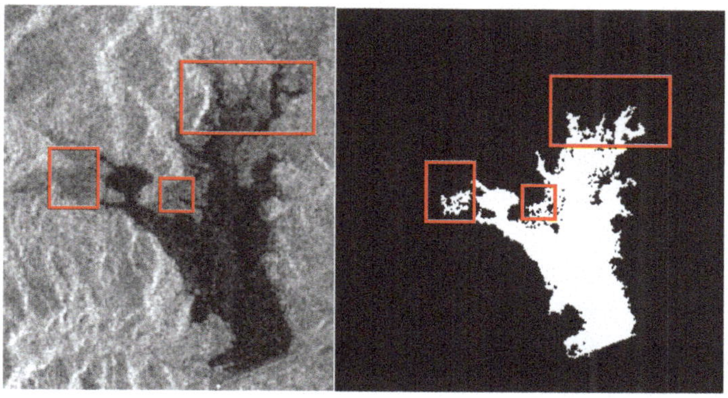

Fig. 11 The misidentification of water surface of Tuong Son reservoir (red rectangulars)

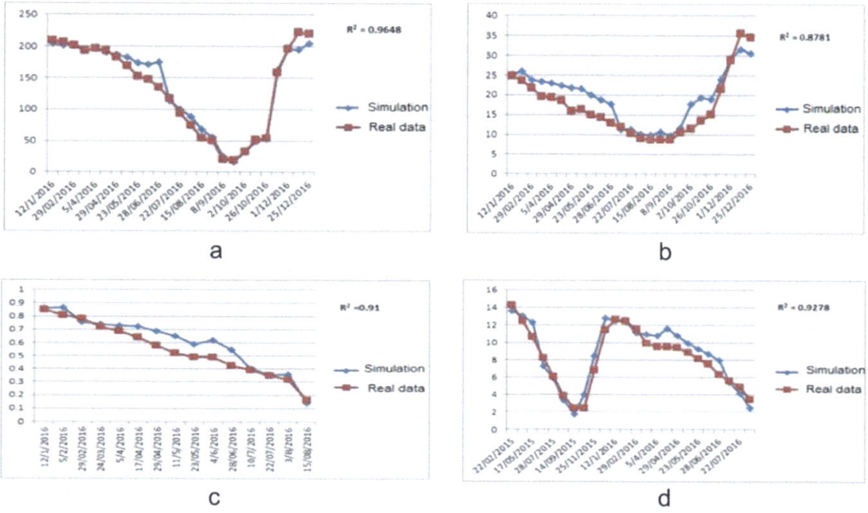

Fig. 12 Validation with observed reservoir volumes **a** Dinh Binh reservoir, **b** Thuan Ninh reservoir (using DEM for reservoir curve), **c** Suoi Duoc reservoir(using DEM for reservoir curve), **d** Suoi Tre reservoir

Overall, water surface area and reservoir curve estimation the results of this study indicated capability of combination between SAR data and DEM extracted—reservoir curves to supervise change of reservoir storage, which is very useful for water accounting, plan to irrigation allocation, and drought management etc. The developed tool also reduced the manual analysis and computation of multiple reservoirs volume in the whole research area. Nevertheless, the method should be studied in more detail to avoid accumulation errors from water surface area and reservoir curve estimation steps.

5 Conclusion

From application of integrated filters consisting Frost filter and Symmetric Triangular Moving Average fuzzy filter for pre-treatment of Sentinel-1 data and use of Otsu thresholding method, the water surface of reservoirs have been successfully detected. The results of water surface area combined with 30 m SRTM DEM—extracted reservoir curves provide estimation of reservoir's storage achieving Nash coefficient in range of 0.7–0.9 in comparison with real measurement data. Also, the coarse resolution of DEM and effect of undepressed noise and the mountain shadow inducing errors require further study. However, the study results show a high fidelity in monitoring reservoirs for drought management in the region. The results of the study could be replicated for the reservoir monitoring supporting drought management in the whole country, and further toward international water bodies.

References

Abileah R, Vignudelli S (2011) A completely remote sensing approach to monitoring reservoirs water volume. Heron I(1):59–72

Amitrano D, di Martino G, Iodice A, Mitidieri F, Papa MN, Riccio D, Ruello G (2014) Sentinel-1 for monitoring reservoirs: a performance analysis. Remote Sens 6(11):10676–10693. https://doi.org/10.3390/rs61110676

Chaki N, Shaikh SH, Saeed K (2014) Exploring image binarization techniques. Stud Comput Intell 560:5–16. https://doi.org/10.1007/978-81-322-1907-1

CSCNDPC (2016) Assessment on drought damages from 2014-2016 by Central Steering Committee for Natural Disaster Prevention and Control

Fang M, Yue G, Yu Q (2009). The study on an application of Otsu method in canny operator. Int Symp Inf 2(4):109–112

Keung H (2003) Fuzzy filters for noise reduction in images. *Springer*

Klemas V, Pieterse A (2015) Advances in watershed science and assessment. In Advances in watershed science and assessment, the handbook of environmental chemistry, vol 33. Springer International Publishing, Switzerland. https://doi.org/10.1007/978-3-319-14212-8

Kupidura P (2016) Comparison of filters dedicated to speckle suppression in SAR images. Int Arch Photogramm Remote Sens Spat Inf Sci ISPRS Arch 41 (July):269–276. https://doi.org/10.5194/isprsarchives-XLI-B7-269-2016

Kwan HK, Cai Y (2002) Fuzzy Filters for image filtering. IEEE 672–675

Magome JUN (2003). Method for satellite monitoring of water storage in reservoirs for efficient regional water management. In Water resources systems-hydrological risk, managemnet and development, pp 303–310

Masoomi A, Hamzehyan R, Shirazi NC (2012) Speckle reduction approach for SAR image in satellite communication. Int J Mach Learn Comput 2(1):62–70. https://doi.org/10.7763/IJMLC.2012.V2.90

Miranda N, Meadows PJ (2015) Radiometric calibration of S-1 Level-1 products generated by the S-1 IPF. European Space Agency. ESA-EOPG-CSCOP-TN-0002

MARD (2016) Assessment of drought impacts and reasons during 2014-2106 International cofference on Drought, Phan Rang-Thap Cham city, Ninh Thuan province, June 2016

Musa ZN, Popescu I, Mynett A (2015) A review of applications of satellite SAR, optical, altimetry and DEM data for surface water modelling, mapping and parameter estimation. Hydrol Earth Syst Sci 19 (9):3755–3769. https://doi.org/10.5194/hess-19-3755-2015

Peng D (2006). Reservoir storage curve estimation based on remote sensing data. J Hydrol Eng 11(2)

Raju KMS, Nasir MS, Devi TM (2013). Filtering techniques to reduce speckle noise and image quality enhancement methods on satellite images. IOSR J Comput Eng (IOSR-JCE) 15(4):10–15

Santoso AW (2016) Comparison of fuzzy filters on synthetic aperture radar image

Van De Ville D, Nachtegael M, Van der Weken D, Kerre EE, Philips W, Lemahieu I (2003) Noise reduction by fuzzy image filtering. IEEE Trans Fuzzy Syst 11(4):429–436. https://doi.org/10.1109/TFUZZ.2003.814830

Santoso AW, Bayuaji L, Pebrianti D, Lateh H (2016a) A fuzzy approach for speckle noise reduction in SAR images. Int J Adv Appl Sci 3(August):33–38

Santoso AW, Bayuaji L, Tien Sze L, Lateh H, Mohamad Zain J (2016b) Comparison of various speckle noise reduction filters on synthetic aperture radar image. Int J Appl Eng Res ISSN 11(15):973–4562

WB8 (2014) World Bank-Project Information Document (Concept Stage)-Vietnam Dam Rehabilitation and Safety Improvement Project-P152309 (English) Project Information Document PIDC13292

World Meteorological Organization (2017) WMO statement on the state of the global climate in 2016. World Meteorological Organization (WMO)

Xu X, Xu S, Jin L, Song E (2011) Characteristic analysis of Otsu threshold and its applications. Pattern Recogn Lett 32(7):956–961. https://doi.org/10.1016/j.patrec.2011.01.021

Farmers Perception and Salinity Driven Fresh Water Scarcity in Coastal Bangladesh

Md. Arman Habib, M. Tauhid Ur Rahman, Jannatul Ferdous, Md. Maruful Hoque, and Md. Rasheduzzaman

Abstract

Agriculture is a major economic activity of rural people in Bangladesh and a fundamental pillar for ensuring food security in the country. Agricultural productivity in coastal zone of Bangladesh has been decreasing in the last decade due to water scarcity caused by groundwater contamination coming from soil salinity. This paper aims at assessing the progress of soil salinity and at identifying the effects of salinity intrusion on agricultural production using the perception of the farmers in the Upazilas (sub-district) of Satkhira district located in the southwest coastal region of Bangladesh. Accordingly, landsat images of the years 2006 and 2016 were analyzed to assess and map the progress of soil salinity. In 2006, soils of none of the study areas had electrical conductivity (EC) beyond 16 dS/m. But in 2016, a total of 5644 ha in the two study areas combined had EC values higher than 16 dS/m. Only a very few varieties of crops can sustain in soils with EC values higher than 16 dS/m. Furthermore, interviews with 300 farmers of 15 randomly selected locations of the two Upazilas were conducted and the results validated by statistical tests and compared with the Landsat Images. Farmers identified water scarcity for irrigation purposes as the most prominent effect of salinity intrusion followed by a decline of prices in the livestock market.

Keywords

Agriculture · Coastal Bangladesh · Farmers · Landsat · Salinity intrusion · Water scarcity

1 Introduction

Salinity intrusion is one of the most common problems in coastal regions, which causes severe land degradation and abandonment of agricultural lands (Li et al. 2011). Salinity intrusion in the coastal region of Bangladesh occurs due to both natural (e.g., flood and storm surge) and human-driven (e.g., excessive use of fertilizer and shrimp culture) reasons (Wu et al. 2008). Coastal districts such as Satkhira, Khulna, Bagerhat, Barguna, Patuakhali, and Barisal are the victims of salinity intrusion, which reduces

Md. A. Habib (✉) · J. Ferdous · Md. M. Hoque · Md. Rasheduzzaman
Climate Change Lab, Department of Civil Engineering, Military Institute of Science and Technology (MIST), Mirpur-1216, Dhaka, Bangladesh
e-mail: m.armanhabib@gmail.com

M. Tauhid Ur Rahman
Department of Civil Engineering, Military Institute of Science and Technology (MIST), Mirpur-1216, Dhaka, Bangladesh

L. Ribbe et al. (eds.), *Towards Water Secure Societies*,
https://doi.org/10.1007/978-3-030-50653-7_13

agricultural productivity remarkably (Disaster Management Bureau 2010). Salt affected areas have been increased by 26.7% from the year 1973 to 2009 in Bangladesh (SRDI 2010). 36.8% of arable lands in the coastal and offshore area are affected by varying degrees of salinity (Asib 2011). Salinity intrusion is one of the most crucial limiting factors for crop development. The rise of the mean sea level caused by erratic climatic patterns would mean more salinity intrusion into the agricultural lands especially in the coastal zones of the world. Salinity is the single most significant problem in the southwestern coastal belt of Bangladesh (Islam et al. 2012).

People of coastal region extract groundwater for irrigation at the beginning of the low flow period, which eventually initiates saltwater intrusion into the coastal aquifers (Murshid 2012). Due to increased saltwater intrusion, farmers are obliged to irrigate their land with saline water (Haider and Hossain 2013). As a result, various agricultural regions have significantly lost their productivity due to soil salinity in last several decades (SRDI 2010). There is also a significant decline in net food production and diversity of food production (Islam et al. 2012). Although it is difficult to measure the scale of damage caused by salinity to agriculture, the effect seems to grow with time. So farmers should adopt some measures to cope with the increasing salinity intrusion.

To know the impact of salinity on the agricultural sector, mapping of saline affected area is an important task, which is very time-consuming with the collection of primary field data. But at present, applications of remote sensing technique make this task more economic and efficient and several remote sensing indices provide a wide range of opportunity to map the saline affected areas. Various studies have been conducted to map the saline affected areas from remote sensing images in different regions of the world (Asfaw et al. 2016; Azabdaftari and Sunar 2016; Elhag 2015; Ivushkin et al. 2016; Madani 2005). Morshed et al. (2016) have developed a multiple regression equation for detecting soil salinity from satellite images studying the southern region of Bangladesh.

This study aims to observe the impacts of salinity on the livelihoods of farmers of Tala and Shyamnagar Upazilas of Satkhira district from field survey and to detect the change in soil salinity of the vegetation areas in the study areas from Landsat images of the year 2006 and 2016. The study concludes with some appropriate adaptation measures to combat the effects of salinity intrusion into the agricultural lands.

2 Materials and Methods

2.1 Study Area

Tala and Shyamnagar Upazilas of Satkhira district in the southwest part of Bangladesh are selected as study area. Eight random locations spread across Shyamnagar named Atulia, Nurnagar, Kaikhali, Shyamnagar, Padmapukur, Ishwaripur, Gabura and Harinagar and Seven random locations in Tala named Nagarghata, Kumira, Patkelghata, Tala, Islamkati, Khalishkhali and Khosra were studied (Fig. 1). The mangrove portion of Shyamnagar Upazila has not been considered in this study, as there is no agricultural activity. Around 46,934 ha area of Shyamnagar (23.4% of total area) has been considered here. The principal source of income in these Upazilas is agriculture; about 70.34% people in Tala and 64.98% in Shyamnagar Upazila are engaged in agricultural occupation (Banglapedia 2014, 2015). Poultry and livestock, wage labors follow after shrimp farming as the major livelihood options in the study area (UNICEF 2014).

Shyamnagar is listed as a salinity prone area along with Satkhira Sadar Upazila (CDMP 2013). Tala is also in close proximity to Satkhira Sadar Upazila. Moreover, two recent devastating cyclones Aila and Sidr and excavated flood control embankment have increased the level of salinity in the study area (Islam et al. 2012).

Salinity intrusion is an issue here and people find it difficult to draw the balance between the needs of irrigation water and drinking water. Almost (99.36%) all the cultivated areas in Shyamnagar has been affected by salinity, among

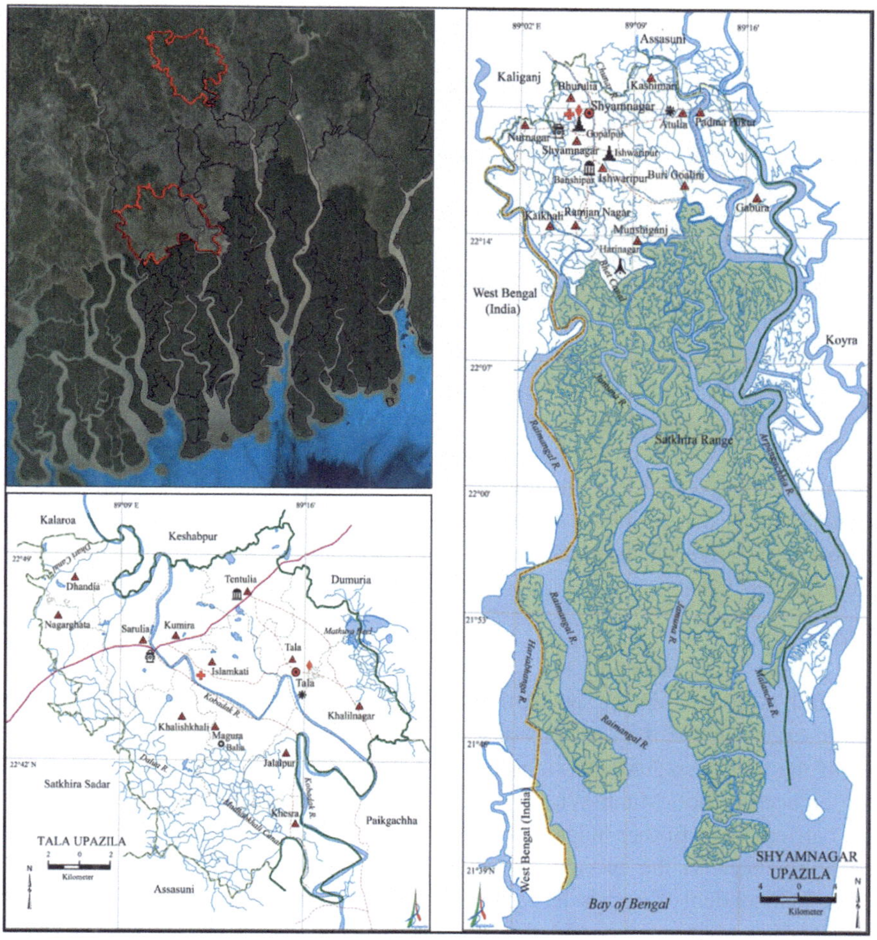

Fig. 1 Location of study area (Shyamnagar and Tala Upazila). *Source* Banglapedia (2014, 2015)

which 37.71% area has salinity level greater than 16 dS/m. On the other hand, 61.13% of cultivated area in Tala has been affected and among them 51.17% has salinity value ranges from 2 to 4 dS/m and remaining had higher salinity value (SRDI 2010).

2.2 Study Methods

Field Survey

The study was based on both primary and secondary data. A field trip to the study areas was organized during March 2016 to collect the primary data. Perceptions of the farmers about the impacts of salinity intrusion were collected by a semi-structured questionnaire. The options in the questionnaire were based on past researches about salinity intrusion and their effects on agriculture. The questionnaire format was kept as simple as possible for better and quick understanding of the farmers keeping their literacy level in concern. Farmers' perceptions were collected on a scale of 1–5 (from very less to very high) on how far they agreed to a predetermined set of possible impacts of salinity intrusion. The set of possible impacts were identified through a focus group discussion prior to preparing the questionnaire.

General information on the type of farmers was collected from Bangladesh Bureau of Statistics (2011) prior to the questionnaire survey to finalize the samples for the questionnaire survey. It was ensured that all types of farmer participated in the questionnaire survey. 95% confidence interval and ±5% level of precision was used for calculating the sample size of the study area and the approximate sample size was 300. Among the respondents, there were 105 owners, 27 owner-cum tenants and 18 tenants in Shyamnagar and 100 owners, 35 Owner-cum tenants and 15 tenants in Tala Upazila.

With the semi-structured questionnaire, 300 farmers of different categories had been interviewed. For an effective and participatory collection of perceptions from the farmers, the local government representatives from Shyamnagar and Tala enlightened the research team and helped with his knowledge of the local communities and agricultural lands.

Statistical Tests
Independent t-test was carried out using IBM SPSS 20.0 software on the perceptions. This was to observe whether the opinions of the respondents of the two areas had any significant statistical differences or not. Mann-Whitney '*U*' test was also carried out to further bolster the fact of statistical differences. If the '*p*-value' of the equal variance independent '*t*' test is larger than the value of 0.05, then it can be said that the perceptions of the respondents in the two Upazilas did not differ significantly. On the other hand, the critical value for the Mann-Whitney '*U*' test is dependent on the number of options (*N*). Critical value is 13 when *N* is 8 and 17 when *N* is 9. If the obtained value is greater than the critical value, then it means that the response of the farmers in the two Upazilas does not vary significantly (The Mann-Whitney test 2011). If the statistical differences are not significant then the risk reduction measures can be applied commonly to both the locations due to the fact that both the areas are exposed to or are suffering from the same type of hazards.

Acquisition and Processing of Satellite Images
To detect the change in soil salinity of the study areas, multi-temporal satellite images (WRS2: 138/44, and 138/45) were collected from the United States Geological Survey (USGS) website. All bands of Landsat Thematic Mapper 5 (TM) (-dated 27 February 2006), and Landsat 8 Operational Land Imager (OLI) (dated 26 March 2016) were collected, and all the spatial data layers were registered to the same Universal Transverse Mercator (UTM) coordinate system and resampled to the same pixel resolution of 30 m. As precipitation is insufficient in the month of February and March in Bangladesh, satellite images for these months are very suitable to detect the soil salinity. Insufficient precipitation cannot leach ions from the soil profile and as a result salts accumulate in the soil (SRDI 2010).

Radiometric correction has been performed in the raw images and the Digital Number (DN) values have been converted into reflectance to avoid the noise from DN value (Huang et al. 2002).

Identification of Common Vegetation Areas of three years through NDVI
The study aims to detect the change in soil salinity of vegetation areas. At first, the vegetation areas have been identified through the Normalized Vegetation Index (NDVI), which has been calculated with the Near infrared and Red bands of Landsat. The areas with NDVI value more than 0.2 has been considered as vegetation area (USGS 2015). Then the common vegetation area of the years 2006 and 2016 has been identified through the intersect tool of ArcGIS. Common vegetation areas are 20,979.50 ha and 9954.81 ha in Tala and Shyamnagar Upazilas, respectively.

Calculation of Soil Salinity Level of Common Vegetation areas and Change Detection
After the radiometric correction of both scenes of Landsat images, the scenes have been mosaicked and the common vegetation areas have been masked from the images. The masked images have been used to compute salinity and to detect changes within the period of 20 years.

Morshed et al. (2016) have developed a multiple regression equation for detecting soil salinity from the Landsat images after computing several remote sensing indices, which can represent the salinity level of soils. OLS, SAR, and SLAG were used as regression models to develop the equation and correlation with the field EC values were the bases of the equation. Following is the multiple

Table 1 Classification of soil salinity level

Salinity class	Salinity value calculated from remote sensing images (EC in dS/m)
Very low saline	<2
Low saline	2–4
Medium saline	4–6
High saline	6–8
Very high saline	>8

Source Morshed et al. (2016)

regression equation, which has been used in this study to detect the soil salinity.

$$\text{Salinity} = 12.53 + 1108.95 * \text{NIR} + 2.73 * \text{NIR}/R - 717.86 * \text{SI2} - 295.7 * \text{SAVI}$$

where

NIR = Near Infrared band of Landsat

R = Red band of Landsat

SI2 = $\sqrt{(G^2 + R^2 + \text{NIR}^2)}$, G is the Green band of Landsat

SAVI = Soil Adjusted Vegetation Index = (default vale of L is 0.5 (Huete et al. 1985)).

The soil salinity of the study area has been detected by using the above equation and the salinity level has been classified into five classes to detect the change. Table 1 shows the classification of soil salinity level.

3 Results

3.1 Nature and Extent of Salinity Intrusion from the Perception of Farmers

Salinity intrusion has been influenced by both man-made and natural factors. The man-made factors include haphazard construction of embankments, sedimentation in the polders mainly, while the natural causes include the effects of climate change such as erratic temperature and rainfall pattern. Farmers' perceptions were noted on how much they believe salinity intrusion is a man-made or a natural hazard. It appears that slightly more percentage of people "agree" to it be termed as a man-made hazard (55%) than a natural hazard (45%) in Tala. The farmers of Shyamnagar also provided similar perceptions that salinity is slightly more of a man-made (60%) hazard than a natural hazard (49%) and that its intensity is on the rise. The 'p-value' (0.971) signifies that the perceptions of the study areas show no statistical differences and can be assumed to be similar. Most of the respondents in Tala (75%) and Shyamnagar (80%) agreed that the intensity/spread of salinity is increasing over time, which is also seen from the analysis of satellite images.

3.2 Impacts of Salinity Intrusion

Figure 2a, b reflects that almost the entire number of respondents (89% in Tala and 86% in Shyamnagar) agreed very strongly to the question of to what extent they thought water scarcity is an impact of salinity. The response received to famine (71% in Tala and 65% in Shyamnagar) and crop failures (45% in Tala and 51% in Shyamnagar) got mixed opinions therefore placed in the medium response category.

The farmers placed the impact of salinity on health issues (e.g., kidney disease) in the medium category as well. Continuous exposure to drinking saline water leads to development of kidney diseases (Naser et al. 2017). Livestock health was found to be affected by salinity intrusion and the response was fairly high as greater proportions of the respondents opted for the 'high' category (32% in Tala and 47% in Shyamnagar). As crop lands are getting depleted, livestock food shortage is persistent in coastal Bangladesh and also pregnant and young livestock are not surviving the high levels of salinity in the drinking water fed to them (Alam et al. 2017).

The increase in food prices in the study area has yet not been very prominent and therefore gathered medium category response. The respondents have identified decrease of livestock prices as an impact of salinity as the largest proportion of them has categorized this impact as high (72% in Tala and 68% in Shyamnagar).

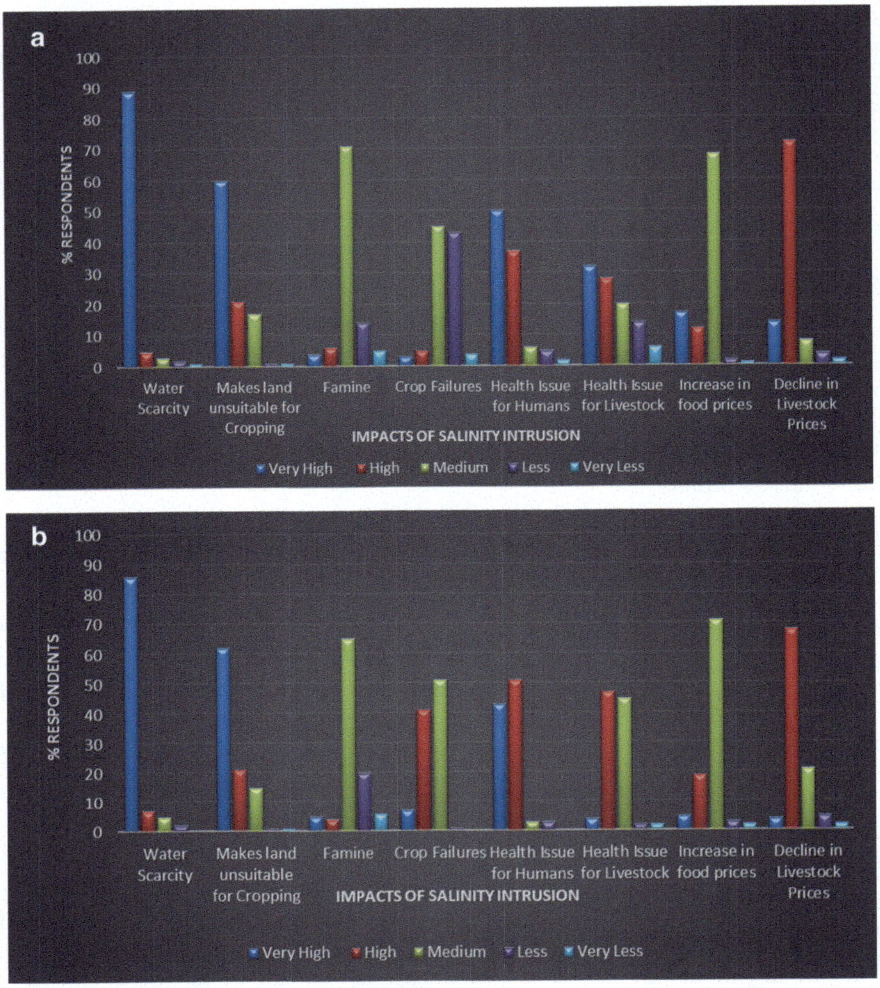

Fig. 2 **a** Perceptions of farmers on impacts of salinity intrusion in Tala. **b** Perceptions of farmers on impacts of salinity intrusion in Shyamnagar

The '*p*-value' (0.972) tells us that the perception of respondents in the two study areas on the impacts of salinity in the eight mentioned categories did not differ much statistically. The Mann-Whitney '*U*' value (24.5) also means that the assessments of the respondents do not differ significantly.

3.3 Change Detection of Soil Salinity from Remote Sensing Images

Tables 2 and 3 show the changing patterns of soil salinity in Tala and Shyamnagar, respectively. For Tala, the levels of soil salinity seem to increase from medium to high to very high from 2006 to 2016 (Fig. 3).

7.5 and 5.5% of vegetation areas have been categorized into medium range salinity having Electrical Conductivity (EC) values from 4 to 6 dS/m in 2006 and 2016, respectively. The medium saline zone areas in Tala were gradually converted into high and very high saline zones. In 2016, however, the percentage of high saline zone area decreased from 50.3 in 2006 to 16.4. But this was only due to the fact that these high saline zones were converted to very high saline zones (EC > 8 dS/m). From 2006 to 2016, the percentage of area falling into very high saline

Table 2 Saline affected vegetation areas in the year 2006 and 2016 in Tala Upazila according to the salinity classes

Salinity class	Area (% of total vegetation area)	
	2006	2016
Very low	0.1	1.2
Low	0.3	1.7
Medium	7.5	5.5
High	50.3	16.4
Very high	41.8	75.2

zones increased rapidly. Alternatively, three fourth of vegetation area of Tala is now a very high saline zone (Table 2 and Fig. 3).

The scenario was somewhat similar in Shyamnagar as well. In 2006, 64.61% area fell into the very high saline category. Compared to Tala, the rate of conversion from medium saline zone to very high saline zone was higher in Shyamnagar. 20% more area was converted into high saline zones in Shyamnagar within 10 years than Tala. However, in Shyamnagar slightly less percentage of area fell into the very high saline zone compared to Tala, about 68.4% (Table 3 and Fig. 4).

Table 3 Saline affected vegetation areas in the year 2006 and 2016 in Shyamnagar Upazila according to the salinity classes

Salinity class	Area (% of total vegetation area)	
	2006	2016
Very low	0.01	3.23
Low	0.02	5.16
Medium	2.06	9.34
High	33.3	13.87
Very high	64.61	68.4

4 Discussions

4.1 Extent and Impacts of Salinity Intrusion: Fresh Water Scarcity

Withdrawal of fresh river water from upstream, irregular rainfall, introduction to brackish water for shrimp cultivation, regular saline tidal water flooding in the unprotected area, the capillary rise of soluble salts, etc. are the main causes of

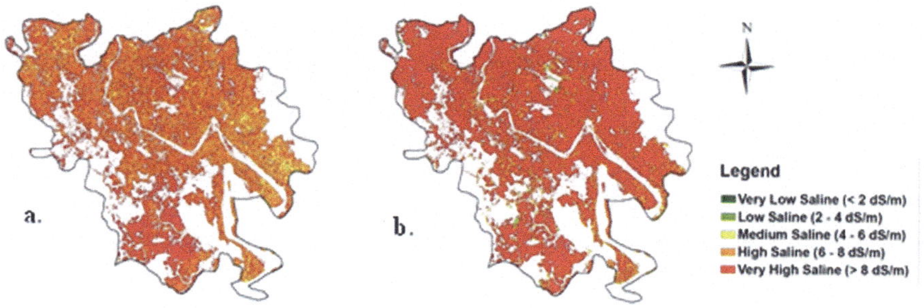

Fig. 3 Spatial pattern of different levels of saline affected soil in Tala Upazila **a** in 2006, and **b** in 2016

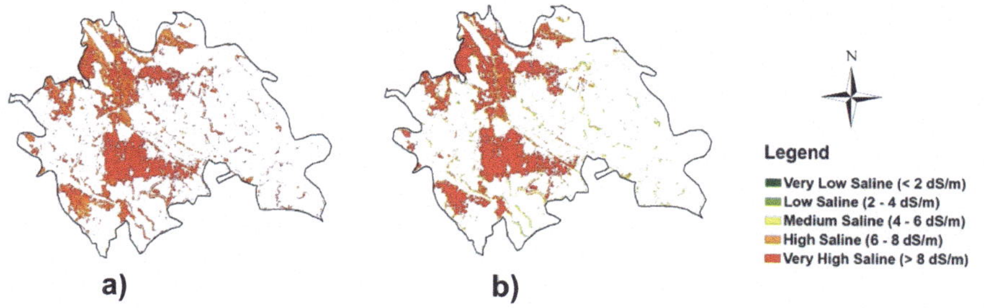

Fig. 4 Spatial pattern of different levels of saline affected soil in Shyamnagar Upazila **a** in 2006 and **b** in 2016

increased soil salinity in the top soils of the coastal region (SRDI 2010).

The farmers' perception also reflected that the spread of salinity is on the rise both in Tala and Shyamnagar Upazila. Figure 2a, b shows some impacts of increasing levels of salinity in the soil of the study area. The effect that gathered the most response was freshwater scarcity, which eventually affects crop production. Freshwater scarcity is the single most severe problem in both Upazilas (Rahman et al. 2016). The spread of salinity into the natural water bodies has further contributed to the water scarcity of the study area. Freshwater scarcity means residents of the study area need to find a balance between drinking water needs and irrigation water. Inadequate irrigation water would make cropping difficult.

Crop production reduced significantly due to the increased influx of salinity in the soils (Kumar et al., 2010). Saline resistant crops may have contributed to decreased number of cases of crop failures. However, this impact gathered medium response from more than half of the respondents in Shyamnagar. The respondents cited that there was a crop failure of huge extent in the year 2009. 30% of the cultivable land of Bangladesh is in the coastal areas and soil salinity is increasing day by day (Clarke et al. 2015). Cultivable lands of the southwestern region would decrease significantly due to sea level rise and the adverse impact of soil salinity

Health issues for humans got a varied response from very high to less response due to the enhanced medical facilities over the years in the study area. Livestock health is a matter of concern as the shortage of fresh drinking water also affects the livestock's normal livelihood and continuous exposure to salty drinking water imposes health effects on them. Food prices are expected to be on the rise when agricultural production is hindered and the opinions of the farmers make no difference to it.

4.2 Agricultural Adaptation Measures

To establish the need for agricultural adaptation measures, farmers must first realize the risks associated with agriculture. In the study area, this risk is primarily from the effects of salinity intrusion. The farmers' perception on salinity intrusion has established the need for adaptation measures to be utilized. Agricultural adaptation measures have therefore become mandatory for the area to prevent a collapse in the crop productivity of the area and to change the increasing rate of land use shifting from agricultural land to shrimp farms. The study area is losing agricultural lands at an alarming rate. In the present scenario, eventually, the farmers will lose their jobs and will have to look forward to food supply from the other parts of the country.

The modern phenomenon to tackle the impact of hazards is the practice of adaptation measures. Some adaptation measures have already been practiced in the study area. The farmers who own their own land are better suited to switch to appropriate adaptation measures at a fast rate than the owner-cum-tenant farmers and tenant farmers, as they tend to have better control on the choices for their respective agricultural lands. This is a positive indication for both Tala and Shyamnagar as over 50% farmers in the study area own agricultural land. Researchers had found the advisory board regarding various agricultural adaptation measures in one of the study places. The advisory board tells about practicing retention irrigation, alley cropping, multi-cropping/intercropping, reusing of sullage water, and mulching technique for reducing the effects of soil salinity as adaptation measures. The agricultural extension department, Shyamnagar branch advises the following adaptation techniques for the area.

(i) Digging of tiny ponds in the farmland to retain water: This practice will help retaining irrigation water for the drier periods of the season. The department advises that farmers start digging the ponds right after the end of monsoon season.

(ii) Alley Farming: Farmers are advised to utilize the alley lands between the main farmlands with crops such as potato, wheat, maze, mustard, cereals, and onion.

(iii) Multi-cropping: Farmers are encouraged to harvest more than one crop in one farmland and if possible also construct poultry farms as well.
(iv) Mulching: Applying mulching practices in the salt drying fields is one of the adaptation techniques that can be followed by the farmers.
(v) Reuse of sullage water: This practice is aimed at conserving irrigation water.

4.3 Integrated Agricultural Adaptation Measures

The results revealed the level of perception gap or insufficiency in incorporating the adaptation measures as most of the measures gathered medium to low response from the farmers. The study recommends some suggestions for integrated adaptation measures in the study areas:

- The course of actions of the farmers and the policy-makers such as local and national government should complement each other and lessoned by the local agriculture departments such as agricultural extension services and the NGOs. Researchers should engage in assessing, monitoring, and modifying, if necessary, the adaptation measures.
- As the study area is composed of mainly low-income groups, so community-based adaptation measures should be encouraged so that any form of investment is distributed in a good manner.
- Green energy technologies such as solar power panels can be installed for small groups of farmers to power the irrigation pumps and conserve electricity or reduce the dependency on electricity. Such initiatives were found in the study area, but were limited and isolated.
- Low water consuming crops variety, salinity resistant crops variety should be a very effective adaptation measure in the study area.
- Crops having a short harvesting period can be cultivated so that the harvesting period does not fall into the crisis times.

5 Conclusions

Problems in agricultural practice due to salinity are the main focus of the study and freshwater scarcity has been identified as the most striking effect of salinity through farmers' perception. Statistical tests also suggested no major differences in the perceptions of the study areas concluding that the crisis scenario is similar for both the areas. Salinity propagating through the soil layers is contaminating both the groundwater and surface water. Mapping of soil salinity from Landsat images has revealed the fast conversion of low class soil saline zones to medium and higher classes gradually. Both the study areas now have negligible extent of area that fall into the low saline zones.

Agricultural practices have been affected significantly due to scarcity of irrigation water free of salinity. To sustain agriculture in these crisis-prone areas, studies on farmers' perception are recommended to formulate an integrated and well-coordinated agricultural adaptation measures.

References

Alam MZ, Boggs LC, Mitra S (2017) Effect of salinity intrusion on food crops, livestock, and fish species at Kalapara coastal belt in Bangladesh. J Food Qual 2017. https://doi.org/10.1155/2017/2045157

Asfaw E, Suryabhagavan KV, Argaw M (2016) Soil salinity modeling and mapping using remote sensing and GIS: the case of Wonji sugar cane irrigation farm. J Saudi Soc Agric Sci, Ethiopia. https://doi.org/10.1016/j.jssas.2016.05.003

Asib A (2011) Some of the major environmental problems relating to land use changes in the coastal areas of Bangladesh. J Geogr Reg Plann 4(1):1–8

Azabdaftari A, Sunar F (2016) Soil salinity mapping using multi-temporal Landsat data. Paper presented at the international archives of the photogrammetry, remote sensing and spatial information sciences, vol XLI-B7, XXIII ISPRS Congress, 12–19 July 2016,

Prague, Czech Republic. https://doi.org/10.5194/isprs-archives-XLI-B7-3-2016

Bangladesh Bureau of Statistics (BBS) (2011) Population and Housing Census 2011 Statistics and informatics division (SID), Ministry of planning, Government of the People's Republic of Bangladesh. http://bbs.dhaka.gov.bd/sites/default/files/files/bbs.dhaka.gov.bd/law_policy/6ed6b42c_2015_11e7_8f57_286ed488c766/SocioEconomic%20and%20demographic%20Report%202012.pdf. Accessed on 19 June 2016

Banglapedia (2014) Tala Upazila. http://en.banglapedia.org/index.php?title=Tala_Upazila. Accessed on 20 June 2016

Banglapedia (2015) Shyamnagar Upazila. http://en.banglapedia.org/index.php?title=Shyamnagar_Upazila. Accessed on 20 June 2016

CDMP (2013) Annual progress report, comprehensive disaster management programme phase 2010–2014 (CDMP II). https://info.undp.org/docs/pdc/Documents/BGD/CDMP-II%20Annual%20Progress%20Report%20-%202013.pdf. Accessed on 19 June 2016

Clarke D, Williams S, Parks K, Jahiruddin M, Salehin M (2015) Projections of on-farm salinity in coastal Bangladesh. J Environ Sci Processes Impacts. https://doi.org/10.1039/C4EM00682H

Disaster Management Bureau (2010) National plan for disaster management 2010–2015. http://www.ddm.gov.bd/. Accessed on 21 June 2016

Elhag M (2015) Evaluation of different soil salinity mapping using remote sensing techniques in arid ecosystems, Saudi Arabia. J Sens 2016 (Article ID 7596175). http://dx.doi.org/10.1155/2016/7596175

Haider MZ, Hossain MZ (2013) Impact of salinity on livelihood strategies of farmers. J Soil Sci Plant Nutr 13(2):417–431

Huang C, Yang L, Homer C, Wylie B, Vogelman J (2002) University of Nebraska—Lincoln Digital Commons@University of Nebraska—Lincoln. At satellite reflectance: a first order normalization of LANDSAT 7 ETM+ images. https://digitalcommons.unl.edu/cgi/viewcontent.cgi?article=1108&context=usgspubs. Accessed on 22 June 2016

Huete AR, Jackson RD, Post DF (1985) Spectral response of a plant canopy with different soil backgrounds. J Remote Sens Environ 17:37–53. https://doi.org/10.1016/0034-4257(85)90111-7

Islam MS, Rahman MA, Sultana N, Nath B, Paul A (2012) Using geospatial techniques to assess the salinity impact on agricultural landuse: a study on Shyamnagar Upazila, Satkhira. J Agric Environ Int Dev 106(2):157–169. http://dx.doi.org/10.12895/jaeid.20122.91

Ivushkin K, Bartholomeus H, Bregt AK, Pulatov A (2016) Satellite thermography for soil salinity assessment of cropped areas in Uzbekistan. J Land Degrad Dev 28(3):870–877. https://doi.org/10.1002/ldr.2670

Kumar U, Baten MA, Masud AA, Osman KS, Rahman MM (2010) Cyclone Aila: one year on, natural disaster to human sufferings. Unnayan Onnesan the innovators, Dhaka, Bangladesh. http://www.unnayan.org/reports/climate/ailareport_final.pdf. Accessed on 17 May 2016

Li DJ, Chun WM, Tiyip T (2011) Study on soil salinization information in arid region using remote sensing technique. J Agric Sci China 10(3):404–411. https://doi.org/10.1016/S1671-2927(11)60019-9

Madani AA (2005) Soil salinity detection and monitoring using landsat data: a case study from Siwa Oasis, Egypt. J GISci Remote Sens 42(2):171–181. https://doi.org/10.2747/1548-1603.42.2.171

Morshed MM, Islam MT, Jamil R (2016) J Environ Monit Assess 188:119. https://doi.org/10.1007/s10661-015-5045-x

Murshid SM (2012) Impact of sea level rise on agriculture using groundwater in Bangladesh. Dissertation, TU Delft

Naser AM, Rahman M, Unicomb L, Doza S, Matin Ahmed K, Nasir Uddin M, Selim S, Gribble MO, Anand S, Clasen TF, Luby SP (2017) Drinking water salinity and kidney health in southwest coastal Bangladesh: baseline findings of a community-based stepped-wedge randomised trial. J Lancet 389(Supplement 2):S15. https://doi.org/10.1016/S0140-6736(17)31127-3

Rahman MTU, Rasheduzzaman M, Habib MA, Ahmed A, Tareq SM, Muniruzzaman SM (2016) Assessment of fresh water security in coastal Bangladesh: an insight from salinity, community perception and adaptation. J Ocean Coast Manag 137:68–81. https://doi.org/10.1016/j.ocecoaman.2016.12.005

SRDI (2010) Saline soils of Bangladesh. Soil Resources Development Institute, Ministry of Agriculture, SRMAF Project, Ministry of Agriculture. http://srdi.portal.gov.bd/sites/default/files/files/srdi.portal.gov.bd/publications/bc598e7a_df21_49ee_882e_0302c974015f/Soil%20salinity%20report-Nov%202010.pdf. Accessed on 25 May 2016

The Mann-Whitney test (2011) Graham hole research skills Mann-Whitney test handout version 1.0. Page 1. http://users.sussex.ac.uk/~grahamh/RM1web/MannWhitneyHandout%202011.pdf. Accessed on 15 June 2016

UNICEF (2014) Local capacity building and community empowerment (LCBCE) program: district equity Profile-Satkhira

USGS (2015) NDVI, the foundation for remote sensing phenology. https://phenology.cr.usgs.gov/ndvi_foundation.php. Accessed on 25 May 2016

Wu J, Vincent B, Yang J, Bouarfa S, Vidal A (2008) Remote sensing monitoring of changes in soil salinity: a case study in Inner Mongolia, China. J Sens 8(11):7035–7049. https://doi.org/10.3390/s8117035

Remote Sensing for Monitoring Suspended Sediment Concentration in the Lower Ganges (Padma) River

Mashrekur Rahman, G. M. Tarekul Islam, and Md. Munsur Rahman

Abstract

Suspended Sediment Concentration (SSC) is a significant parameter in the hydrologic, morphologic, and ecosystem studies of large alluvial rivers of the Ganges-Brahmaputra-Meghna (GBM) delta. Traditional in situ measurement of SSC in the Lower Ganges River is challenging in terms of time, cost, and spatial coverage. This study investigated the applicability of open-access Landsat Enhanced Thematic Mapper (ETM+) images in estimating the SSC of the Lower Ganges. Multiple-temporal Landsat 7 ETM+ images were processed to extract Digital Numbers (DN) of pixels corresponding to Bangladesh Water Development Board (BWDB)'s river measurement station, Mawa (SW93.5L). The DNs were converted to radiance and ultimately to top-of-atmosphere (ToA) reflectance. The ToA values for Landsat-7 bands 1–4, which sense electromagnetic radiation of 0.45–0.52, 0.52–0.60, 0.63–0.69, and 0.76–0.90 μm, respectively, were combined with corresponding measured values of SSC, between the years 2000 and 2012 for determination of statistical relationship between them. R^2 for bands 1, 2, 3, and 4 were 0.64, 0.51, 0.44, and 0.67, respectively; showing that Coefficient of Determination (R^2) value of band 4 presented the best relationship—therefore chosen as the best SSC indicator. Scatter plot of predicted SSC values from a polynomial equation based on band 4 against in situ values of SSC with 1:1 fit line generated strong positive coefficient of determination of 0.89 and Root Mean Square Error of 88.3 ppm. Using a polynomial model based on the band 4 data, spatial distribution maps and temporal variation of SSC, between the years 2000 and 2010, for monsoon and post-monsoon seasons were demonstrated and river cross-sectional analysis was performed.

Keywords

Landsat · Sediment · Remote Sensing · Ganges · GIS · Alluvial · River · GBM · Delta

1 Introduction

The quantification of suspended sediments and their transport mechanism is essential in understanding riverine, estuarine, and coastal processes. SSC is required to estimate and predict

M. Rahman (✉) · G. M. T. Islam · Md. M. Rahman
Institute of Water and Flood Management, Bangladesh University of Engineering and Technology, Dhaka, Bangladesh
e-mail: mrahman18@crimson.ua.edu

G. M. T. Islam
e-mail: tarek@iwfm.buet.ac.bd

Md. M. Rahman
e-mail: mmrahman@iwfm.buet.ac.bd

L. Ribbe et al. (eds.), *Towards Water Secure Societies*,
https://doi.org/10.1007/978-3-030-50653-7_14

soil erosion and sediment transport caused by changes in land use patterns (Collins and Walling 2004). Suspended sediments transported in rivers is an important parameter to consider in river engineering, sustainable water resources, and environmental management (Halbe et al. 2013; Kisi 2008; Melesse et al. 2011). Suspended sediment is a dominant factor in the Ganges and Brahmaputra Rivers. For Brahmaputra, the ratio of the bedload discharge to the suspended sediment discharge is 10–25 (Okada et al. 2016). The task of obtaining reliable and constant spatio-temporal SSC data of rivers in Bangladesh is severely limited. These limitations include size and extent of rivers, financial and economic constraints, lack of experienced personnel. Although in situ measurement techniques of suspended sediment are continually changing, it is generally accepted that they do not completely satisfy requirements. Instruments used traditionally to evaluate SSC have been limited to single location, and therefore variations in SSC are obtained by profiling or by using a fixed vertical array of measuring devices. Conventional techniques are often constrained in their capability to provide comprehensive spatial and temporal data of SSC (Crawford and Hay 1993; Sheng and Hay 1988; Thorne and Hardcastle 1997; Thorne et al. 1993).

2 Application of Satellite Remote Sensing in the Monitoring of SSC

Multiple studies have demonstrated that water reflectance in the NIR and visible spectral regions (bands 1–4) of Landsat images correlate quite strongly with SSC (Ritchie and Schiebe 2000). Therefore, reflectance is usable as an indicator of SSC for inland and coastal waters. Miller and McKee (2004) used the moderately high resolution of MODIS 250 m data for mapping small-scale features of TSM concentration in different inland and coastal waters. Digital Numbers (DN)s were directly used in past studies to retrieve SSC values from water using satellite remote sensing. Digital spectral data from 14 Landsat MSS scenes of Moon Lake in Coahoma County, Mississippi were analyzed and compared with ground measurements of total solids and suspended sediments in the lake surface water for the period between January 1983 and May 1985. Coefficients of determination of greater than 0.81 were calculated between MSS Band 2 (0.6–0.7 μm) or Band 3 (0.7–0.8 μm) and suspended sediments or total solids. Coefficients of determination for multiple regression using three or four MSS bands were greater than 0.90 (Ritchie et al. 1987). Physically quantified values of spectral radiance, which is converted from DN through in-flight calibration sensor parameters, were also used to indicate SSC values for coastal waters. Aranuvachapun and LeBlond (1981) used digital data from Landsat's Multiple Spectral Scanners (MSS) to calculate the amount of radiation received from sea level at the position of the satellite—after assessing the atmospheric effects, on the radiance data, the amount of radiance was then related to sediment concentration.

However, reflectance values, especially exoatmospheric or top-of-atmosphere reflectance are the most widely used for indicating SSC; for example, Band 3 (Red) of SPOT-XS2 demonstrated strong correlation with SSC within a range of 0–10 ppm. Similarly, MODIS Terra Band 1 showed significant correlation with SSC range of 0–60 ppm. Reflectance in the Band 2 and Band 4 of Landsat ETM+ have also been used in the past to retrieve SSC values (Ouillon et al. 2004). Ekercin (2007) conducted a study on a river in Istanbul, Turkey using High Resolution Ikonos Multispectral Imagery to retrieve Total Suspended Solids (TSS). Few studies also used algorithms to define the relationship between in situ SSC and corresponding spectral radiance or reflectance to indicate SSC directly using satellite images (Pavelsky and Smith 2009; Reddy and Srinivasulu 1994; Xia 1993). Most of these studies investigated relations for water bodies such as lakes, reservoirs, and coastal regions; mainly waters with very low to moderate turbidity. Also, many of their concentrations were on water quality (Zhang et al. 2003) aspects rather than morphological aspects associated with rivers, especially alluvial rivers.

Highly turbid large rivers have been largely ignored when it comes to the study of remote sensing of suspended sediments. Wang et al. (2009) used multi-temporal Landsat ETM+ to remotely sense SSC for the Yangtze River in China. Wang and Lu (2010) estimated SSC of the Yangtze at Jiujiang using time-series satellite data of high temporal resolution Terra MODIS, and found differences in water reflectance between Band 2 and Band 5 could provide relatively accurate SSC estimation even when in situ atmospheric conditions were unknown. Fleiflea (2013) assessed the applicability of Landsat ETM+ data for monitoring suspended sediment load (SSL) in the Mekong River over temporal and spatial dimensions. Spectral mixture analysis from Landsat MSS and TM images have been used in the past to estimate the concentration of suspended sediment in surface waters of the Amazon River wetlands (Mertes et al. 1993). Islam et al. (2001) explored the satellite remote sensing of suspended sediments in the Ganges and Brahmaputra Rivers using older Landsat TM and AVHRR data. Furthermore, Islam et al. (2002) focused on the distribution of suspended sediment in the coastal sea off the Ganges–Brahmaputra River mouth. However, a comprehensive study regarding the remote sensing of SSC in the Padma River has not yet been published.

3 The Ganges–Brahmaputra Rivers System

The Ganges–Brahmaputra basin (Fig. 1) also represents the world's largest sediment dispersal system and has formed the world's biggest submarine fan, known as the Bengal fan (Kuehl et al. 1989). River Ganges enters into the Bengal basin from the northwest after draining the Himalayas and about 2500 km through India. The Ganges divides into two distributaries: the River Lower Ganges flows southeast toward the confluence with the River Brahmaputra in Bangladesh (Mukherjee et al. 2009). The Brahmaputra channel belt shows a rapid lateral migration of as much as 800 m/year (Allison 1998). Brahmaputra can be classified as braided. The detailed processes responsible for the formation of a braided river are still poorly understood. Many published works deal specifically with the process of braiding, but little agreement can be found, and the hydraulic parameters of a braided stream are extremely complex (Coleman 1969). Ganges is a sediment laden, wide, meandering river. It's annual mean predicted sediment load is 498.8×10^6 tons (Islam and Jaman 2006).

The Padma River in Bangladesh, also known as the Lower Ganges starts after the Ganges and the Brahmaputra Rivers confluence, flowing toward southeast until it confluences with the Meghna River near the Bay of Bengal. Secondary data procured from Bangladesh Water Development Board (BWDB) corresponding to their gauging station at Mawa SW 93.5L (23°28′ N & 90°15′ E) were used in this study. This station is the only reliable historical measured data point source currently available in Bangladesh for the Lower Ganges River; therefore, specifically chosen for this study.

4 Data Collection

In situ data corresponding to Mawa 93.5L (Fig. 2) gauging station was procured from BWDB and maximum sediment concentration was chosen for this study. 26 Landsat 7 Enhanced Thematic Mapper (ETM+) images were selected as the remotely sensed data. Landsat 7 captures visible bands in the spectrum of blue, green, red, near-infrared, and mid-infrared (bands 1–5 & 7). It is also equipped with thermal infrared channel (band 6) with a spatial resolution of 60 m. Landsat 7 data were glitch-free up to 2003, then there was a mechanical disturbance in the Scan Line Corrector (SLC); therefore, images resulted in partially missing data. However, the SLC error did not have any significant implication regarding this study, and therefore correction was not necessary. Water level, discharge, flow velocity, and cross-sectional data corresponding to sediment data were also procured from the Bangladesh Water Development Board through appropriate acquisition procedure. These data sets were used for morphological analyses in this study. 26 Landsat

Fig. 1 The Ganges–Brahmaputra basin

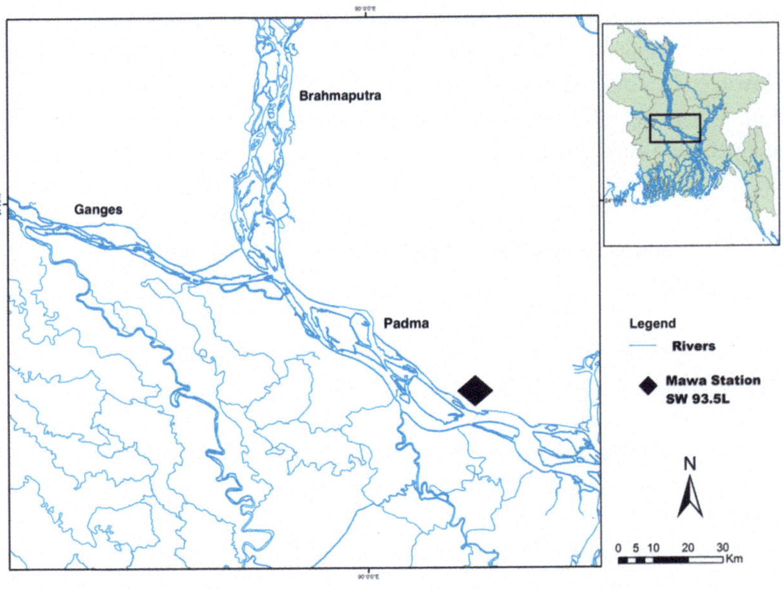

Fig. 2 Location of in situ data collection point

ETM+ L1G products between the years 2000 and 2012 were processed for analysis in this study. These images were selected based on two broad temporal categories, i.e., monsoon and post-monsoon season images. The months of June to September were considered to be inside monsoon, whereas the months of October and November were deemed to be in post-monsoon season. These images were processed using open-source GIS softwares.

5 Satellite Image Processing

Digital Numbers (DN) corresponding to bands 1–4 of Landsat 7 ETM+ images were extracted. Raw Landsat ETM+ Digital Numbers were converted to physical values of radiance using the following formula (1) (Fleiflea 2013).

$$L\lambda = ((L_{max} - L_{min})/255)D + L_{min} \quad (1)$$

where D is the DN value, $L\lambda$ is the spectral radiance (mW cm^{-2} sr^{-1} μm^{-1}), L_{max} is the radiance measured at detector saturation (mW cm^{-2} sr^{-1} μm^{-1}), and L_{min} is the lowest radiance (mW cm^{-2} sr^{-1} μm^{-1}).

Top-of-Atmosphere or Exoatmospheric Reflectance corrects for the earth–sun distance and solar angle errors. Reflectance is the proportion of the radiation striking a surface to the radiation reflected off of it. Thus to increase the accuracy of results, it was preferable to convert those radiance values into ToA reflectance values. The following formula (2) (Fleiflea 2013) was used for conversion.

$$p\lambda = \left(\pi d^2 L\lambda / E_o\lambda \cos\theta_s\right) \quad (2)$$

where $p\lambda$ is the ToA or Exoatmospheric Reflectance as a function of the band width λ, d^2 is the factor which corrects the variation in solar irradiances due to varying earth–sun distance at different times of the year. The value of d for each image was acquired from a table provided by USGS, which consists of d values corresponding to the day of year. $E_o\lambda$ is the mean solar Exoatmospheric Irradiance (W μm^{-2}); θ_s represents the Solar Zenith Angle in radians which was obtained by subtracting the Solar Elevation Angle from 90°.

6 Development of SSC—Spectral Reflectance Relationships

We investigated which band or what combination of bands of Landsat ETM+ provides the strongest correlations with SSC. For the purpose of analysis, the ToA Reflectance values were converted to ToA % values. Thereafter XY scatter plots were drawn for each of the four bands. For each scatter plot, a best-fit polynomial curve was plotted and its Coefficient of Determination (R^2) was found out (Fig. 3).

Although reflectance at band 1 shows significantly strong relationship with measured SSC, band 4 (R^2= 0.6729) demonstrated the strongest statistical relationship among the four, which is consistent with previous studies (Sterckx et al. 2007; Wass et al. 1997). Therefore band 4 (Near Infrared) was chosen as the indicator for SSC. Near-Infrared spectrum is sensitive to SSC and relative to shorter wavelength bands, also less influenced by bottom reflectance in environments with shallow water (Tolk et al. 2000). Owing to the non-linear nature of the data, curvilinear polynomial Eq. (3) was chosen. The polynomial relationship obtained from the best-fit curve can be stated as,

$$\text{SSC} = 22.565\,b^2 - 549.27b + 3616.7 \quad (3)$$

where 'SSC' is the Suspended Sediment Concentration (ppm) and 'b' is the Band 4 ToA reflectance percentage. Polynomial relationship, which provided the best coefficient of determination values, was chosen over linear, exponential and log formulations, after applying them similarly to examine the relationship between SSC and reflectance percentage values. In order to validate the regression relation between SSC and band 4 reflectance percentage values, scatter plot of predicted values of SSC from Eq. (3) versus measured values of SSC was drawn and the Root Mean Square Error (RMSE) was extracted (Fig. 4) and the p-value of t was calculated as a function of the correlation coefficient found between ToA reflectance and SSC, and the total number of images used in the regression analysis. Thereafter the residuals were also plotted against measured SSC (Fig. 5). Relative error percentage was also demonstrated in an individual plot (Fig. 6).

Scatter plot of predicted SSC values from the polynomial Eq. (3) against in situ values of SSC

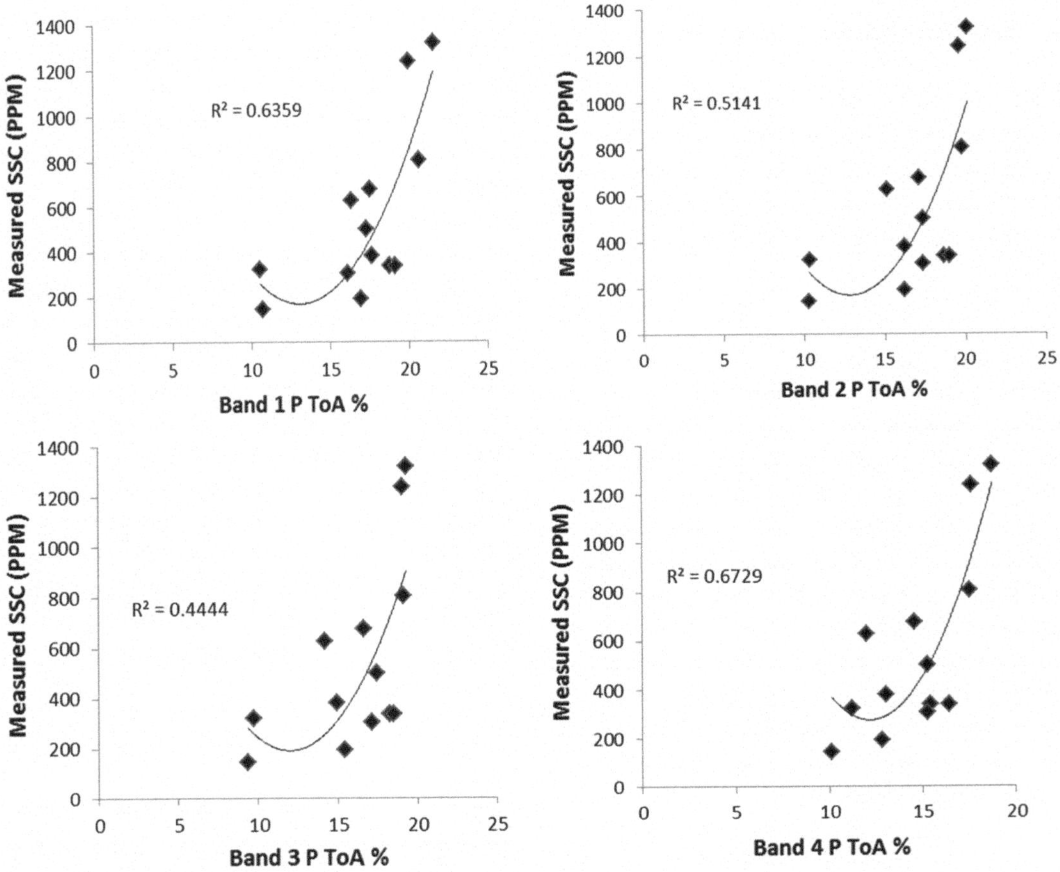

Fig. 3 Scatter plot of measured SSC versus ToA reflectance percentage of bands 1–4

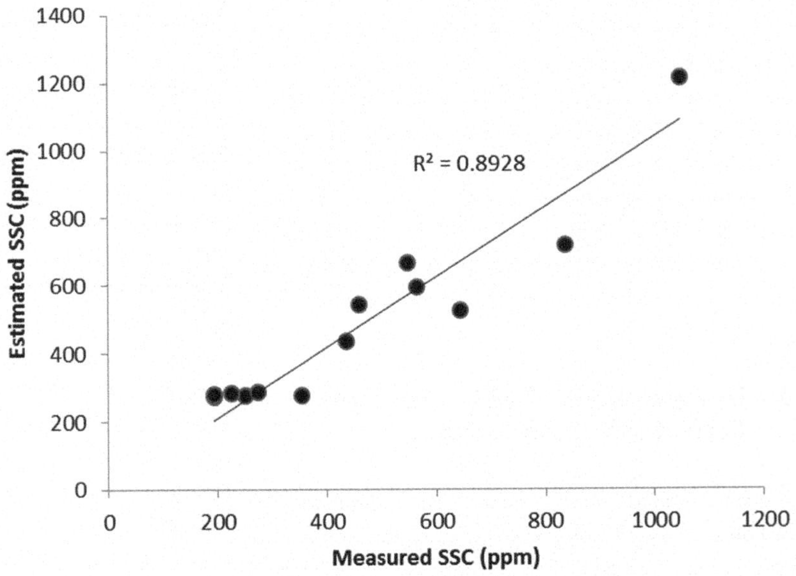

Fig. 4 Scatter plot of estimated SSC values by the polynomial model based on band 4 (Near Infrared) ToA reflectance and measured SSC data with 1:1 fit line ($R^2 = 0.89$, $n = 13$)

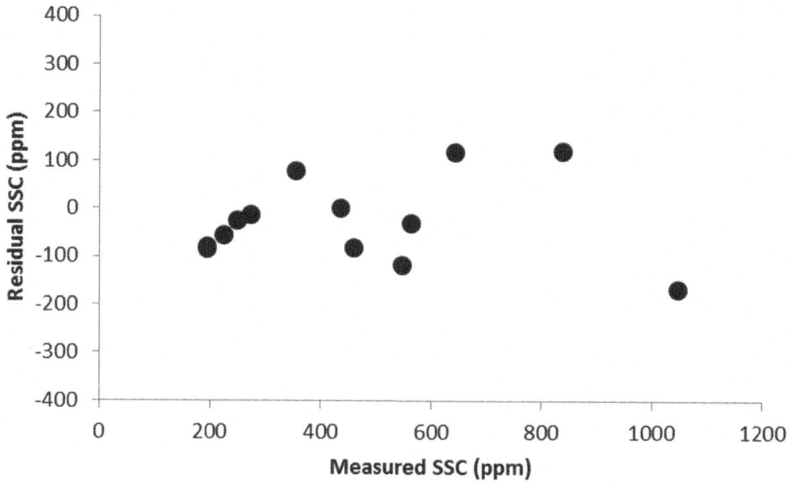

Fig. 5 Residue of SSC versus measured SSC

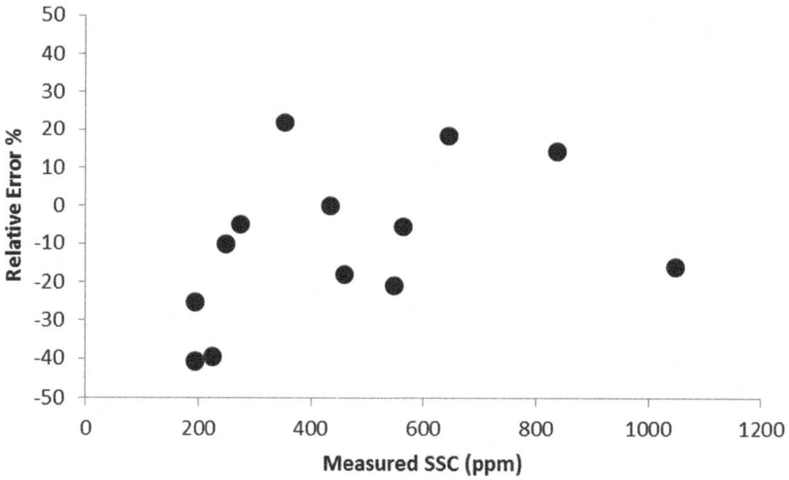

Fig. 6 Scatter plot of relative error percentage of estimated SSC from measured SSC (average relative error is ±18.11%)

with 1:1 fit line generated strong positive coefficient of determination of 0.89, Root Mean Square Error (RMSE) of 88.3 ppm and the p-value of t was found to be 0.005. The residual values of SSC were calculated by subtracting estimated values of SSC from the measured data. RMSE was extracted by calculating the square root of average squared residual. Therefore, predicted values and measured values of SSC are in strong statistical agreement (Taylor 1990), and that correlation does imply causation for the polynomial relationship. It is apparent that this model's prediction accuracy is better for higher values of SSC, i.e., above 200 ppm to approximately 800 ppm. From the measured values of SSC, it is observed, the Lower Ganges River's SSC values usually range above 200 ppm for most of the time during monsoon and even during post-monsoon season; thus, affirming this model's applicability. The average relative error is ±18.11%.

7 Assessment of Spatio-Temporal Variation of SSC in the Lower Ganges River

To demonstrate the spatial and temporal variations of SSC in the Lower Ganges River, the previously developed polynomial relationship (3) was applied, and using a GIS software, the spatial

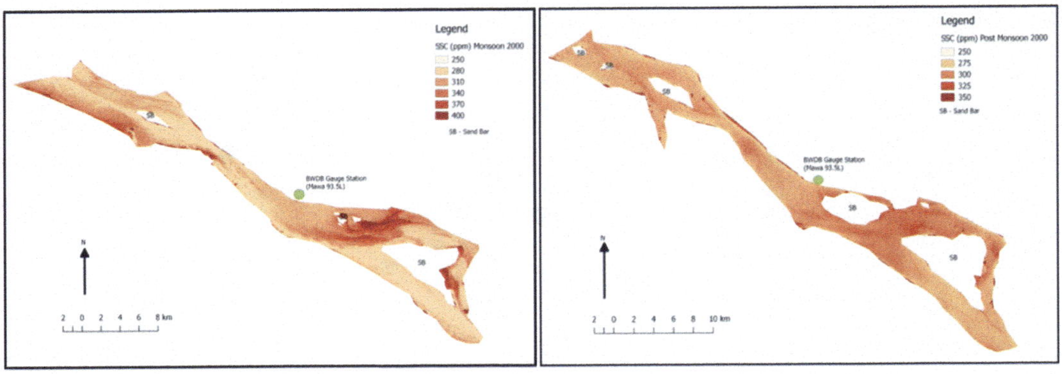

Fig. 7 Spatial distribution map of SSC in the Lower Ganges River for monsoon and post-monsoon of 2000

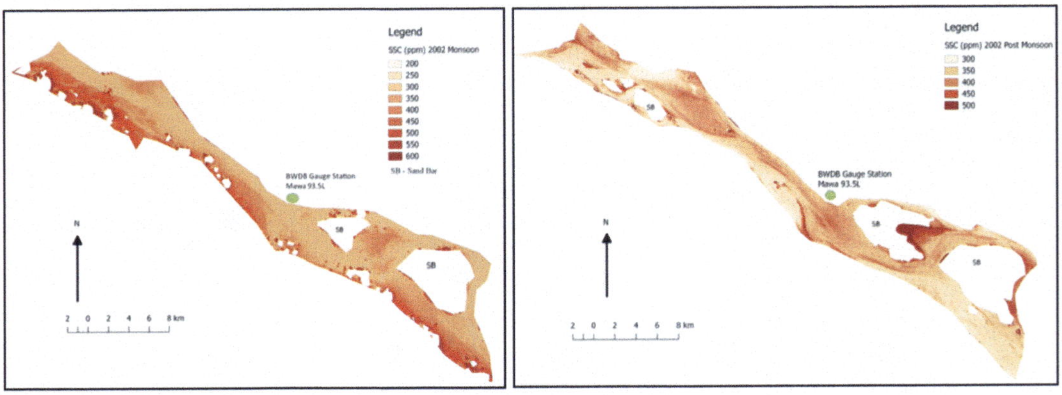

Fig. 8 Spatial distribution map of SSC in the Lower Ganges River for monsoon and post-monsoon of 2002

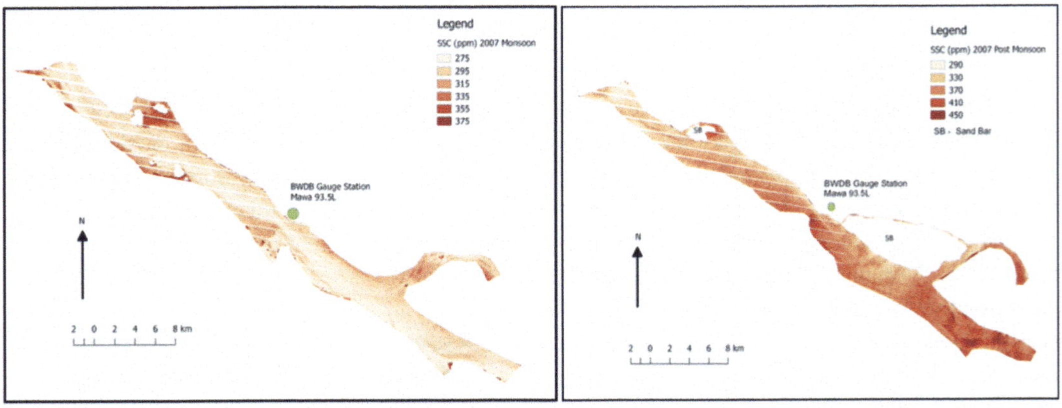

Fig. 9 Spatial distribution map of SSC in the Lower Ganges River for monsoon and post-monsoon of 2007

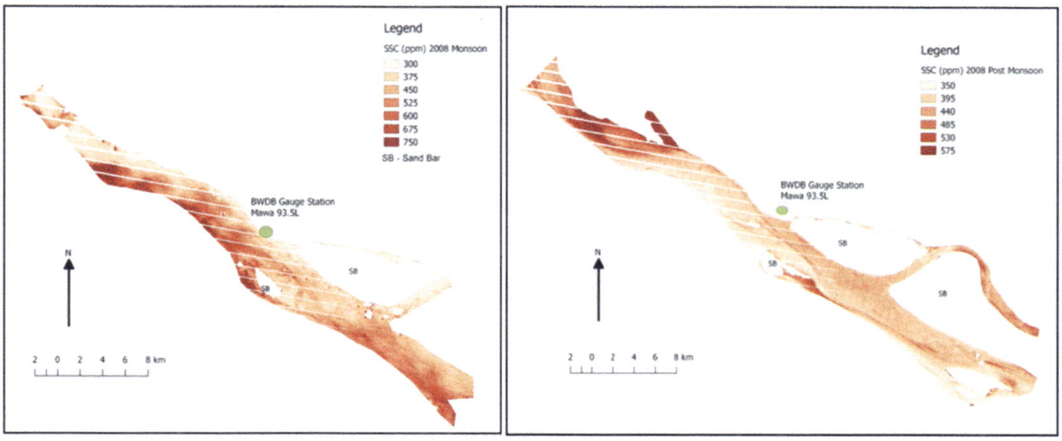

Fig. 10 Spatial distribution map of SSC in the Lower Ganges River for monsoon and post-monsoon of 2008

distribution maps of SSC were generated. Landsat ETM+ images from both monsoon and post-monsoon seasons, between years 2000 and 2010, were used for this purpose.

The SSC values for monsoon of 2000 (Fig. 7) ranged between 250 and 400 ppm. There was no significant cloud patch over the water body of the river. The sand bars excluded from the map have been marked appropriately. A notable rise of suspended sediment concentrations can be observed immediately downstream of sand bars; an observation which may be attributed to the erosive effect of the river, water level rise, and other unknown effects. Another interesting observation in this map is the sand bar, which apparently is situated at the margin where water flows from the Ganges and Brahmaputra conflate, was not subjected to the erosive effect of the river, contrary to the observed behavior of sand bars downstream. The detailed processes responsible for the formation of a braided river are still poorly understood. Many published works deal specifically with the process of braiding, but little agreement can be found, and the hydraulic parameters of a braided stream are extremely complex (Coleman 1969). Compared to the monsoon image, slight decline in SSC levels can be observed for the post-monsoon image; SSC levels varied between 250 and 350 ppm. Also, in contrast with the previous map, the sand bars are not facing erosion apparently. The average measured discharge of Lower Ganges River at Mawa for monsoon season of 2000 was 63,519 m^3/s, and the average measured discharge for the corresponding post-monsoon season was 38,028 m^3/s. The average water flow velocity declined from 3.06 m/s in monsoon to 2.30 m/s in post-monsoon season. These declines in discharge and flow velocity could be driving factors behind the decline of sand bar erosion. There was also a decline of measured average water level at Mawa 93.5 L of more than 1 m between monsoon and post-monsoon seasons, also possibly influencing erosion and increases the area above water surface. The decline of SSC levels may be attributed to declines in discharge, water level, and flow velocity (Bogárdi and Szilvássy 1974).

It must be noted that in the monsoon map of 2002 (Fig. 8), excluding the designated sand bars, the missing patches toward are bank are cloud covers. A significant rise of SSC level can be seen for 2002 monsoon. This rise may be due to increased discharge, water level, or flow velocity considering the prevalence of a monsoon flood in 2002. However, there is no definitive data available on the aforementioned parameters for this map. In this map again, the flow from the Ganges and the Brahmaputra can be delineated based on their sharp difference in SSC levels. The Ganges seemed to be contributing higher levels suspended sediments into

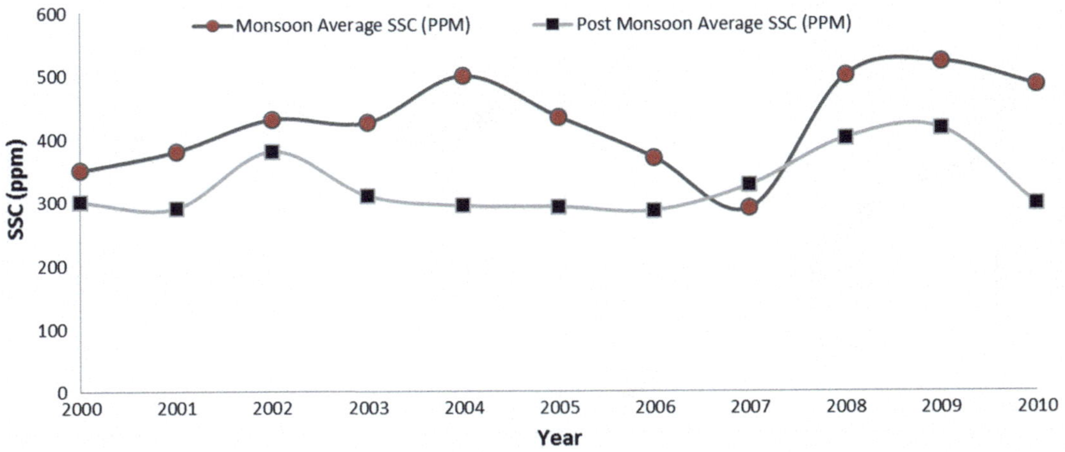

Fig. 11 Time-series variation of SSC

the Lower Ganges compared to the Brahmaputra. Although the SSC levels in this map are higher than the post-monsoon seasons of previous years, a significant amount of sand bar erosion can be observed. Since there is no reliable in situ data available corresponding to this map, the erosive behavior cannot be exactly attributed to certain hydro-morphological aspects.

For 2007 (Fig. 9), the monsoon SSC values ranged between 275 and 375 ppm. Here again, there are no in situ data of corresponding water level, discharge or flow velocity available for this image. However, compared to previous monsoons, SSC values were considerably lower for this particular image. Contrary to previous observations, the post-monsoon image has higher SSC range compared to the monsoon season. Even though there are no corresponding in situ discharge, water level and flow velocity data available for this image, the rise in post-monsoon sediment transport may be possibly due to increased flow in delayed monsoon (Mirza 2002).

For 2008 (Fig. 10), SSC range varied between 300 and 750 ppm for the monsoon image. This increase is expected due to increase in average discharge, flow velocity, and water level compared to the previous monsoon (Bogárdi and Szilvássy 1974). Owing to decreases in water level in the post-monsoon image, discharge and flow velocity, the maximum SSC also declined compared to corresponding monsoon season (Bogárdi and Szilvássy 1974). Some differences in upstream and downstream SSC levels can be seen. These observations may possibly be due to variations in sediment settling rate and boundary layer shear stress (Van Rijn 1993).

Time-series Variation of SSC

To investigate how average monsoon and post-monsoon SSC values varied at the Mawa, SW 93.5L station at the Lower Ganges River, the temporal variation of SSC between the years 2000 and 2010 was plotted (Fig. 11).

Relationship between SSC and Cross-section of River

To investigate any existing relationship between spatial variations of SSC and measured cross section of the Lower Ganges River at Mawa, SSC values corresponding to cross-sectional distance were recorded and plotted. The results are shown in Figs. 12, 13, 14 and 15.

The cross-sectional data acquired for 2001 post-monsoon (Fig. 12) is from BWDB's River Maintenance Project (RMP)-6. For 2002 Monsoon (Fig. 13), the cross-sectional data was recorded during RMP-4, and for post-monsoon of 2005 (Fig. 14) and 2008 (Fig. 14), cross-sectional data was recorded during RMP-7. It can be observed from all the four plots that SSC rises, albeit through variable range, whenever the Reduced Level (RL) value rises. In other words, SSC tends to increase with depth, and when bed

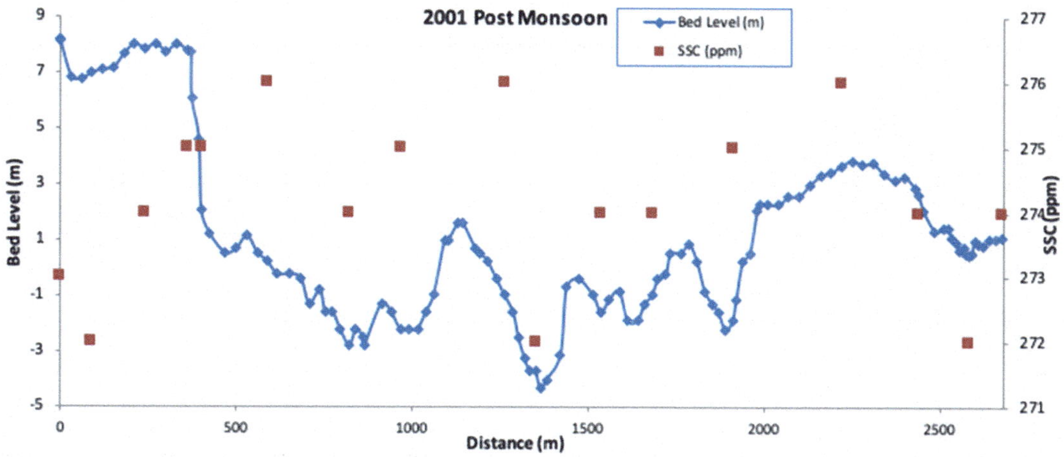

Fig. 12 Variation of cross section and corresponding SSC for post-monsoon of 2001

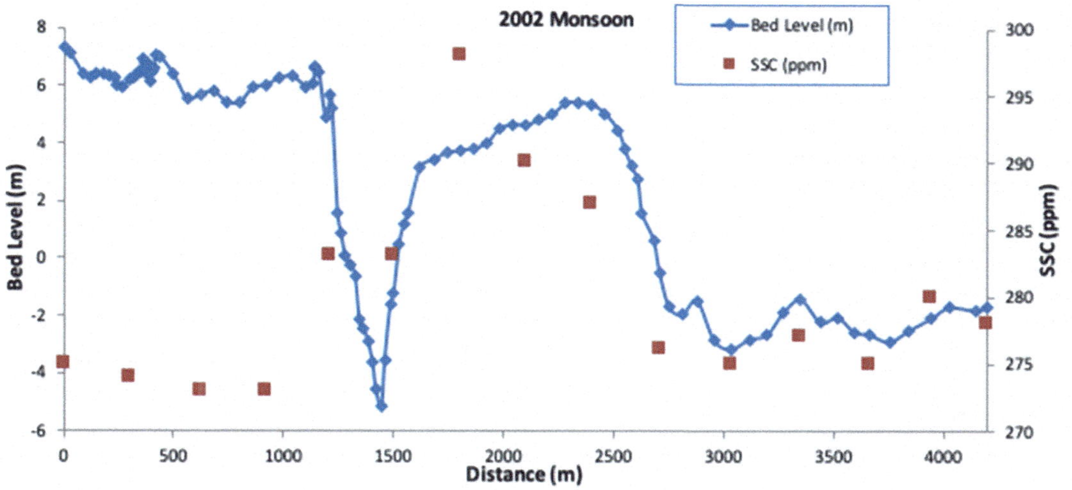

Fig. 13 Variation of cross section and corresponding SSC for monsoon of 2002

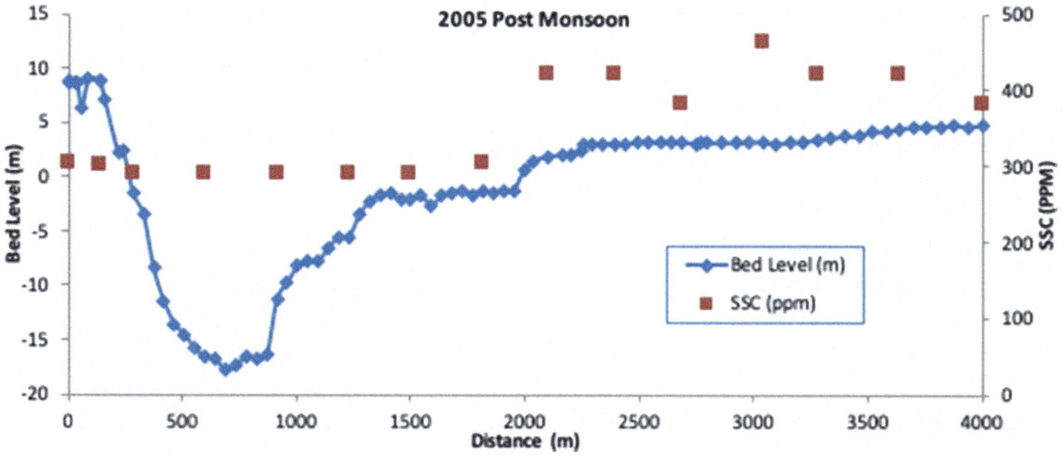

Fig. 14 Variation of cross section and corresponding SSC for post-monsoon of 2005

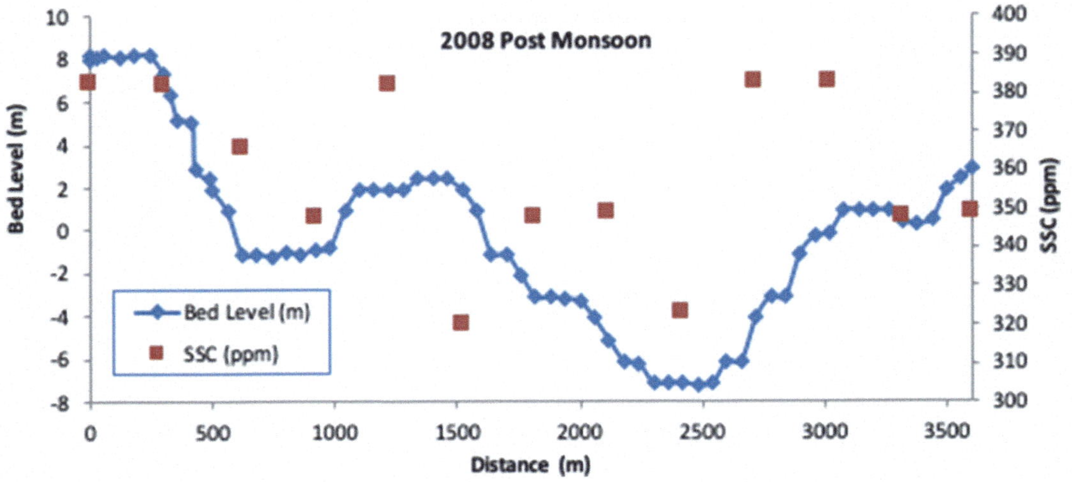

Fig. 15 Variation of cross section and corresponding SSC for post-monsoon of 2008

level rises, the corresponding SSC also tends to increase (Okada et al. 2016). Although near-infrared wavelengths of band 4 of Landsat 7 ETM+ is less susceptible to be influenced by shallow water environments (Tolk et al. 2000), there may still be some degree of influence during changes in water depth. This phenomenon maybe a limitation when remotely sensed SSC is used to assess changes in water depth (Gao 2009).

8 Conclusions

In this study, Landsat ETM+ images were used in combination with available in situ values of SSC to develop a polynomial model for retrieving temporal values of SSC at the Lower Ganges River in Bangladesh. For validation of the model involving SSC and Band 4 Reflectance Percentage values, scatter plot of predicted values of SSC from the model versus measured values of SSC was drawn and the Root Mean Square Error (RMSE) was extracted. Scatter plot of predicted SSC values from the polynomial Eq. (3) against in situ values of SSC with 1:1 fit line generated strong positive coefficient of determination (R^2) of 0.89 and Root Mean Square Error (RMSE) of 88.3 ppm.

Because predicted values and measured values of SSC were in strong statistical agreement, polynomial model based on the band 4 data was selected to estimate monsoon and post-monsoon spatial variations of SSC for the years 2000–2010. Spatial distribution maps of SSC, between years 2000 and 2010 were observed. Among the notable observations, overall SSC levels appeared to be generally higher in monsoon and flood seasons compared to post-monsoon season. However, there were exceptions in this observation too. Rise in discharge, water level, and flow velocity increased the overall SSC. During cross-sectional analysis, it was generally observed that rise in bed level also caused small jumps in corresponding SSC.

References

Allison MA (1998) Historical changes in the Ganges-Brahmaputra Delta Front. J Coast Res 14:1269–1275

Aranuvachapun S, LeBlond PH (1981) Turbidity of coastal water determined from Landsat. Remote Sens Environ 11:113–132. https://doi.org/10.1016/0034-4257(81)90013-4

Bogárdi J, Szilvássy Z (1974) Sediment transport in alluvial streams. Akadémiai Kiadó Budapest, Hungary

Coleman JM (1969) Brahmaputra River: channel processes and sedimentation. Sed Geol 3:129–239

Collins AL, Walling DE (2004) Documenting catchment suspended sediment sources: problems, approaches and prospects. Prog Phys Geogr 28:159–196. https://doi.org/10.1191/0309133304pp409ra

Crawford AM, Hay AE (1993) Determining suspended sand size and concentration from multifrequency acoustic backscatter. J Acoust Soc Am 94:3312–3324. http://dx.doi.org/10.1121/1.407237

Ekercin S (2007) Water quality retrievals from high resolution ikonos multispectral imagery: a case study in Istanbul, Turkey. Water Air Soil Pollut 183:239–251. https://doi.org/10.1007/s11270-007-9373-5

Fleiflea AE (2013) Suspended sediment load monitoring along the Mekong river from satellite images. J Earth Sci Climatic Change

Gao J (2009) Bathymetric mapping by means of remote sensing: methods, accuracy and limitations. Prog Phys Geogr 33:103–116. https://doi.org/10.1177/0309133309105657

Halbe J, Pahl-Wostl C, Sendzimir J, Adamowski J (2013) Towards adaptive and integrated management paradigms to meet the challenges of water governance. Water Sci Technol 67:2651–2660

Islam G, Jaman S (2006) Modelling sediment loads in the Lower Ganges, Bangladesh. In: Proceedings of the Institution of Civil Engineers-Water Management, vol 2. Thomas Telford Ltd., pp 87–94

Islam MR, Yamaguchi Y, Ogawa K (2001) Suspended sediment in the Ganges and Brahmaputra Rivers in Bangladesh: observation from TM and AVHRR data. Hydrol Process 15:493–509

Islam MR, Begum SF, Yamaguchi Y, Ogawa K (2002) Distribution of suspended sediment in the coastal sea off the Ganges-Brahmaputra River mouth: observation from TM data. J Mar Syst 32:307–321. https://doi.org/10.1016/S0924-7963(02)00045-3

Kisi Ö (2008) Constructing neural network sediment estimation models using a data-driven algorithm. Math Comput Simul 79:94–103. https://doi.org/10.1016/j.matcom.2007.10.005

Kuehl SA, Hariu TM, Moore WS (1989) Shelf sedimentation off the Ganges-Brahmaputra river system: evidence for sediment bypassing to the Bengal fan. Geology 17:1132–1135

Melesse A, Ahmad S, McClain M, Wang X, Lim Y (2011) Suspended sediment load prediction of river systems: an artificial neural network approach. Agric Water Manag 98:855–866

Mertes LAK, Smith MO, Adams JB (1993) Estimating suspended sediment concentrations in surface waters of the Amazon River wetlands from Landsat images. Remote Sens Environ 43:281–301. https://doi.org/10.1016/0034-4257(93)90071-5

Miller RL, McKee BA (2004) Using MODIS Terra 250 m imagery to map concentrations of total suspended matter in coastal waters. Remote Sens Environ 93:259–266

Mirza MMQ (2002) Global warming and changes in the probability of occurrence of floods in Bangladesh and implications. Glob Environ Change 12:127–138

Mukherjee A, Fryar AE, Thomas WA (2009) Geologic, geomorphic and hydrologic framework and evolution of the Bengal basin, India and Bangladesh. J Asian Earth Sci 34:227–244. https://doi.org/10.1016/j.jseaes.2008.05.011

Okada AY, Koseki H, Kudo S, Muraoka K (2016) Comprehensive measurement techniques of water flow, bedload and suspended sediment in large river using acoustic doppler current profiler. In: Wieprecht et al (eds) River sedimentation, pp 1274–1280

Ouillon S, Douillet P, Andréfouët S (2004) Coupling satellite data with in situ measurements and numerical modeling to study fine suspended-sediment transport: a study for the lagoon of New Caledonia. Coral Reefs 23:109–122. https://doi.org/10.1007/s00338-003-0352-z

Pavelsky TM, Smith LC (2009) Remote sensing of suspended sediment concentration, flow velocity, and lake recharge in the Peace-Athabasca Delta, Canada. Water Resour Res 45

Reddy MA, Srinivasulu S (1994) Comparison of IRS-IB LISS-IIA pixel array sizes for estimating suspended solids concentration in Hussain Sagar Lake, Hyderabad, India—a statistical approach. Int J Remote Sens 15:3693–3706

Ritchie JC, Schiebe FR (2000) Water quality. In: Remote sensing in hydrology and water management. Springer, Berlin, pp 287–303

Ritchie JC, Cooper CM, Yongqing J (1987) Using Landsat multispectral scanner data to estimate suspended sediments in Moon Lake, Mississippi. Remote Sens Environ 23:65–81

Sheng J, Hay AE (1988) An examination of the spherical scatterer approximation in aqueous suspensions of sand. J Acoust Soc Am 83:598–610

Sterckx S, Knaeps E, Bollen M, Trouw K, Houthuys R (2007) Retrieval of suspended sediment from advanced hyperspectral sensor data in the Scheldt estuary at different stages in the tidal cycle. Mar Geodesy 30:97–108

Taylor R (1990) Interpretation of the correlation coefficient: a basic review. J Diagn Med Sonography 6:35–39

Thorne PD, Hardcastle PJ (1997) Acoustic measurements of suspended sediments in turbulent currents and comparison with in-situ samples. J Acoust Soc Am 101:2603–2614. http://dx.doi.org/10.1121/1.418501

Thorne PD, Hardcastle PJ, Soulsby RL (1993) Analysis of acoustic measurements of suspended sediments. J Geophys Res Oceans 98:899–910. https://doi.org/10.1029/92JC01855

Tolk BL, Han L, Rundquist D (2000) The impact of bottom brightness on spectral reflectance of suspended sediments. Int J Remote Sens 21:2259–2268

Van Rijn LC (1993) Principles of sediment transport in rivers, estuaries and coastal seas, vol 1006. Aqua Publications, Amsterdam

Wang J-J, Lu X (2010) Estimation of suspended sediment concentrations using Terra MODIS: an example from the Lower Yangtze River, China. Sci Tot Environ 408:1131–1138

Wang J-J, Lu XX, Liew SC, Zhou Y (2009) Retrieval of suspended sediment concentrations in large turbid rivers using Landsat ETM+: an example from the

Yangtze River, China. Earth Surf Proc Land 34:1082–1092. https://doi.org/10.1002/esp.1795

Wass PD, Marks SD, Finch JW, Leeks GJL, Ingram JK (1997) U.K. fluxes to the North Sea, Land Ocean Interaction Study (LOIS) Rivers Basins Research, the first two years monitoring and preliminary interpretation of in-river turbidity and remote sensed imagery for suspended sediment transport studies in the Humber catchment. Sci Tot Environ 194:263–283. http://dx.doi.org/10.1016/S0048-9697(96)05370-3

Xia L (1993) A united model for quantitative remote sensing of suspended sediment concentration. Int J Remote Sens 14:2665–2676

Zhang X, Zhang Y, Wen A, Feng M (2003) Assessment of soil losses on cultivated land by using the 137Cs technique in the Upper Yangtze River Basin of China. Soil Tillage Res 69:99–106

Conclusion

Justyna Sycz

The studies selected for this contribution form part of the inter- and transdisciplinary exchange conducted in the scope of the International Conference Series: Water Security and Climate Change (WSCC). In this chapter, we summarize the main messages emerging from the contributions published in this volume and discussed in the scope of the WSCC conference series.

Water security for sustainable development: Since all life and socioeconomic activities rely on water, we consider water and the progress of the Sustainable Development Goal 6 (SDG 6) "Clean water and sanitation" as a critical determinant for achieving almost all other SDGs set up in 2015 by the United Nations. "Zero hunger" (SDG 2) is inconceivable without water for agriculture. "Improving human health" (SDG 3) depends on progress in drinking water and sanitation. Carrying water especially by women and girls consumes time for attending schools influencing quality of education (SDG 4) and gender equity (SDG 5). Secure water supply for industry promotes decent work and economic growth (SDG 8). Clean energy (SDG 7), responsible consumption (SDG 12), life below water (SDG 14), and life on land (SDG 15) also depend on water availability (UN 2015; World Bank 2016; BMZ 2017; Karnib 2019).

J. Sycz (✉)
Faculty of Spatial Development and Infrastructure Systems, TH Köln - University of Applied Sciences, Cologne, Germany
e-mail: justyna.sycz@th-koeln.de

L. Ribbe et al. (eds.), *Towards Water Secure Societies*,
https://doi.org/10.1007/978-3-030-50653-7_15

However, achieving water security for sustainable development in the face of rapid changes—in demography, economies, geopolitics, and climate is one of the major challenges for societies today, at both global and local level.

1 Next Steps for Water Action

Considering already taken initiatives and the current research status to deal with the challenges related to ensure "the reliable availability of an acceptable quantity and quality of water (…) coupled with an acceptable level of water-related risks" (UN-Water 2013), we have identified the following gaps and recommendations for action.

1.1 Water Management in the Anthropocene Requires an Interdisciplinary Approach

We are living in the Anthropocene where the water availability and demand are affected by a strong tandem interaction between the hydro-climatic and socioeconomic systems (Savenije et al. 2014; Rockström et al. 2014; Graham et al. 2000). Therefore, there is a huge request for interdisciplinary research. This research should bring together socioeconomic and hydro-climatic knowledge and data to seek a profound understanding of the interactions between the two systems. Such knowledge is crucial to support evidence-based water management for which we propose a simple conceptual framework that comprises four steps:

1. Answer the questions: How much water is available? What is the water demand of all sectors including the ecosystems? = > **2.** Based on the result of point 1 answer the question: What are the challenges?: Too much water; Too little water to cover the demand; Sufficient but poor quality water; **3.** What are the reasons for the challenges in 2?: (socioeconomic, governance failure, hydro-climatic conditions) = > **4.** Formulate interventions to deal with the identified challenges.

1.2 New Technologies for Evidence-Based Water Management in the Anthropocene

The recent developments in digital technologies like data analytical capacities (big data, machine learning, and artificial intelligence), data storage and sharing capabilities, hydrological modeling or remote sensing open up a range of possibilities for gathering information necessary for answering the questions of: *How much water is available?, What is the water demand? and What are the challenges?* and, accordingly, enable an evidence-based water management in the Anthropocene.

In order to systematically use the potential of these emerging technologies, a range of new competencies is required by professionals working in the water sector. Hence, the development of digital competencies should receive more attention in capacity-building and in university curricula.

1.3 Straighten the Role of Local Universities as Knowledge Hubs for Sustainable Development

Capacity building, education, and technological innovation have already been acknowledged as fundamental to achieving water security and climate resilience, (UN 2020). However, not always clear is by whom these actions should and could be implemented. We believe that as research and education centers universities are the actors with huge potential to act as engines of progress toward not only water secure societies but toward delivering all sustainable development goals at regional and local scales (Hoffmeister et al. 2016). We urge the local and national governments as well as the financial sector to empower universities for acting as hubs of knowledge and to consider them as important consultation bodies in creating "the future we want" (UN 2015).

1.4 Coordinators of the Science-Policy Dialogue

However, knowledge creation alone is insufficient. Crucial for making knowledge and innovations working for sustainable development is their implementation into decision-making processes. This requires an effective science-policy dialogue. Nonetheless, such a dialogue is extremely challenging. Scientific knowledge is not always easily accessible or comprehensible for decision-makers. Scientists and politicians have different objectives and responsibilities. In order to facilitate this dialogue and mutual understanding, we suggest educating professionals with capacities to understand both perspectives and act as coordinators between science and policy. Universities and institutions of public administration would be adequate places to establish positions for such science-policy coordinators.

1.5 Water Management Requires Local Solutions

Challenges of water management are always determined by local hydro-climatic and socioeconomic conditions (FAO 2012; Jaeger et al. 2013). Hence all activities related to water management ranging from capacity building, investments, infrastructure construction, implementing new technologies to gather information about water demand and availability should always consider and be adapted to the local conditions.

1.6 Establish a Global Database of Local Water Projects

There are thousands of water-related projects funded by international and national development organizations spread around the globe. However, the experience shows that gathering information about existing projects in a certain region to ascertain what has been already implemented, what gaps remain, and how to create synergies by developing new initiatives requires extremely time-consuming research work. Even local stakeholders often lack information about regional initiatives. The creation of a Global Database of Local Water Projects would provide a comprehensive list of existing projects and facilitate coordination across different initiatives. Such a database could be implemented into the World Bank Water Data Portal and include different search functions such as by region, project, funding volume or thematic scope. International and national funding institutions could assume the responsibility for entering of new projects into the database.

1.7 Improve the Access of the Water Sector to Climate Finance

Water security and climate change are urgent and interconnected challenges currently facing humankind. Climate change is expected to significantly impact the predictability of water availability and affect water quality requiring urgent adaptation actions (UN 2020). Conversely, sustainable water management has a huge potential to support climate change mitigation, e.g., by using CO_2 emission-reducing technologies for wastewater treatment (UN 2020), or by protecting peatlands, which are the second natural carbon store after oceans (Harenda et al. 2016). However, as reported by the Climate Policy Initiative, in 2016, only 2.6% of the climate finance went directly to water management (CPI 2018). Accordingly, we urge considering the demand of the water sector for climate change adaptation and the potential to reduce greenhouse emissions and call to strengthen the inclusion of water in climate actions and negotiations.

1.8 Water Management as a Crucial Factor for Progress Toward Achieving a Sustainable Future

We have already acknowledged water as a critical determinant for achieving almost all SDGs. Additionally, as mentioned above, sustainable water management has also a large potential to mitigate the emissions of greenhouse gases and is essential for reducing the occurrence and impacts of water-related disasters such as e.g. floods and droughts. This makes the water sector an important factor for reaching the targets of the

Sendai Framework for Disaster Risk Reduction, adopted by the UN Member States and for the implementation of the Paris Agreement. We urge national and local governments to consider water by distributing investments and designing development plans as an important determining factor for progress toward achieving a sustainable future.

1.9 Finally: Governance Matters

Even if socioeconomic and climatic changes pose a huge challenge for water managers to "safeguard sustainable access to adequate quantities of and acceptable quality water with an acceptable level of water-related risks (…)" (UN-Water 2013) the current technological progress and available infrastructures enable adapting to these challenges. Good governance and political willingness are crucial for ensuring water security.

References

Climate Policy Initiative (2018) Global climate finance: an updated view 2018. CPI. https://www.climatepolicyinitiative.org/publication/global-climate-finance-an-updated-view-2018/

FAO (2012) Coping with water scarcity. An action framework for agriculture and food security. http://www.fao.org/3/i3015e/i3015e.pdf

Federal Ministry for Economic Cooperation and Development (BMZ) (2017) BMZ water strategy. A key contribution to implementing the 2030 Agenda and Paris Agreement. BMZ Paper 08/2017. https://www.bmz.de/en/news/publications/publikationen-reihen/strategiepapier390-08-2017-23756

Graham NT, Hejazi MI, Chen M, Davies EGR, Edmonds JA, Kim SH et al (2020) Humans drive future water scarcity changes across all Shared Socioeconomic Pathways. Environ Res Lett 15:014007. https://doi.org/10.1088/1748-9326/ab639b

Harenda KM, Lamentowicz M, Samson M, Chojnicki BH (2018) The role of peatlands and their carbon storage function in the context of climate change. In: Zielinski T, Sagan I, Surosz W (eds) Interdisciplinary approaches for sustainable development goals. GeoPlanet: earth and planetary sciences. Springer, Cham. https://doi.org/10.1007/978-3-319-71788-3_12

Hoffmeister V, Averill M, Huq S (2016) The role of universities in capacity building under the Paris agreement. Policy Brief. http://icccad.net/wp-content/uploads/2016/08/Capacity-Building-Policy-Brief-July-4.pdf

Jaeger WK, Plantinga AJ, Chang HK, Dello G, Grant D, Hulse J, McDonnell J, Lancaster S, Moradkhani H, Morzillo AT, Mote P, Nolin A, Santelmann M, Wu J (2013) Toward a formal definition of water scarcity in natural-human systems. Water Resour Res 49:4506–4517. https://doi.org/10.1002/wrcr.20249

Karnib A (2019) Water as crosscutting factor in the SDGs under review at the high-level panel forum for sustainable development (HLPF) 2019 in the Arab States. http://www.unesco.org/new/fileadmin/MULTIMEDIA/HQ/SC/images/HLPF_2019_Arab_States_correction_long.pdf

Rockström J, Falkenmark M, Allan T, Folke C, Gordon L, Jägerskog A, Kummu M, Lannerstad M, Meybeck M, Molden D, Postel S, Savenije HHG, Svedin U, Turton A, Varis, O (2014) Ecohydrology bearings—invited commentar. The unfolding water drama in the Anthropocene: towards a resilience-based perspective on water for global sustainability. Ecohydrology 7:1249–1261. https://doi.org/10.1002/eco.1562

Savenije HHG, Hoekstra AY, van der Zaag P (2014) Evolving *water* science in the *Anthropocene*. Hydrol Earth Syst Sci 18:319–332. https://doi.org/10.5194/hess-18-319-2014

United Nations—Water (2013) Water security & the global water Agenda: a UN-Water analytical brief. https://www.unwater.org/app/uploads/2017/05/analytical_brief_oct2013_web.pdf

United Nations General Assembly (2015) Transforming our world: the 2030 Agenda for sustainable development. https://sustainabledevelopment.un.org/post2015/transformingourworld.

United Nations (UN) World Water Development Report (2015). Water for a sustainable world. Paris, UNESCO. https://www.unwater.org/

UN World Water Development Report (2020) (WWDR) Water and climate change. Paris, UNESCO. https://www.unwater.org/

World Bank (2016) High and dry: climate change, water, and the economy. World Bank, Washington, DC. https://openknowledge.worldbank.org/handle/10986/23665

World Bank Water Data Portal. https://www.worldbank.org/en/news/feature/2020/10/26/data-data-everywhere-new-world-bank-water-data-portal

The manufacturer's authorised representative in the EU is Springer Nature Customer Service Centre GmbH, Europaplatz 3, 69115 Heidelberg, Germany. If you have any concerns regarding our products, please contact ProductSafety@springernature.com

Printed and bound by CPI Group (UK) Ltd, Croydon, CR0 4YY
30/09/2026
02225691-0014